Oceanography and Sailing in the 19th Century: The Water World

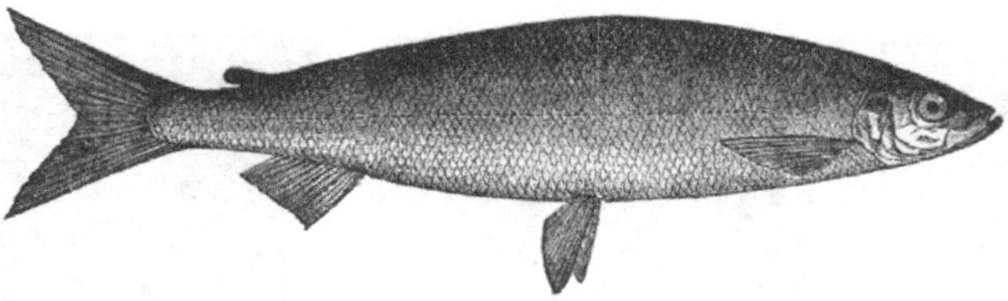

ENLARGED ILLUSTRATED SPECIAL EDITION

J.W. Van Dervoort

Oceanography and Sailing in the 19th Century:
The Water World Enlarged Illustrated Special Edition

by J.W. Van Dervoort
Originally published in 1884

Cover Design by Mark Bussler

www.CGRpublishing.com

FIRING THE LYLE GUN.

THE WATER WORLD;

A Popular Treatise

ON THE BROAD, BROAD OCEAN. ITS LAWS; ITS PHENOMENA; ITS PRODUCTS AND ITS INHABITANTS; GRAPHICALLY DESCRIBING ITS CURRENTS, TIDES, WAVES; ITS WHIRLPOOLS, WATER-SPOUTS, TYPHOONS AND TRADE WINDS; ITS CORAL REEFS, PEARLS, SHELLS, SPONGES, FISHERIES; ITS ANIMAL LIFE, MINUTE AND MAMMOTH, FROM THE BUTTERFLIES OF SUB-MARINE FORESTS AND MEADOWS, TO SHARKS, WHALES AND SEA DRAGONS; WITH CHAPTERS ON STEAMSHIPS, LIGHT-HOUSES, LIFE SAVING SERVICE &C., &C., &C.

BY

PROF. J. W. VAN DERVOORT.

EDITOR OF

THE TWO HEMISPHERES.

Profusely Illustrated.

SOLD ONLY BY SUBSCRIPTION.

1884.

Earth has not a plain
So boundless or so beautiful as thine;
The eagle's vision cannot take it in;
The lightning's glance too weak to sweep its space,
Sinks half-way o'er it, like a wearied bird;
It is the mirror of the stars, where all
Their hosts within the concave firmament,
Gay marching to the music of the spheres,
Can see themselves at once.

—CAMPBELL.

INTRODUCTION.

VERY little, comparatively, has heretofore been written on this subject. This is most singular when it is considered that this majestic ocean, "whose mighty heart throbs in sympathy with the pulse of God," covers more than three-fourths of the entire globe.

We write about the ocean because it is a subject of universal interest, and for the reason that a knowledge of its laws, phenomena, and inhabitants is conducive to right-living and enjoyment.

It has been our aim to write for the people and by avoiding technical terms clothe the subject in such language as shall make it interesting and easily comprehended by all.

We have endeavored to fill these pages, not with dry and uninteresting facts compiled from the cyclopædia, but with living, breathing thoughts, which, if rightly entertained, will lessen some of the weariness of daily life, give a greater impulse to right living, and cause us to revere and adore a Creator who has multiplied, everywhere in nature, countless objects and future well-being.

Of our fifty millions of people, many will live and die without ever having heard the voice of the sea. All want to see it; all are interested in its majestic power and the life with which it teems. To those who are denied the privilege of witnessing it for themselves, as well as to the dwellers on its border, do we send this pen picture. "God gave this beautiful world—the *whole of it*—to subdue and enjoy." We have denied ourselves a great boon and benefit by heretofore confining our research too exclusively to the narrow earth. We see the sun rise, approach its meridian, and decline in the West; we are interested in the changing seasons, to know when to sow, to reap and to rest; we are interested in the beautiful flowers, growing shrub, fruitful vine, and majestic forests; we are interested in the animal species, both wild and domestic; and this is right and reasonable, for such knowledge is enjoined and is necessary for our well-being and prosperity.

Moisture is as necessary to animal and vegetable life as is the heat radiated from the sun. The rain and the dew come not by accident. Over the whole world the rain-fall is about the same, year in and year out. So interesting and wonderful is the machinery that pumps out of the ocean, day by day, all the waters conveyed to it by the rivers, and distributes it over the land back to the sources of the rivers again, that the reader can but enjoy and be profited by a contemplation of the causes that produce such marvellous effects. This subject, together with the Gulf Stream,—that wonderful equalizer of terrestrial climate—all the currents, aerial and oceanic; the tides, trade winds, typhoons, monsoons, saltness and specific gravity of the sea, and other laws and phenomena, will be found described and treated in detail in their proper places in this book. Modern research has defined absolutely many things that heretofore have been but imperfectly understood. Arduous and protracted though the labor has been, we have carefully examined and weighed the conclusions of the late scientists, and therefore write with more confidence than as if we had depended entirely upon our own unaided observations.

Under the divisions of Laws and Phenomena, quotations and conclusions will be found from a book entitled the "Physical Geography of the Sea," by the late Capt. Maury. For this kindly courtesy we are indebted to Colonel Richard L. Maury, of Richmond, Va., son and legal representative of the late Capt. Maury.

We have drawn to some extent from other sources for many of the incidents so graphically illustrating some of the subjects considered.

In the chapter entitled "The Frozen Ocean," will be found something new respecting late expeditions, and much that is of varied and thrilling interest. In the consideration of the subject of "Deep Sea Dredging," and the "Beds of the Waters," we have taken advantage of researches prosecuted by late Government expeditions, in arriving at the conclusions presented to the reader.

We should be negligent did we not call attention to the subject of "Life in the Ocean," minute and mammoth.

We have called attention to those that must be objects of universal interest, from the tiny polyp, the wonderful rock-builder of the ocean, to the "monarch of the deep." Neither have we contented ourselves by giving a mere barren technical description, but we have alluded to their habits, their uses in creation, methods, dangers and exciting incidents relative to their capture. The forests of the deep are infinitely more densely inhabited than the mountains of the earth. God has created nothing in vain. Each being has its use in the scale of creation, and nowhere can we see His works in such perfection as in the vast deep. Here we see perfect and marvelous adjustment; we see such exquisite care displayed in the sustenance of an infinitesimal creature; we may also see what important agents are these little creatures in preserving the equilibrium of the ocean. Surely such contemplations must cause man to think more and better of himself, for he realizes that God thinks much of him. There is reverence in such thoughts.

We believe we have not exceeded the license of an author in the preparation of this volume. We trust the reader may be as much entertained and benefited by the perusal as has been the author in the preparation of this book. If so, "The Water World" shall not have been written in vain, but may go forth on its humble mission of exalting the handiwork of the Creator..

J. W. V.

MOUNT VERNON, N. Y.

CONTENTS.

CHAPTER I.

THE OCEAN—ITS LAWS AND ELEMENTS.

CHAPTER II.

THE FROZEN OCEAN.

CHAPTER III.

ICEBERGS.

CHAPTER IV.

LIFE IN THE OCEAN.

CHAPTER V.

MINUTE ANIMAL LIFE.

CHAPTER VI.

CORAL—THE ROCK BUILDERS.

CHAPTER VII.

PEARLS.

CHAPTER VIII.

SPONGES.

CHAPTER IX.

SEALS.

CHAPTER X.

WHALES THE MONARCHS OF THE OCEAN.

CHAPTER XI.

THE WHALE FISHERY AND ITS PERILS.

CHAPTER XII.

SHARKS THE PIRATES OF THE OCEAN.

CHAPTER XIII.

SEA-HORSES AND NARWAHLS.

CHAPTER XIV.

NAUTILI THE FLOATING NAVIGATORS OF THE OCEAN.

CHAPTER XV.

MODES OF FISHING IN VARIOUS COUNTRIES.

CHAPTER XVI.

ODDS AND ENDS ABOUT FISHES.

CHAPTER XVII.

SHELLS.

CHAPTER XVIII.

SEA BIRDS.

CHAPTER XIX.

SUPERSTITIONS CONNECTED WITH THE OCEAN.

CHAPTER XX.

MARINE PRODIGIES.

CHAPTER XXI.

MONSTERS OF THE DEEP—SEA DRAGONS.

CHAPTER XXII.

SUBMARINE SCENERY—ANIMAL AND VEGETABLE.

CHAPTER XXIII.

THE BED OF THE OCEAN. DEEP SEA SOUNDINGS.

CHAPTER XXIV.

PHENOMENA OF THE OCEAN.

CHAPTER XXV.

OCEAN STEAMSHIPS.

CHAPTER XXVI.

THE SIGNAL SERVICE.

TO

MY ESTEEMED FRIEND,

MR. JAMES MONTEITH,

AS A

TESTIMONIAL OF HIS SUCCESS IN POPULARIZING THE SCIENCE OF GEOGRAPHY IN THE UNITED STATES,

THIS VOLUME IS RESPECTFULLY DEDICATED.

SHIP UNDER FULL SAIL.

CHAPTER I.

THE OCEAN—ITS LAWS AND ELEMENTS.

"IN the beginning," the sacred historian informs us, "God created the heavens and the earth: and the earth was without form and void, and darkness was upon the face of the deep, AND THE SPIRIT OF GOD MOVED UPON THE FACE OF THE WATERS."

How wondrously solemn and grand are these inspired and holy words! What human imagination can fully realize their sublimity? In a few plain but soul-stirring sentences the great mystery of creative power is unfolded, and the mind gets bewildered in the contemplation of such vastness, beauty, and beneficence. We may exclaim with the royal psalmist, "Thou, even Thou, art Lord alone; Thou hast made heaven, the heaven of heavens, with all their host; the earth, and all things that are therein; the seas, and all that are therein; and Thou preservest them all."

On the second day, or generation, uprose progressively the fine fluids or waters of the firmament, and filled the blue ethereal void with a vital atmosphere. The third day, or generation, the waters more properly so called, or the grosser or more compact fluids of the general mass, were gathered together into the vast bed of the ocean, and dry land began to make its appearance.

No subject, surely, could be more delightful than the study of the "world of waters" and its strange inhabitants, and there is none upon which the mind of man has been more absorbed in inquiry and research.

We never tire of the sea; it is a laboratory in which delightful processes are continually being wrought out for our admiration and use. Its flora and its fauna, its waves and its tides, its salts and its currents, all afford grand and profitable themes of study and thought. But, as interesting as they are separately, and as wonderful, too, they are not half so marvelous as the offices which, with their aid, the sea performs in the physical economy of our planet.

Viewed in this light, the ocean, its inhabitants, and its vapors, is a mechanism constructed by the All Wise, of perfect workmanship.

It is so fixed and true in its work that nothing can throw it out of gearing; and yet its returns are so delicate that the task of preserving them is allotted to the minutest of sea dwellers, and to agents apparently the most subtle and fickle.

They preserve its intricate relations, making its adjustments, in beauty and sublimity of effect, to vie with the heavens. These marvelous wonders proclaim, in songs divine, that they, too, are the work of holy fingers.

We may but imperfectly represent this great body of water and the many wonderful objects it contains, but any deficiencies may be supplied later, when the open book of nature is read by thoughtful minds eager for knowledge.

The ocean is essential to the existence of man and of all vegetation; it is the great moderator and equalizer of terrestrial climates, purifying the atmosphere that we breathe, and sending off a perpetual supply of vapors, which condense into clouds, and are the sources of moisture and fertility to the soil. We must also think of the facilities afforded for an intercourse with distant nations. It has been remarked that contact with the ocean has unquestionably exercised a beneficial influence on the cultivation of the intellect and formation of the character of many nations, on the multiplication of those bonds which should unite the whole human race, on the first knowledge of the true form of the earth, and on the pursuit of astronomy, and of all the mathematical and physical sciences.

Since Columbus was sent to unbar the gates of ocean, man has boldly ventured into intellectual as well as geographical regions before unknown to him. How perfect, O Infinite One, are all thy works, and how shortened our aspirations!

If the existing waters were increased only one-fourth of their present area, they would drown the earth, with the exception of some high mountains. If the volume of the ocean were augmented only by one-eighth, considerable portions of the present continents would be submerged, and the seasons would be changed all over the face of the globe. Evaporation would be so much extended, that rains would fall continually, destroy the harvests, fruits, and flowers, and overturn the whole economy of nature.

There is, perhaps, nothing more beautiful in our whole system than the process by which the fields are irrigated from the skies, the rivers are fed from the mountains, and the ocean restrained within bounds which it never can exceed so long as that process continues on the present scale. The vapor raised from the sea by the sun floats wherever it is lighter than the atmosphere; condensed, it falls upon the earth in water. And what is *water?* It is composed of two important gases—oxygen and hydrogen—these being, probably, the two most abundant and essential substances in nature, as regards ourselves and our earth.

These, when combined, become converted into vapor, many gallons of them in this state forming one small drop of fluid water. It is the simplest of combinations, and the compound most resembling a simple element; the most universal solvent at all temperatures; the most widely distributed substance in nature; the most powerful agent; the most perfect representation of perpetual motion, penetrating everything, passing everywhere, always present, in sight or out of sight, and everywhere producing a marked effect. When it is remembered that a very large proportion of the weight of every living being, animal or vegetable, consists of water, and that for life to continue at all, an incessant supply of fresh fluid is required, the necessity of water will be fully understood.

The *Saltness* which distinguishes the waters of the ocean is explained by the circumstance that chloride of sodium (common salt) and other dissolvable salts, which form essential ingredients of the earth, are being constantly washed out of the soil and rocks by rain and springs, and carried down by the rivers; and as the evaporation which feeds the rivers carries none of the dissolved matter back to the land, the tendency is to accumulate in the sea. We know that beds of rock-salt, of enormous thickness, form part of the crust of the globe; and we may infer that immense banks of salt exist in the bed of the deep. The uniformity of this saltness is preserved by the constant movement of the waters, caused by the regular and perpetual action of the winds. It has been said that if all the salts of the sea were spread equally over the northern half of this continent, it would cover the ground to the depth of one mile! What force could move such a mass of matter on dry land? Yet, the machinery of the ocean, of which it forms a part, is so wisely, marvelously, and wonderfully compensated, that the most gentle breeze that plays on its bosom—the tiniest insect that secretes solid matter for its sea-shell—is capable of putting it instantly in motion. Still, when solid and placed in a heap, all the mechanical contrivances of mankind, aided by the tremendous forces of all the steam and water power of the world, could not move so much as an inch in centuries of this matter, which the sunbeam, the zephyr, and the infusorial insect keep in perpetual motion and activity.

Why was the sea made salt? If the sea were not made salt, the rays of the sun could not so readily penetrate it. This penetration of the waters by the sun's rays produces expansion. The force or dynamical power resulting from this expansion, or the spreading up and outward of the waters, increases the circulation of the currents. Were the waters of the sea fresh instead of salt, we should probably have no such thing as a Gulf Stream nor marine climate; the torrid zone would have been hotter and the frigid zone colder; and the climate of England would have vied with Labrador for inhospitality: all for the lack of the watery circulation. With no salts in the seas, evaporation, volume of our rivers, and the quantity of rain, would all have been different. The thunderbolt of the heavens, the sheet lightning of the clouds, and the fitful flashes of the storm, all have their beginning principally in the salts of the sea.

With a few exceptions, such as the Red Sea, Great Salt Lake, etc., the salts of the sea are everywhere the same. They could not be made so, were they not well shaken together. The circulation of the currents of the sea is quite as perfect and wonderful as the circulation of the blood in our bodies. Evaporation in some waters is more rapid than in others. Water can hold only a given amount of salt in solution. We cannot *see* that the quantity of salt deposits is increasing. It reasonably follows from all this that there must be a system of circulation in the waters, whereby an equilibrium is produced, making each and all of the waters of the same degree of saltness. The currents which produce these results do not flow from chance, but in accordance with physical laws, assisting to maintain the order and preserve the harmony which is so apparent in every department of God's handiwork.

The coral islands of the Pacific were built up of matter which a certain kind of animal quarried from the ocean. These rivers of the sea become the hod-carriers of the little animal.

If the currents of the sea were not employed to carry off from this animal the waters that have been emptied by it of their lime, and to bring to it others supplied with more, it is apparent that it would have died for want of something to eat long before its work was completed. But for the benign current, the emptied drop of water would have remained, not only as the grove of the little builder, but as a monument recording a monstrous failure in the beautiful system of terrestial adaptations.

It may be reasonably concluded that the marine animals, whose secretions are so constituted as to alter the specific gravity of the water, to disturb its equilibrium, to originate currents in the ocean, and to control its circulation, are not in any place nor doing this work by accident. Nature is sublime and perfect in adaptation through all her domain.

Currents, which exercise so great an influence on the circulation of the waters, and in producing remarkable changes in the form of coasts, are described as constant, periodical, and variable; the two latter classes being determined chiefly by the winds and tides. The first motion of the ocean waves is derived either from the attraction of the sun or moon, or from winds which blow over the surface of the waters; the second arises from the sun, which directly through its heat, and indirectly by scorching dry winds, produces evaporation, to a great extent, of the parts most exposed to its influence; and by its similar action on the atmosphere, causes a transference of this vapor to remote latitudes, where it descends as rain, and by destroying the equilibrium of the ocean, gives rise to currents. The principal currents of the ocean are four, two warm, and two cold; these originate, the former among the islands of the Archipelago and in the Gulf of Mexico, and the latter in the Arctic and Southern Oceans.

The most important and best known of ocean currents, the *Gulf Stream*—the river in the ocean, one of the most marvelous things in this world of waters—derives its name from the Gulf of Mexico. The general direction of this stream is in the arc of a great circle, towards England, by which it is divided; one branch, passing to the west and north, reaches the coast of Norway, and can be perceived on the southern borders of Iceland and Spitzbergen. The waters are of a deep indigo blue, and are so distinctly marked that their line of junction with the common sea-water may be traced by the eye.

The existence of the Gulf Stream can also be readily ascertained by means of a thermometer, the temperature being so elevated. It is this warmth which tempers and softens the climate of all Western Europe. It is the influence of the Gulf Stream upon the climate that makes Ireland the Emerald Island of the sea, and clothes the shores of England with evergreen robes; while in the same latitude, on our side of the Atlantic, the shores of Labrador are fast bound in fetters of ice. How wonderful is this beneficent operation of Providence, when we think that this warm stream felt on England's shores, which are thus bathed with water heated under a tropical sun, comes from a distance of four thousand miles! Nor is its influence thus circumscribed. In mid-winter, off the inclement coasts between Cape Hatteras and New Foundland, ships, when beaten back from their harbors by fierce north-westers, loaded down with ice, and in danger of founding, turn their prows to the east, and seek relief and comfort in the Gulf Stream. In high northern latitudes, after having run three thousand miles towards the north, it still preserves even in winter the heat of summer. With this temperature, it spreads itself out for thousands of square miles over the cold waters around, and covers the ocean with a mantle of warmth that serves so much to mitigate in Europe the rigors of winter.

With a breadth of about fifty miles in its narrowest portions, the Gulf Stream has a velocity, at times, of five miles an hour, pouring on like an immense torrent.

The cause of these phenomenal ocean river currents, up to the present time, is only conjectured, and the nature and extent of this work will hardly warrant any extended theoretical discussion.

Each current seems to have a circulation of its own, *i.e.*, an upper and lower stratum. In the warm currents, the upper portion only is warm, while beneath runs a counter cold current. In the cold currents, the order of stratums is reversed.

There is a constant tendency of polar waters toward the tropics, and of tropical waters toward the poles.

It is a custom often practiced by seafaring people to throw a bottle overboard, having inside a paper, stating the time and place at which it is done. These minute little voyagers leave no trace behind them, and therefore their routes cannot be exactly ascertained, though we can approximate closely, knowing where they were cast and where they were found. Charts have been prepared showing the routes of over one hundred bottles, by drawing straight lines from the starting to returning point, with the time elapsed. From this it appears that the waters from every quarter of the Atlantic tend towards the Gulf of Mexico and its stream. Good circumstantial evidence exists to prove that bottles cast overboard in the Gulf Stream have performed the entire tour of that current.

Other currents as well as the Gulf Stream are utilized in a similar manner. As an instance of this we quote the instructions of Mr. G. M. Robeson, Secretary of our Navy, to Capt. Francis Hall, of the *Polaris* expedition in 1869, as follows: "To keep the Government as well informed as possible of your progress, you will, after leaving Cape Dudley Digges, throw overboard daily, as open water or drifting ice may permit, a bottle or small copper cylinder, closely sealed, containing a paper, stating date, position, and such other facts as you may deem interesting. For this purpose you will have prepared papers, containing a request, printed in several languages, that the finder transmit it by the most direct route to the Secretary of the Navy, U. S. of A."

Toward the end of the seventeenth century, a Dutch brig, pursued by the French corsair, Phœnix, was overhauled between Tangier and Tarifa, and seemed to be sunk by a single broadside; but in place of going down, the brig, being freighted with a cargo of oil and alcohol, floated between the two currents, and, drifting toward the west, finally ran aground in the neighborhood of Tangier, more than twelve miles from the spot where she had disappeared under the waves. She had therefore floated that distance, driven by the action of the under current, in a direction opposite to that of the surface current.

Recent invention has wrought out an improved plan of warming houses in winter by hot water. A furnace heats the water; this heated water and steam is conveyed by pipes to the place to be warmed.

This convenient mode of warming our offices and dwellings was probably suggested by two things: first, because of its utility; and, second, the fact that we have a similar heating apparatus in nature, in the warm waters of the Gulf of Mexico. The heater is the torrid zone; the Gulf and the Caribbean Sea are the boilers; the Gulf current is the means of conveyance; from New Foundland to Europe is the reservoir. According to *Maury*, "the quantity of heat discharged over the Atlantic by the Gulf Stream in a winter's day would be sufficient to raise the whole column of atmosphere that rests upon France and the British Islands from the freezing point to summer heat."

How benign is the influence of this wonderful stream, and how a contemplation of it leads one to revere its creator.

"He treadeth upon the waves of the sea, and is seen in

the waters of the deep. Yea, He calleth for its waters, and poureth them out on the face of the earth."

Investigation as to the causes of the severe storms raging so frequently in certain portions of the Atlantic, and which have proved so disastrous to navigation, led to the conclusion that they were caused by the irregularity between the temperature of the Gulf Stream and neighboring regions, both in the air and water. "To use a sailor's expression, the Gulf Stream is the great 'weather breeder' of the North Atlantic." It is a "storm fiend" that out-tops the "stormy capes," and out-vies the furious storms of the North Pacific and China Seas.

Storms from the right and left of the Gulf Stream break in upon it; and, turning about, rush along with it, leaving behind a steamy mist, caused by the cold water and warm air coming in contact, to mark its course.

Formerly the Gulf Stream exercised a greater influence npon commerce than it does at the present day. Up to the last century, the navigator *guessed* as much as he *calculated* the place of his ship.

For three centuries navigators had been crossing and recrossing this Gulf Stream daily, without using it as a means of giving them their longitude, or warning them of their approach to this continent.

Before the warmth of the Gulf Stream was known, a voyage from Europe in winter to portions of our own coast was exceedingly perilous. Gales and snow storms would be met which would set at naught the seaman's skill. His vessel becomes incrusted with ice and her crew benumbed and helpless. She remains obedient only to her helm, which almost instinctively guides her to the Gulf Stream. She crosses its magic boundary and is embraced by its healing presence. The ice vanishes from her garments; the weary sailor laves in its healing properties, being invigorated by its genial warmth. He is now ready to make another effort to enter his port, perhaps to be as rudely driven back again. But each breathing spell renews his energies, until at last he may enter his haven in safety, though many, in this terrible contest, may sink to rise no more

Other currents as well as the Gulf Stream materially affect navigation. While an intimate knowledge of them is necessary, in order to avoid the danger of mistaking the true position of a vessel, its progress to port may be facilitated by falling in with a local stream, or steering clear of it, according as its direction is favorable or adverse.

The effect of currents was perceived long before anything was known of their direction and velocity, and Columbus was strengthened in his belief that land might be reached across the Atlantic westward, by substances which had drifted from that quarter. After the commencement of his great undertaking, when, day after day, nothing had been seen but a shoreless horizon, and hope had nearly expired in his own breast, while his crew were on the verge of open rebellion, the effect of the oceanic currents restored his confidence and allayed their clamors. A branch of thorn, with berries on it, appeared; a reed was picked up, and a staff artificially carved—intimations that an inhabited land lay before the adventurers, which was at length revealed to their gaze, and terminated forever the mystery which had rested upon the western flood.

A *Tide* is a wave of the whole ocean, which is elevated to a certain height, and then sinks after the manner of a common wave. The interval between the two positions forms the tide. The principal cause is the attraction of the sun and moon, the latter being the more potent agent. The sea rises or flows, as it is called, by degrees, about six hours; it remains stationary about a quarter of an hour; and then retires or ebbs during another six hours, to flow again after a brief repose. Thus every day, or the period elapsing between successive returns of the moon to the meridian of a place—which is twenty-four hours, fifty minutes and a half—the sea ebbs and flows twice, much less, indeed, towards the poles than within the tropics, where the waters lie under the direct influence of the lunar attraction. It is in the southern hemisphere that the tidal wave originates, and from thence moves northward, influenced in its direction by the motion of the earth. Almost excluded from the Northern Pacific by the barrier of islands and coral reefs which stretch across from Australia nearly to South America, the effect of the tides, excepting on the west coast of that continent, is little felt in that ocean. In the Indian Ocean, compressed between Africa on the north and Australia and Sumatra on the east, it bursts in full strength on the shores of Hindoostan. In the narrow channel of the Atlantic the tidal wave progresses northward with great rapidity, and on the shores both of Europe and America, producing, as in Southern India, the Bore, which is described in the chapter on the "Phenomena of the Ocean,"

The highest floods and the lowest ebbs occur at the period of new and full moon, near the equinoxes, in March and September, when the moon is nearest the earth.

Winds have also a powerful influence over the tidal currents, especially in narrow seas, keeping them back when blowing from an opposite quarter, and quickening their flow when pursuing the same direction; but the motion of the water in the tide-wave is totally unlike that in an ordinary surface-wave, such as the wind produces; and it differs, also, in affecting the whole depth of the ocean equally from the bottom to the surface, while the wind-waves, even in the most violent storms, agitate it to a very trifling depth. In the deep water of the ocean, the tidal-wave does not exceed twelve feet in height.

The ancients knew that the time of high water, and also the height of the tide, were in some way connected with the age of the moon. It was the illustrious Sir Isaac Newton

BOATS STRANDED BY THE TIDE.

SPRING TIDE.

who made the first attempt to explain the phenomena of the tides, on the principle of the influence of gravitation, the grand agent in the movement of the universe.

What are called *wind-waves* are small at their first origin, commencing with a mere ripple, or, as the sailors term it, a "cat's-paw." But each wave, as it advances, acquires increased height by the continued pressure of the wind. Thus it is that the larger waves are not developed in narrow seas, or where the wind blows off the land; they require breadth of water and continued pressure for their formation. The greatest waves known are those off the Cape of Good Hope, under the influence of a north-west gale (the storm-wind of that region), which drifts the swell around the Cape, after traversing obliquely the vast area of the South Atlantic. In such gales, the waves attain a height of above forty feet, so that two ships in the trough of the sea, with such a wave between them, lose sight of one another from their decks. Off Cape Horn, also, the waves reach upwards of thirty feet in height. In our own seas, they rarely exceed eight or nine feet.

The crossing of waves, instead of dividing the water into parallel ridges, causes the pitching and rolling so distressing to passengers and trying to vessels. When more than two series of waves cross one another, they give rise to the term chopping seas.

Whatever relates to the *color* of the ocean is a matter on which many and various opinions have been expressed. Very curious is the statement of Martyn, one of the early voyagers, attributing these changes in the sea to the color of the skies: "If," he says, "the sky be clear, the sea looks as blewe as saphire; if it is covered somewhat with clouds, the sea is as greene as an emeralde; if there be a foggy sunshine, it looketh yellow; if it be quite darke, like unto the color of indigo; in stormy and cloudy weather, like blacke sope, or exactly like unto the color of blacke leade."

The Greenland sea varies in color from ultramarine blue to olive green, differences which have been found, on examining the water, were due to the presence of innumerable minute animals. The red, brown, and white patches of the Pacific and Indian Oceans, are attributed to the presence of swarms of animalculæ, and the colors of the Red and Yellow Seas to matters of vegetable origin. On both sides of the island of Ceylon, during the south-west monsoon, a broad expanse of the sea assumes a red tinge, considerably brighter than brick-dust; and this is confined to a space so distinct, that a line seems to separate it from the green water which flows on either side. On examining some of this water with a microscope, it proved to be filled with animalculæ, probably similar to those which have been noticed near the shores of South America, and whose abundance has imparted a name to the Vermillion Sea off the coast of California.

Captain Kingman passed through a tract of water twenty-three miles in breadth, and of unknown length, so full of minute (and some not very minute) phosphorescent animal organisms, as to present the aspect at night of a boundless plain covered with snow. Some of the animals were serpents six inches in length, of a transparent jelly-like nature. This appearance is noticed by Dr. Collingwood as a "milky sea," the whole surface composed of a white fluid-like milk. The contrast of the ocean, thus colored, with the dark sky, is very striking.

This proceeds from a great variety of marine organisms, some soft and gelatinous, and some minute shelly animals. They mostly shine when excited by a blow or by agitation of the water, as when a fish darts along or oar dashes, or, in the wake of a ship, when the water closes on its track. In the latter case are often seen what appear to be lamps of light rising from under the keel, and floating out to the surface, apparently of many inches in diameter. One of the most remarkable of these luminous creatures is a species of shell animals with muff-shaped bodies upwards of an inch in length, which, when thrown down on deck, burst into a glow so strong as to appear like lumps of white-hot iron.

There are few subjects of study more interesting than the luminous appearance presented by the sea under various circumstances. That the sea, the great extinguisher of fire, should be turned into flame—that the darkness of night should be illuminated by the luminous glow which bathes every ripple and breaks over every wave—that globes of light should traverse the ocean, or that lightning flashes should coruscate no less in the billows of the sea than in the clouds of the air—are all facts which seize on the imagination. Nor is the interest lessened by the knowledge that all these phenomena are produced by animals whose home is in the great waters; that not only do the fiery bodies of large animals give out steady patches of light, but that of the myriad animalculæ with which the sea teems, like motes in a sunbeam, each contributes its tiny scintillation, the aggregate forming a soft and lovely radiance.

A vivid description of a luminous sea is given by an eminent French naturalist, as follows:

"It exhibited to us in all its splendor the glorious phenomena of its phosphorescence. For more than an hour the waters around us seemed to be kindled into a blaze of light, as if they had borrowed some of the hidden fires of Stromboli. The waves, as they broke along the rocky shores of Sicily, encircled it with a glowing band of light, while every projecting cliff was circled with a wreath of fire. Our boat seemed as if it were opening for itself a passage through some glowing and fused liquid, while in its wake it left a long track of light, each stroke of the oar brightening the bosom of the waves with a broad silver gleam. The water that was taken up in a bucket presented the appearance of molten lead, as it was poured back into the sea. Everywhere over this brilliant surface of calm light, myriads of dazzling green sparks and globes of fire were flashing, quivering, and dying amidst the undulations of the waves, and these sparks and globes of fire were so many living beings. At certain times of the year these microscopical beings acquire the property of emitting light at each muscular contraction; and hence every movement in these animalculæ is made apparent by a luminous flash."

Mr. Edmonds alludes to the luminous waters frequently witnessed in Mount's Bay:

"On these occasions, particularly when the night is dark, if a fish rise from the calm water, a most brilliant and beautiful effect is produced. Were you, from a boat, to look down into the sea while fishes were darting to and fro, their paths would be luminous. and the deep would be traversed by streams of light as bright and beautiful as those of stars shooting through the sky. If you draw in your fishing-line, it will appear as a line of fire, and the fish at the end of it like a ball of fire coming near you. A net suspended in the sea appears 'like a brilliant lacework of fire,' and the fishes may be seen carefully avoiding it. When fishermen by night wish to know whether any fish are near, they stamp on the bottom of the boat, and instantly, if there are any beneath, they will be seen darting away in all directions."

To these observations may be added the interesting description of this phenomenon, as witnessed in the vicinity of the Plata by the distinguished Darwin:

"One very dark night the sea presented a very beautiful and singular appearance. There was a fresh breeze, and every part of the surface which, during the day, is seen as foam, now glowed with a pale light. The vessel drove before her bows two billows of liquid phosphorus, and in her wake she was followed by a milky train. As far as the eye reached, the crest of every wave was bright, and the sky above the horizon, from the reflected glare of these livid flames, was not so utterly obscure as over the vault of the

LUMINOSITY OF THE SEA.

heavens. Farther south, the sea is seldom phosphorescent, probably owing to the scarcity of organic beings in that part of the ocean. The same torn and irregular particles of gelatinous matter seem, in the Southern as well as in the Northern Hemisphere, to be the common cause of this phenomenon. The particles were so minute as easily to pass through fine gauze, yet many were distinctly visible by the naked eye. The water, when placed in a tumbler and agitated, gave out sparks, but a small portion in a watch-glass scarcely ever was luminous. All these particles retain a certain degree of irritability. My observations gave a different result. Having used the net one night, I allowed it to become partially dry, and twelve hours after, having occasion to use it again, I found the whole surface sparkle as brightly as when first taken out of the water. It does not appear probable, in this case, that the particles could have remained so long alive. When the waves scintillate with bright green sparks, it is generally owing to minute shell-covered animals; but there can be no doubt that very many other pelagic animals, when alive, are phosphorescent. The phenomenon is the result of the decomposition of the organic particles, by which process the ocean becomes purified."

Having briefly glanced at some of the most important features of the world of waters, it may not be amiss to call attention to some of its principal divisions, and these are five: the Atlantic, Pacific, Indian, Arctic, and Antarctic Oceans. Although no one portion is completely set off from the rest, it has been found desirable to arrange it into these divisions.

The extreme breadth of the Atlantic system is about five thousand miles, and its narrowest part about sixteen hundred miles. The extent of its shores is immense—above fifty thousand miles—several thousand more than the Pacific and Indian Oceans combined. The Atlantic, from its relation to civilized countries, and as the most frequented highway of communication for commerce, is regarded as the most important, and is, consequently, much better known than the Pacific. Its waters wash the eastern coasts of North and South America, and the western coasts of Europe and Africa. Its northern and southern extremities are the Polar waters.

The Mediterranean Sea, one of the arms or tributaries of the Atlantic, with which it is connected by the Straits of Gibraltar, is one of the greatest inland seas of the world. Its shores were the successive seats of the governments of the earth for thousands of years. It was the central ocean of the Ancients, on which all the early discoveries and hardships of navigation were experienced.

The Pacific was discovered by Balboa, in 1513, not quite four hundred years ago. The causes that led up to this important discovery, and the effect it produced upon what was then called the Old World, are matters of common history, and need not be related nor discussed here. As a highway of commerce, it does not compare with its sister, the Atlantic, though each decade increases its importance in this respect; for the light of the Gospel, and the rigor of modern research, and commercial enterprise, is gradually but surely opening up a lively correspondence and communication between the civilized inhabitants of North and South America bounding its eastern shores, and the benighted hosts of Asia on its west.

The Indian Ocean, an arm of the Pacific, and embraced by Africa on the west, Asia on the north, and Australia on the east, possesses a remarkable interest, inasmuch as the earliest voyage on record, made by the navy of Solonic, was taken on its romantic waters.

CHAPTER II.

THE FROZEN OCEAN.

THOSE of us who pass our days in a sun-favored and temperate portion of the earth, with every comfort we could desire around us, the green face of nature only covered at brief wintry intervals with a mantle of snow, and a wide-spread fertility attesting the bounty of an indulgent Providence, cannot realize the dark and repelling picture of the frozen North.

We can only fancy, with a shudder, a winter of nine months reigning over the boundless regions of ice; and we might wonder how human nature is able to support such an intensity of cold with its attendant privations, did we not know that the inhabitants of this bleak climate, accustomed to hardships which we could not endure, pursue an existence which *we* might consider miserable, but which *they*, active, self-reliant, and with but few wants to satisfy, except the cravings of hunger, are contented with, and would not, probably, exchange for what we might consider a happier lot.

It is astonishing what amount of cold can be endured by the human frame. Dr. Kane, one of the Arctic navigators, records, 7th of February, 1851, a frost three degrees below the freezing-point of mercury! Only a few degrees above this, the crew of the ship engaged in the expedition performed a farce, called "The Mysteries and Miseries of New York." One of the sailors had to perform the part of a damsel with bare arms, and when a cold flat-iron, which was employed in the play, touched his skin, the sensation was like that of burning with a hot iron. On the 22d of the same month (Washington's birthday), there was another theatrical performance. "The ship's thermometer *outside* was at 46°; inside, the audience and actors, by aid of lungs, lamps, and hangings, got as high as 30°, *only* sixty-two degrees below the freezing-point, perhaps the lowest atmospheric record of a theatrical representation. It was a strange thing altogether. The condensation was so excessive, that we could barely see the performers; they walked in a cloud of vapor. Any extra vehemence of delivery was accompanied by volumes of smoke. Their hands steamed; when an excited Thespian took off his coat, it smoked like a dish of potatoes."

As another instance of extreme cold in these fearful regions, it may be mentioned how, under a temperature of 15° below zero, Captain M'Clure, one of the most adventurous of Arctic explorers, spent the night of the 13th of October, 1851, on the ice, amid prowling bears, and that without food or amunition, his only guide being a pocket compass, which, however, the darkness, aided by mist and drift, rendered useless. He, nevertheless, wiled away the time by sleeping three hours on "a famous bed of soft dry snow," and by wandering ten miles by the crow's flight, over a surface so rugged with ice and snow as to endanger his limbs. It was at the close of a walking expedition of nine days, on a very short allowance of food and water, he accomplished his desire of reaching the winter quarters of the expedition, so as to ensure a warm meal ready for his men when they arrived at their destination.

Edward Parry mentions his experience of Arctic rigors thus: "Our bodies appeared to adapt themselves so readily to the climate, that the scale of our feelings was soon reduced to a lower standard than ordinary, so that after being some days in a temperature of 15° or 20°, it felt quite mild and comfortable when the thermometer rose to zero—that is, when it was 32° below the freezing-point!" One of Dr. Kane's crew put an icicle into his mouth to crack it, when the thermometer was at 28°; one fragment stuck to his tongue, and two to his lips, each taking off a bit of skin, *burning* it off, if this term might be used in an inverse sense. The same writer observes, "that at 25° the beard, eyebrows, eyelashes, &c., acquire a delicate, white, and perfectly enveloping cover of venerable hoar-frost. The moustache and under-lip form pendulous beads of dangling ice. Put out your tongue, and it instantly freezes to this icy crusting, and a rapid effort and some hand-aid will be required to liberate it. Your chin has a trick of freezing to your upper jaw by the biting aid of your beard. My eyes have often been so glued as to show that even a wink may be unsafe."

One day Dr. Kane walked himself into "a comfortable perspiration" with the thermometer *seventy* degrees below freezing-point! A breeze sprang up, and instantly the sensation of cold was intense. His beard, coated before with icicles, seemed to bristle with increased stiffness, and an unfortunate hole in the back of his mitten "stung like burning coal." On the next day, while walking, his beard and moustache became one solid mass of ice. Inadvertently he put out his tongue, and it instantly froze fast to his lip. This being nothing new, costing only a smart pull and a bleeding afterwards, he put up his mittened hands to "blow hot," and thaw the unruly member from its imprisonment. Instead of succeeding, his mitten was itself a mass of ice in a moment; it fastened on the upper side of his tongue, and flattened it out like a batter-cake between the two disks of a hot griddle. It required all his care with the bare hands to release it, and then not without laceration.

Such is the relation of the rigors experienced by Arctic navigators in the frozen regions. The Esquimaux, on the approach of winter, cut the hard ice into tall square blocks, with which they construct their dwellings. They pass their nights covered with bear and seal skins, near a stove or lamp, every portion of the hut being closed against the piercing cold. Their provisions are often frozen so hard as to require to be cut with a hatchet. The whole of the inside of the hut sometimes becomes lined with a thick crust of ice; and, if a window is opened for a moment, the moisture of the confined air it immediately precipitated in the form of a shower of snow.

Without interest and adventure to stimulate the energies and excite the curiosity of mankind, these gloomy regions might not, probably, have been penetrated by the brave seamen who have imperilled their lives amidst those icy waters or on the inhospitable coasts, and whose explorations have developed and tested more heroism and skill than, perhaps, the exploration and discovery of all the rest of the world since the age of Columbus. But for these Arctic voyagers, we should have been ignorant of the strange and wonderful countries of the North, and their inhabitants. These voyages originated in an attempt to discover a shorter passage to India across the Northern seas. In 1553, an expedition of three vessels for this purpose left England. The results to two of these ships were most disastrous; the crews, seventy in number, and the commander of the expedition, Sir Hugh Willoughby, being frozen to death. Since this period, upwards of a hundred expeditions have been made in search of the North-west Passage—that is, a navigable channel from the Atlantic to the Pacific Ocean, round the northern margin of America. Among the heroic leaders of these expeditions are the conspicuous names of Parry, John and James Ross, Back, Franklin, Beecher, Austin, Kellett, Osborne, Collinson, M'Clure, Rae, Simpson, M'Clintock, Hayes, Kane, Hall, and other famous men.

The fate of the unfortunate Sir John Franklin, one of the

bravest and boldest of the Arctic explorers, is well known: how, in 1845, when nearly sixty years of age, he started on his last and fatal voyage to the frozen regions, with the ships "Erebus" and "Terror." The vessels were seen three months afterwards, but for eleven years their fate remained a mystery, although twenty expeditions were sent, at the cost of a million sterling, to discover traces of the missing crews. In 1857 the "Fox," commanded by the gallant M'Clintock, was fitted out, at the expense of Lady Franklin, on the same mission; and in 1859, the sad end of Franklin and his associates was ascertained. The "Erebus" and "Terror" had been beset by ice and abandoned in 1848; the commander himself had died the year previously (11th of June), and was thus spared the agony of witnessing and sharing the sufferings of his crews, all of whom had, it is presumed, perished on those fearful shores. Many sad and interesting relics of the Franklin expedition were recovered and brought home. The discoverers obtained their information in a remarkable manner: lying amongst some stones, which had evidently fallen off from the top of a pillar, was a small tin case, deposited on this spot by the crews of the abandoned vessels and containing a record of the long-lost expedition.

It was in one of the attempts in search of Franklin and his companions that the discovery of the North-west Passage was effected in 1850, by the successful though perilous exertions of Captain M'Clure, who had shared in the Arctic expedition of Captain Back in 1836, and in the voyage of James Ross in 1848. Captains M'Clure and Collinson were sent out in the "Investigator" and the "Enterprise." The course of the latter vessel was chiefly in open waters, close to our shores; but M'Clure steered in a more northern route, and encountered fearful perils from the ice in those storm-bound regions. During four years he underwent trials and exposures, which would have daunted many a navigator, however accustomed to these dangers. His vessel, several times beset by ice, was at length so firmly locked in, that M'Clure, seeing no hope of release, decided upon sending thirty of his crew to make their way homewards; some by way of North America, up the Mackenzie River, and the others by Cape Spencer, Beechey Island; while he himself, with the remainder of the officers and crew, would stay by the ship, spend a *fourth* winter in those dreary regions, and then, if not relieved, endeavor to retreat upon Lancaster Sound. Such was the arrangement, when an incident occurred that thrilled their hearts with joy. The captain and his first lieutenant were walking near the ship conversing, when they perceived a figure rapidly approaching them from the rough ice at the entrance of the bay. When about a hundred yards from them, he shouted and gesticulated, but without enabling them to guess who he might be. At length he approached, and to their astonishment thus announced himself: "I am Lieutenant Pym, late of the 'Herald,' and now in the 'Resolute.' Captain Kellett is in her at Denby Island." Lieutenant Pym had come from Melville Island, in consequence of one of Captain Kellett's parties having discovered an inscription left by M'Clure on Parry's famous sandstone rock in Winter Harbor.

The ship was abandoned, and the commander and his crew, released from a very perilous position, returned to England in 1854. Although he was obliged to leave his ship blocked in mountains of ice, and had to walk and sledge over hundreds of miles of ice, to reach other ships which had entered the frozen regions in the opposite direction, still, he had *water under him* all the way, and was thus the first commander of a vessel who really solved the problem of the famous North-west Passage.

The Arctic and Antarctic Circles are the boundaries which separate the frigid and temperate zones. At the poles themselves there is only one day of six months, during which the sun never sets, and one night of six months, when the sun never rises. At the Arctic Circle the greatest length of continuous light is twenty-four hours, at the summer solstice or Midsummer's day; while, at the same time, at the Antarctic Circle, the sun is twenty-four hours below the horizon, and the reverse at the opposite seasons of the year.

The coldness of the Polar regions arises from the fact of the rays of the sun striking the earth obliquely, as, at the equator, the heat is produced by the sun's rays falling upon the earth vertically. In the Arctic Ocean—that part of the universal sea which surrounds the North Pole—lie the most fearful dangers which can beset the seaman on his perilous course, arising from floating ice, the ship being frozen in, the fogs, the blinding snow, the darkness, the storms, and the tides and currents, comparatively unknown, which he has to encounter.

The following thrilling incident, described in the Westminster Review, is one of the most fearful histories that have been recorded:

"One serene evening in the middle of August, 1775, Captain Warrens, the master of a Greenland whale-ship, found himself becalmed among an immense number of icebergs, in about 77° of north latitude. On one side, and within a mile of his vessel, these were of an immense height and closely wedged together, and a succession of snow-covered peaks appeared behind each other as far as the eye could reach, showing that the ocean was completely blocked up in that quarter, and that it had probably been so for a long period of time. He did not feel altogether satisfied with his situation; but, there being no wind, he could not move one way or the other, and he therefore kept a strict watch, knowing that he would be safe as long as the icebergs continued in their respective places. About midnight the wind rose to a gale, accompanied by thick showers of snow, while a succession of thundering, grinding, and crashing noises gave fearful evidence that the ice was in motion. The vessel received violent shocks every moment, for the haziness of the atmosphere prevented those on board from discovering in what direction the open water lay, or if there was actually any at all on either side of them. The night was spent in tacking as often as any case of danger happened to present itself, and in the morning, the storm abating, he found, to his great joy, that his ship had not sustained any serious injury. He remarked with surprise that the accumulated icebergs, which had the preceding evening formed an impenetrable barrier, had been separated and disengaged by the wind, and that in one place a canal of open sea wound its course among them as far as the eye could discern.

It was two miles beyond the entrance of this canal that a ship made its appearance about noon. The sun shone brightly at the time, and a gentle breeze blew from the north. At first some intervening icebergs prevented the captain from distinctly seeing anything but her masts; but he was struck by the strange manner in which her sails were disposed, and with the dismantled aspect of her yards and rigging. She continued to go before the wind for a few furlongs, and then grounding upon the low icebergs, remained motionless. His curiosity was so much excited that he immediately leaped into his boat, with several seamen, and rowed towards her.

On approaching, he observed that her hull was considerably weather-beaten, and not a soul appeared upon the deck, which was covered with snow to a considerable depth. He hailed her crew several times, but no answer was returned. Previous to stepping on board, an open port-hole near the main chains caught his eye, and on looking in he perceived a man reclining on a chair, with writing materials before him; but the feebleness of the light made everything indistinct. The party went upon the deck, and having

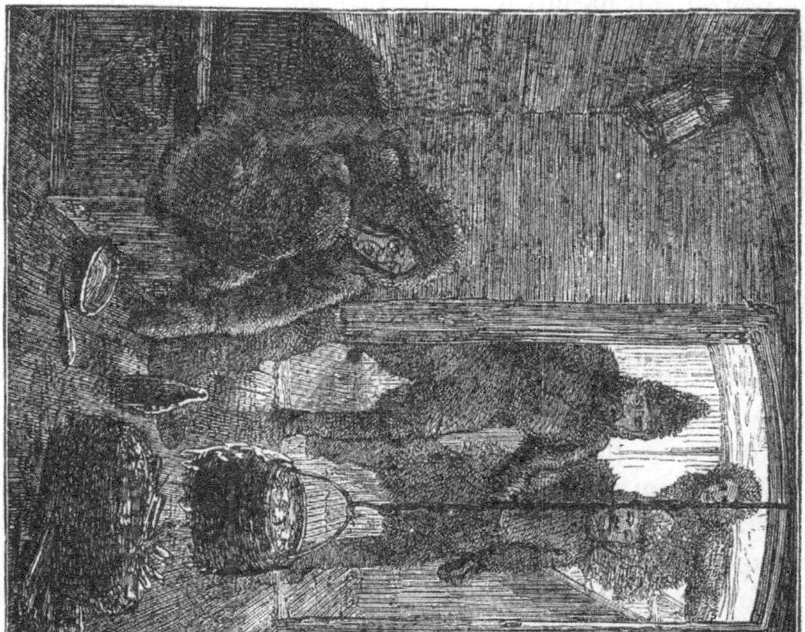

FROZEN TO DEATH

uncovered the hatchway, they descended below to the cabin which the captain had viewed through the port-hole. A tremor seized him as he entered it. Its inmate retained his former position, and seemed to be insensible to the presence of the strangers. He was found to be a corpse, and a green damp mold had covered his cheeks and forehead, and veiled his eye-balls He had a pen in his hand, and a log-book lay before him, the last sentence in whose unfinished page ran thus: "—November 11th, 1762. We have now been enclosed in the ice seventeen days. The fire went out yesterday, and our master has been trying ever since to kindle it again, but without success. His wife died this morning. There is no relief."

The captain and his men hurried from the spot without uttering a word. On entering the principal cabin, the first object that attracted their attention was the dead body of a female, reclining on a bed in an attitude of deep interest and attention. Her countenance retained the freshness of life, and a contraction of the limbs alone showed that her form was inanimate. Seated upon the floor was the corpse of an apparently young man, holding a steel in one hand and a flint in the other, as if in the act of striking fire upon some tinder which lay beside him. In the forepart of the vessel several sailors were found lying dead in their berths, and the body of a boy was found crouched at the bottom of the gangway stairs.

Neither provisions nor fuel could be discovered anywhere; but Captain Warrens was prevented, by the superstitious prejudices of his seamen, from examining the vessel as minutely as he wished to have done. He therefore carried away the log-book already mentioned, and returning to his own ship, immediately steered to the southward, deeply impressed with the awful example which he had just witnessed of the danger of navigating the Polar seas in high northern latitudes. On returning to England, he made various inquiries respecting vessels that had disappeared in an unknown way; and by comparing these results with the information which was afforded by the written documents in his possession, he ascertained the name and history of the imprisoned ship and of her unfortunate master, and found that she had been frozen in thirteen years previous to the time of his discovering her imprisoned in the ice."

One of the most successful Polar expeditions was that of the late Capt. Francis Hall, ship *Polaris*.

The Geographical Society of Paris voted, at its session, April 21, 1875, the biennial prize, a gold medal, devoted to Arctic explorations, to Capt. Hall. In giving a brief account of his explorations, we shall quote a few extracts from the report of the Geographical Society, both as a deserved tribute to the memory of Capt. Hall, and as helping to furnish the reader with a description of the voyages undertaken by him.

"The Prize Commission has before it several enterprises, which have had for their object either Smith's Sound, Eastern Greenland, Spitzbergen, or Nova Zembla. All of these deserve our tribute of praise; but especially that of the *Polaris*, the ship in which the American Francis Hall, passed beyond Smith's Sound and Kennedy's Channel, as far as 82° 16″—that is to say, the nearest to the pole that any vessel has reached *under sail*—has particularly commended the unanimous vote of the Commission."

Capt. Hall was a veteran in arctic explorations. In 1850 he was seized with the desire to take part in the expedition sent out in search for Franklin. Laying aside his graving-tools, he devoted all his leisure hours to the study of the polar regions of America. He designed taking part in the McClintock expedition, failing in which, he resolved to organize a new expedition. He succeeded in interesting in his enterprise, Mr. Henry Grinnell and other distinguished philanthropists; he left New London, Conn., in 1860, in the whale ship *George Henry*. The loss of his own boat prevented him from completing his expedition; but he satisfied himself, among other geographical determinations, that what on previous charts had been marked as Frobisher's straits is a long open bay, without any communication with the bay of Hudson.

On his return here, in 1862, he published the results of his researches, in a work entitled, "Life with the Esquimaux." In 1864, he returned to the Polar regions with his faithful companions, Joe and Hannah. The five succeeding years he spent in these regions in explorations. Sharing the daily life of this rude people, he made himself thoroughly acquainted with their language, customs and traditions, and thus was prepared on his return to this country, in 1869, for his great expedition to the Pole—the final object of all his efforts.

He busied himself very promptly in organizing it, appealing to Congress for assistance, and while awaiting its action, sustained himself and his dusky friends, by lectures upon his preceding voyages. He met with many hindrances, but finally obtained the use of a tug of 400 tons, which he admirably fitted up for its rough navigation in the ice, significantly naming her the *Polaris*.

The following is an extract quoted from a letter written by Capt. Hall, in 1869: "There is a great sad blot upon the present age, which ought to be wiped out, and this is the blank on our maps from about the parallel of 80° North up to the North Pole. I, for one, hang my head in shame, when I think how many thousands of years ago it was that God gave to man this beautiful world—the whole of it—to subdue; and yet that part of it which must be most interesting and glorious, at least to me, remains as unknown to us as though it had never been created. Neither glory nor money has caused me to devote my very life and soul to Arctic explorations."

The *Polaris* sailed from New London, July 3, 1871. Capt. Hall died, November 8, 1871. Capt. Budington then took charge of the expedition.

The voyage from this time on, and until most of its survivors providentially returned to their homes, is very sad, though full of heroic endurance. The sad tale has been read in most of our homes with moist eyes and aching hearts; how the *Polaris* left Thank God Harbor, drifted south and west, sprung a leak, requiring the most constant efforts to keep her from going down; how, on that terrible stormy night of October 15, 1872, it was thought the vessel must sink, and orders were given to take to the ice. Instruments, charts, boats, etc., were hurriedly transferred to the floe; but the drift changes its direction, the *Polaris* is released from her grim pressure, the floe parts assunder, and the vessel, breaking from her moorings, drifts away in the darkness and howling tempest, leaving Capt. Tyson and eighteen of the crew on the ice. "Several men were seen hurrying toward the ship as she was leaving, but they failed to reach her. The voice of the steward, John Herron, was heard calling out, 'Good-bye, *Polaris!*'"

We will not attempt to picture the consternation of the separated voyagers, nor try to describe their after adventures; suffice it to say, that most of them marvelously escaped the thousand dangers incidental to their perilous position.

In concluding this meager description of the *Polaris* expedition, we quote from the closing paragraphs of the report of the Paris Geographical Society:

"In consideration of these results, your Prize Commismission has judged it their duty to award to Captain Francis Hall, the promoter and chief of the Polaris expedition, that which is otherwise due him for his previous labors, the gold medal of the Roquette Foundation. But Francis Hall, like his fellow countryman Kane, seventeen years before him, has fallen a victim to his sufferings, and it is on a tomb that we must once more deposit a crown. If we are denied the gratification of giving to Francis Hall the medal which we have awarded him, we will have at least the consolation of transmitting it to his family. It will bear witness across the seas that death itself cannot prevent the just tribute of your gratitude for services rendered to geographical science. The Prize Commission awards this year the gold medal of the Roquette Foundation to the Arctic Explorer, Francis Hall, a medal which will be sent to the family of the unfortunate explorer."

The Jeannette Expedition.—On July 8, 1879, about ten years after Capt. Hall started out on his voyage of discovery with the *Polaris*, Mr. James Gordon Bennett sent out an Arctic expedition from San Francisco, commanded by Lieutenant De Long, U. S. N. The details respecting route and purpose were withheld from the public. The *Jeannette* was early caught in the ice drift, having no volition of her own, and being driven hither and thither by the mechanical impulse of the pack for nearly two years, when she was finally overwhelmed, and went down, June 11, 1881. The crew escaped in three boats, two of which reached the coast of Siberia, September, 1881. The deplorable news of the loss was received at New York in December of the same year. The boats were separated by a gale of wind before leaving the Arctic Ocean; the one containing Engineer Melville and others landed at the mouth of the Lena River. The boat containing De Long and his party landed a little north of that point. The cutter, in which was Lieutenant Chipp and the remainder of the ship's crew, has never been heard of since the separation. Melville's party found shelter and succor at Irkutsk. Two of Commander De Long's party succeeded in escaping, none of the rest ever being found alive. Lieutenant Danenhower, with a party of the *Jeannette* crew, reached New York in May, 1881, Engineer Melville staying behind to try to recover the lost bodies. When the sad tidings reached this country, great interest and pity was felt. Congress generously appropriated $175,000 to fit out a vessel with which to prosecute a diligent search for the lost parties. The *Rogers*, Commander Lieut. Berry, reached Wrangle Land, August 24, 1881. Three parties were immediately detailed from the ship to continue the search already so anxiously begun. On January 1, 1882, the *Rogers* was destroyed by fire, her crew taking refuge on the shore in hastily constructed tents and huts. As soon as the information reached Washington, the ship *Corwin* was immediately fitted out, and sent to pick up the officers and crew. Whatever is known respecting the awful suffering and unflinching adherence to duty of the *Jeannette* crew has been sufficiently detailed by the daily press. We need not further dwell upon it here. Owing to the unfortunate accident attending this expedition so early in its course, but few new geographical determinations or scientific knowledge was attainable. In any event, the result is not yet sufficiently a matter of written history to be judged without prejudice.

Every decade of modern history witnesses renewed attempts to draw aside the veil that seems to shroud the beginning of earth's distance in impenetrable mystery. Does the magic circle encompass vast treasures hoarded by nature, that God in His own good time will reveal to man for his admiration and use? Will the North Pole ever be discovered? Will such discovery explain the attraction of the magnetic needle, and tell us what *is* electricity? Will gravitation cease to be a law, and a new law take its place that will be the basis of new departures and inventions? These queries remain to be answered, and the solution of the enigma is engaging the best thought of the world.

CHAPTER III.

ICEBERGS.

"These are
The palaces of Nature, whose vast walls
Have pinnacled in clouds their snowy scalps,
And throned eternity in icy halls,
Of cold sublimity."

BYRON.

AMONG the most imposing and grand of the many wonders of the ocean world, are the fixed and floating icebergs, the "palaces of nature," which assume extraordinary and fantastic shapes, and more than realize the most sublime conceptions of the imagination. Well indeed may the mind become awe-struck and the heart almost cease to beat as the lips exclaim, "Wonderful Thou art in all Thy works! Heaven and earth are full of the majesty of Thy glory!" on beholding these mighty and surpassing works of the great Creator. East and west, north and south, the Arctic regions present a picture of grandeur and magnificence nowhere to be equalled—great beyond conception—impossible to be portrayed.

These icebergs are described by Arctic navigators as imitating every style of architecture on earth; cathedrals with pillars, arches, portals, and towering pinnacles, overhanging cliffs, the ruins of a marble city, palaces, pyramids, and obelisks; castles with towers, walls, bastions, fortifications, and bridges; a fleet of colossal men-of-war under full sail; trees, animals, and human beings: one is described as an enormous balloon lying on its side in a collapsed state.

A number of icebergs seen at the distance of a few miles presented the appearance of a mountainous country, deceiving the eyes of experienced mariners.

These icebergs differ somewhat in color, according to age, solidity, or the atmosphere. A very general appearance is that of cliffs of chalk, or of white-gray marble. A few have a blue or emerald-green tint. The sun's rays reflected from them give a glistening appearance to their surface, like that of silver. In the night, they are readily distinguished in the distance by their natural effulgence, and, in foggy weather, by a peculiar blackness of the atmosphere.

A writer thus describes the strange and sudden transformations and the changing tints of icebergs. "One resembled, at first, a cluster of Chinese buildings, then a Gothic cathedral of the early style. It was curious to see how all that mimicry of a grand religious pile was soon to change to another like the Coliseum, its vast interior now a delicate blue, and then a greenish white. It was only necessary to run on half a mile to find this icy theatre split asunder. An age of ruin seemed to have passed over it, leaving only to view the inner cliffs, one a glistening white, and the other blue, soft and airy as the July heavens." Another berg shone like polished silver, dripping with dews, the water streaming down in all directions in little rills and falls, glistening in the light like molten glass. Veins of gem-like transparency, blue as sapphire, crossed the mass.

"Solomon, in all his glory," was not clothed like the flowers of the field. Would you behold an iceberg appareled with a glory that eclipses all floral beauty, and makes you think not only of the clouds of heaven at sunrise and sunset, but of heaven itself, you must come to it at sunrise and sunset. Lofty ridges of the shape of flames have the tint of flames; out of the purity of the lily bloom the pink and the rose. We will not say cloth of gold drapes, but water of gold *washes*—water of green, orange, scarlet, crimson, and purple wash—the crags and steeps; strange metallic tints gleam in the shaggy caverns, copper, bronze, and gold: endless grace of form and outline.

These icebergs—so beautiful in summer, so grand and awful under a wintry aspect—project above the surface of the sea like high hills composed of rugged and steep rock. Navigators have frequently stated that they have seen them rising from four to five hundred feet above the water, and extending more than a mile in length. A Danish navigator examined an iceberg on the eastern coast of Greenland, and estimated its circuit, at its base, at four thousand feet. In height it was one hundred and twenty feet above the sea-level. He calculated that its contents amounted to upwards of nine millions of cubic feet.

The reader may be interested to know the origin of these stupendous floating bergs, whence they come, how they are formed, and their ultimate destination. It has been ascertained, beyond all doubt, that they originate in the land, being nothing more than fragments of glaciers—a name given to immense masses of ice, or appendages to snow mountains. By far the larger number of these are formed on the coast of Greenland. The mountains are always covered with snow; the valleys between them are filled with ice, derived from the higher portions of the mountains, and are thus converted into enormous glaciers. If the extent of all the shores of Greenland, in which the glaciers advance to the very sea, were put together, it is probable they would constitute a coast-line exceeding six hundred miles in length. These are the birth-places of the icebergs. The average height or depth of the ice at its free edge, or seaward, in these valleys is about twelve or fifteen hundred feet. As the glaciers advance farther into the sea, the rise and fall of the tide undermine the base, and enormous masses become detached and fall into the sea with a

AURORA BOREALIS—NORTHERN LIGHTS.

crash like thunder. The icebergs thus formed—vast moving mountains or islands—are drifted along, some finding their way to the North Atlantic—a distance of more than two thousand two hundred miles from the place of departure—brought down by a strong current which appears to originate under the immense masses of ice which surround the Arctic Pole.

Fearfully appalling are the dangers arising from these icebergs on their floating voyages, and we cannot wonder at the terror excited by their appearance among the early navigators among these ice-bound seas. In the expedition of Captain James Hall, under Danish auspices, for exploring Greenland, the sailors were in sight of the south point of that country, and, to avoid the ice, which encompassed the shore, they stood to the westward, and fell in with "mighty islands of ice, being very high, like huge mountains of ice, making a hideous and wonderful noise," and on one of them was observed "a huge rockstone of the weight of three hundred pounds or thereabouts." Finding nothing but ice and fog from the 1st to the 10th of June, the *Lion's* people hailed the admiral, "calling very fearfully, and desiring the pilot to alter his course, and return homeward." The alarm spread to the admiral's ship, and they had determined to put about, had not Cunningham (the captain) protested he would stand by the admiral, "as long as his bloode was warme, for the good of the Kinge's majestie." This pacified the seamen for a moment, but the next floating island of ice renewed the terrors of those on board the *Lion*, who, having fired a piece of ordnance, stood away to the southward.

All later voyagers in the Arctic Seas describe the sublimity of these moving mountains and islands of ice, and the fearful perils encountered among them. The following thrilling instance of hairbreadth escape is related: "It was awful to behold the immense icebergs, working their way to the northeast from us, and not one drop of water to be seen; they were working themselves right through the middle of the ice. The dreadful apprehensions that assailed us yesterday, by the near approach of the iceberg, were this day awfully realized. About three P. M. the iceberg came in contact with our floe, and in less than one minute it broke the ice we were frozen in quite close to the shore; the floe (similiar to field ice, but smaller, as its extent can be seen), was shivered to pieces for several miles, causing an explossion like an earthquake, or one hundred pieces of cannon fired at the same moment. The iceberg, with awful but majestic grandeur (in height and dimensions resembling a vast mountain), came almost to our stern, and every one expected it would have run over the ship. The intermediate space between the berg and the vessel was filled with heavy masses of ice, which, though they had been previously broken by the immense weight of the iceberg, were again formed into a solid body by its pressure. The iceberg was drifting at the rate of about four knots an hour,—and by its force on the mass of ice, was pushing the ship before it, and, as it seemed, to inevitable destruction. A gracious Providence ruled this otherwise: the iceberg, that so lately threatened destruction, was driven completely out of sight to the northeast."

It has been supposed that the unfortunate steamship the *President*, which left England for New York in 1841, was crushed to pieces between icebergs. In the year that this magnificent vessel was lost, the Atlantic Ocean was more thickly beset with icebergs, and at an earlier season, than commonly occurs. This is ascertained from a report of the *Great Western* steamer, which was published in New York. This vessel left England about the middle of April in the same year, and encountered an ice-field, which extended far more than a hundred miles, and along the southern edge of which she proceeded. This edge was lined by

BREAKING UP OF ICEBERGS.

a broad border of loose ice, consisting of numerous floes and icebergs, and a considerable quantity of floating ice. To make way between these masses, the steamer was compelled frequently to change her course, for fear of coming in contact with them. The number of icebergs which were in sight of the vessel amounted to three hundred, and the largest was three-fourths of a mile long, and about a hundred feet high. A similar calamity to that which is supposed to have befallen the *President* is said to have well-nigh occurred to the brig *Anne*, of Poole, which, in a voyage from Newfoundland to England, was so completely beset by ice that no means of escape were visible. The ice in its whole extent rose fourteen feet above the surface of the water. It drifted toward the southeast, and bore the ship along with it for twenty-nine successive days. An opening most providentially occurred, by which the vessel became disengaged.

The *President* in 1841, the *City of Glasgow* in 1854, the *Pacific* in 1856, and, later, the *City of Boston*, have disappeared, from, it is supposed, their contact with icebergs.

Captain Ross draws a vivid picture of what a vessel is exposed to in sailing amidst these moving hills. He reminds his readers that ice is like stone, as solid as if it were granite, and he bids them "imagine these mountains hurled through a narrow strait at a rapid rate, meeting with the noise of thunder, breaking from each other's precipices huge fragments, or rending each other asunder, until, losing their former equilibrium, they fall over headlong, lifting the sea around in breakers, and whirling it in eddies. There is not a moment in which it can be conjectured what will happen in the next; there is not one which may not be the last."

It is generally found that a strong current runs along the sides of an iceberg, and a vessel approaching too near is viovently forced against the mass, and dashed to pieces.

Another source of danger arises from mooring vessels to

icebergs, which is frequently done for shelter in strong adverse winds, or when the vessel is rendered unmanageable by the accumulation of drift-ice around; but there is this danger: the icebergs are very nicely poised; if a large piece of ice breaks off from one side, the whole mass is suddenly and rapidly turned over, by which vessels have often been wrecked or destroyed, while boats have been upset, even at a considerable distance, by the vast waves produced by the sudden change of position of an iceberg.

An incident is related of two sailors who were attempting to fix an anchor to an iceberg. They began to hew a hole in the ice, but scarcely had the first blow been struck, when suddenly the immense mass split from top to bottom, and fell asunder, the two halves falling in contrary directions with a prodigious crash. Fortunately the men escaped.

Sometimes vessels moor to icebergs when in want of water, and obtain it from the deep pools which, in the summer season, are found on the depressed surface of some bergs, or from the streams running down their sides; but if, meanwhile, the iceberg should fall to pieces, which is likely at any moment during the summer season to be the case, the vessel is liable to be buried under its icy mooring. The precarious character of these huge mountains of ice will be understood from an anecdote related by Dr. Hayes, the Arctic navigator: "A few years ago, while a French man-of-war was lying at anchor in Temple Bay, Labrador, the younger officers resolved on amusing themselves upon an iceberg a mile or more distant in the straits. They made sumptuous preparations for a picnic upon the very top of it, the mysteries of which they were curious to see. All warnings of the fishermen in the ears of the smartly-dressed gentlemen who 'had seen the world,' were useless. It was a bright summer morning, and the jolly-boat with a showy flag went off to the iceberg. By twelve o'clock the colors were flying from the icy turrets, and the wild young midshipmen were shouting from its walls. For two hours or so they hacked and clambered the crystal palace, frolicked and feasted, drank toasts to the King and the ladies, and laughed at the thought of peril where all seemed so fixed and solid. As if in amazement of such rashness, the grim Alp of the sea made neither sound nor motion. A profound stillness reigned on its shining pinnacles and in the blue shadows of its caves. When the youngsters, like thoughtless children, had played themselves weary, they went down to their boat. As if the time and distance were measured, they were scarcely out of harm's way when the mighty iceberg collapsed and broke into a myriad fragments, which filled the surrounding waters. This was, no doubt, the first and last day of amusement on an ice-berg by the daring young seamen."

Icebergs are not affected by the swell of the sea, which breaks up the largest fields of ice in the space of a few hours; they rise and fall with a tremendous noise, though their size and form remain the same. But, when acted upon by the sun or a temperate atmosphere, they become hollow and fragile. Few icebergs are destroyed in the Northern seas; a large number get as far as the great banks of Newfoundland, which is occasionally crowded with them.

The fields of ice that float in the Polar Seas are often twenty or thirty miles in diameter, and some hundreds of feet in thickness. It is calculated that upwards of twenty-thousand square miles of drifting ice come down every year along the coast of Greenland into the Atlantic, moving on during the winter at the rate of about five or six miles a day. The *Resolute* exploring ship, which was abandoned in Melville's Straits, on account of its being enclosed firmly in a vast field of ice, was afterwards found in Baffin's Bay, having been carried a thousand miles from its former position by the drift of an icefield three hundred thousand square miles in extent and seven feet thick. This will give an idea of the quantity of ice which is carried out of the Polar regions, independent of the icebergs, and drifted into warmer climates.

The formation and destruction of ice within the Arctic Circle is a beautiful provision of Nature for adjusting the inequality of temperature. Had only dry land been thus exposed to the sun, it would, in summer, have been actually scorched by its beams, yet severely pinched during the darkness of the winter by the most intense and penetrating cold. None of the animal or vegetable tribes could have supported such extremes. But in the actual arrangement the surplus heat of summer is spent in melting away the ice. As long as ice remains to thaw or water to freeze, the temperature of the atmosphere can never vary beyond certain limits.

CHAPTER IV.

LIFE IN THE OCEAN.

"THE appearance of the open sea," says Fridol, "far from the shore—the boundless ocean—is to the man who loves to create a world of his own, in which he can freely exercise his thoughts, filled with sublime ideas of the Infinite. His searching eye rests upon the far distant horizon. He sees there the ocean and the heavens, meeting in a vapory outline, where the stars ascend and descend, appear and disappear in their turn. Presently this everlasting change in Nature awakens in him a vague feeling of that sadness, which, says Humboldt, 'lies at the root of all our heartfelt joys.'"

Emotions of another kind are produced by the contemplation and study of the habits of the innumerable organized beings which inhabit this great deep. In fact, that immense expanse of water which we call the sea, is no vast liquid desert; light dwells on its bosom as it does on that of dry land. Here this mystery of life reigns supreme. It is among the most beautiful, the most noble, and the most incomprehensible of His manifestations. Without life, the world would be as nothing. All the beings endowed with it transmit it faithfully to other beings, they, again, to their successors, which will be, like them, the depositories of the same mysterious gift; the marvelous heritage thus traverses years and hundreds of years without losing its powers; the globe is teeming with the life which has been so bounteously distributed over it.

In every living being there are two powers, between which a silent but incessant combat is being carried on—*life*, which builds up; and *death*, which pulls asunder. At first, life is all powerful—it lords it over matter; but its reign is limited.

Beyond a certain point, its physical vigor becomes gradually impaired; with old age, it feebly struggles; and it is finally extinguished with time, when the chemical and physical laws seize upon it, and its organization is destroyed. But, in turn, the very elements, though inert at first, are soon reanimated and occupied with new life. Every plant, every animal, is bound up with the past, and is a part of the future; for every generation which starts into life is only the corollary upon that which is about to be born. Life is the school of death; death is the foster-mother of life.

Life, however, does not always exhibit itself at the actual moment of its formation. It is visible later, and only after other phenomena. In order to develop itself, a suitable medium must be prepared, and other determinate physical and chemical conditions provided.

If we expose a quantity of pure water to the light and air, in the spring-time or summer season, we would soon see it producing minute spots of a yellowish or greenish color. These spots, examined through the microscope, reveal thousands of vegetable forms. Presently thousands of Rhizapods and Infusoria appear, which move and swim about the floating vegetable forms upon which they nourish themselves. Other infusoria then appear, which, in their turn, pursue and devour the first.

In short, life transfers unorganized into organized matter. Vegetables appear first, then come herbivorous animals, and then come the carnivorous. Life maintains life. The death of one provides food and development to others; for all are bound up together, all assist at the metamorphosis continually occurring in the organic as in the inorganic world, the result being general and profound harmony—harmony always worthy of admiration. The Creator alone is unchangeable, omnipotent, and permanent; all else is transition.

The inhabitants of the water are at least as numerous as those of the solid earth. "Upon a surface less varied than we find on continents," says Humboldt, "the sea contains in its bosom an exhuberance of life, of which no other portion of the globe could give us any idea. It expands in the north as in the south; in the east as in the west. The seas, above all, abound with this life; in the bosom of the deep, creatures corresponding and harmonizing with each other sport and play. Among these the naturalist finds instruction, and the philosopher subject for meditation. The changes they undergo only impress upon our minds more and more a sentiment of thankfulness to the Author of the universe."

Yes, the ocean, in its profoundest depths—its plains and its mountains, its valleys, its precipices—is animated and beautified by the presence of innumerable organized beings. Among these we find the Algæ, solitary or social, erect or drooping, spreading into prairies, grouped in patches, or forming vast forests in the ocean valleys. These submarine forests protect and nourish millions of animals, which creep, which run, which swim among them; others, again, sink into the sands, attach themselves to rocks, or lodge themselves in their crevices; these construct dwellings for themselves; they seek or fly from each other; they pursue or fight, caress each other lovingly, or devour each other without pity. Our terrestial forests do not maintain nearly as many living beings as those which swarm in the bosom of the sea.

The sea influences its numerous inhabitants, animal or vegetable, by its temperature, by its density, by its saltness, by its bitterness, by the never-ceasing agitation of its waves, and by the rapidity of its currents.

When the tide retires from the shore, the sea leaves upon

the coast some few of the numberless beings which it carries in its bosom. In the first moments of its retreat, the naturalist may collect a crowd of substances, vegetable and animal, of various characteristic color and properties. The inhabitants of the coast may find there their food, their commerce, their occupations.

At low water, the nearest villages and hamlets send their contingents, old and young, to gather the riband seaweed, a source of great wealth to the dwellers by the sea, being much used in making kelp; others gather the small shells left on the sand; boys mount upon the rocks in search of whelks and of mussels, and detach limpets from the rocks to which they attach themselves.

On some coasts, shells are sought for their beauty. By turning the stones, or by sounding the crevices of the rocks with a hook at the end of a pole, cuttles and calmars are sometimes surprised, sometimes even a young conger eel which has sought refuge there; while the pools, left here and there by the retiring tide, are dragged by nets of very small mesh, in which the smaller crustacea, mollusks, and small fish are secured.

In the Mediterranean and other inland seas, where the tide is almost inappreciable, there will be found to exist a great number of animals and Algæ belonging to the deep sea, which the waves or currents very rarely leave upon the sea-shore. There are others again so fugitive, or which attach themselves so firmly to the rocks, that we can watch them only in their habitats. It is necessary to study them, floating on the surface of the waves, or in their mysterious retirements.

" We find in the sea," says Lacepede, " unity and diversity, which constitutes its beauty; grandeur and simplicity, which give it sublimity; puissance and immensity, which command our wonder."

SUB-MARINE SCENERY.

CHAPTER V.

MINUTE ANIMAL LIFE.

"Oh, what an endless work hath he in hand
Who'd count the sea's abundant progeny;
Whose fruitful seed far passeth that on land,
And also them that roam the azure sky,
So fertile be the floods in generation,
So vast their numbers, and so numberless their nation."
—SPENCER.

TRUE and just are the words of the British poet; though the surface of the ocean is less rich in animal and vegetable forms than that of continents, still, when its depths are searched, perhaps no other portion of our planet presents such fullness of organic life. It has been said that our land forests do not harbor so many animals as the low-wooded regions of the ocean, where the sea-weeds, rooted to the shoals, or long branches detached by the force of waves and currents, and swimming free, upborne by air-cells, unfold their delicate foliage. The microscope still further increases our impression of the profusion of organic life which pervades the recesses of the ocean, since throughout its mass we find animal existence, and at depths exceeding the height of our loftiest mountain chains. Here swarm countless hosts of minute animals, which, when attracted to the surface by particular conditions of weather, convert every wave into a crest of light. The abundance of these minute creatures, and of the animal matter supplied by their rapid decomposition, is such, that the sea-water itself becomes a nutritious fluid to many of the large inhabitants of the ocean.

Even in the bleak and dreary regions of the Northern world the wintry seas are filled with a profusion of animal life. The smaller species, of which the herring may be taken for an example, are found amidst the depths of the Arctic zone in immense shoals; countless millions of creatures, sometimes known as sea nettles, a genus of *Acalephæ*, signifying "nettles" (so named from the stinging power which many of them possess), of higher organization than the *Medusæ*, or jelly-fish, exist here, with globular or oval bodies of a delicate or jelly-like substance, strengthened by bands which are covered with rows of large *cilia* (a peculiar sort of moving organs resembling microscopic hairs), the motion of which is extremely rapid, and is evidently controlled by the will of the little animal. *Jelly-Fish*, *Zoophytes*, etc., swarm also to such an extent as to convert the surface water in some places almost into a kind of soup, which furnishes food not only to small fish, but to whales and animals of the largest growth. Even the color of the ocean is influenced by the enormous quantity of the organic life it sustains. The application of the microscope—for by far the most numerous of the animalculæ can only thus be traced—shows them to be the cause of a peculiar tinge observed over a great extent of the Greenland Sea. This color is olive-green, and the water is dark and dense compared to that which bears the common cerulean hue. The portion of the ocean so distinguished amounts to not less than twenty thousand square miles, and hence the number of animalculæ which that space contains is far beyond human calculation.

Some of the calculations of an ingenious and clever writer are very curious and instructive. In a drop of water there were fifty of these animalculæ, on an average, in each square of the micrometer-glass of an eight hundred and fortieth of an inch; and as the drop occupied a circle on a plate of glass containing five hundred and twenty-nine of these squares, there must have been in this single drop of water—taken out of the yellowish-green sea, in a place by no means the most discolored—about twenty-six thousand four hundred and fifty of these animalculæ! Hence, reckoning sixty drops to a dram, there would be a number in a gallon of water exceeding, by one-half, the population of the whole globe! It gives a wonderful conception of the minuteness and vastness of creation, when we think of more than twenty-six thousand animals—living, obtaining subsistence, and moving perfectly at their ease, without annoyance to one another—*in a single drop of water!*

The diameter of the largest of these animalculæ was only the two-thousandth part of an inch, and many only the four-thousandth. The army which Bonaparte led into Russia in 1812, estimated at five hundred thousand men, would have extended—in a double row, or two men abreast, with two feet three inches space for each couple of men—a distance of one hundred and six and a half English miles; the same number of these animalculæ, arrayed in a similar way in two rows, but touching one another, would only reach *five feet two and a half inches!* A whale requires an ocean to sport in, but about one hundred and fifty millions of these animalculæ would have abundant room in a tumbler of water! What a stupendous idea is thus afforded of the immensity of creation, and of the bounty of Divine Providence, in furnishing such a profusion of life in regions so remote from the habitations of men! Even if we consider the number of animals in a space of two miles square as great, what must be the amount requisite for the discoloration of the sea through an extent of, perhaps, twenty or thirty thousand square miles!

If we turn from the Arctic seas to the warmer regions of the ocean, we find the same wonderful profusion of animal life existing in minute forms of infinite variety: small Molluska (soft animals inhabiting shells); Crustacea (with articulated limbs and hard coverings), and luminous creatures, as *Salpæ*, of which vast gelatinous shoals are met with at sea, associated in a round mass like a chain, transparent, and of beautiful colors, of which, we are told, that during a journey of nearly eight hundred miles, they were thickly abundant throughout the track of the ship in the ocean. Each portion of the vast masses of floating seaweed consists—when carefully examined—of a little densely populated world, being crowded with living beings, all active and full of bustling animation—strange-shaped little fishes, bright sea-slugs, tiny shells of the nautilus tribe, grotesque sea-spiders, and whole gangs of odd crabs, jelly-fish, and transparent shrimps.

"The number of living creatures of all orders," observes Darwin, "whose existence intimately depends on the kelp (marine plants) is wonderful. A great volume might be written describing the inhabitants of one of these beds of seeweed. Almost all the leaves, excepting those on the surface, are so thickly encrusted with corallines as to be of a white color. We find exquisitely delicate structures, some inhabited by simple hydra-like *Polypi*, others by more organized kinds and beautiful compound *Ascidiæ* (from the Greek *askos*, a bottle or pouch, these little molluscs resembling sacs everywhere closed, except at two orifices.) Innumerable crustacea frequent every part of the plant. On shaking the great entangled roots, a pile of small fish, shells, cuttle-fish, crabs of all orders, sea-eggs, star-fish, and animals of a multitude of forms all fall out together. Often as I recurred to a branch of the kelp, I never failed to discover animals of new and curious structures. I can only compare these great aquatic forests of the Southern Hemisphere with the terrestrial ones in the intertropical regions. Yet if in any country a forest were destroyed, I do not believe nearly so many species of animals would perish as would here from the destruction of the kelp. Amidst the leaves of this plant numerous species of fish live, which nowhere else could find food or shelter; with their destruction, the many cormorants and other fishing birds, the otters, seals, and porpoises, would soon perish also."

How elevating is the thought that amidst all this prodigious variety and profusion, the boundless extent of which no human mind can conceive, yet the minutest animated particle that is revealed by the microscope is governed by the same laws that regulate the highest objects in creation!

> "Each moss,
> Each shell, each crawling insect, holds a rank
> Important in the scale of Him who framed
> This scale of beings; holds a rank which, lost,
> Would break the chain, and leave a gap behind,
> Which Nature's self would rue."

Very interesting is the study of those curious inhabitants of the ocean, constituting what are termed by naturalists *Acalephæ*, as has been previously mentioned, but which are more commonly known by such names as jelly-fish, sea-blubber, etc., and are sometimes called sea-nettles. Most of them were included in the Linnæan genus Medusa, and the name *Medusæ* is still frequently applied to them. They abound in all parts of the ocean, although some are tropical and others belong to cold latitudes. Some are of a large size, reaching two feet in diameter, and others are very small. They are of an extremely soft jelly tissue, which in most of them, and all in the true Medusæ, is unsupported by any harder substance. The latter comprise various species that shine with great splendor in the water. The South Atlantic abounds with them, and much amusement may be derived in a long sea voyage by observing these beautiful organisms, for endless are the moulds in which prolific Nature has cast them. Some are shaped like a mushroom, others are like ribbons, or globular, flat or bell-shaped; others again resemble a bunch of berries. Their motions are generally slow, their sensations dull and directed

FLEET OF MEDUSÆ.

entirely to the procuring of food. They often float without any apparent animation, trusting in the winds and waves to waft them about, and to carry them their food; some keep a little beneath the surface, and propel themselves by contracting their pellucid disks. They have been termed the "living jellies of the deep," and some are endowed with an acrid secretion, which irritates the skin, and has thus caused them to be termed sea-nettles.

> "Those living jellies which the flesh inflame,
> Fierce as a nettle, and from that the name;
> Some in huge masses, some that you may bring
> In the small compass of a lady's ring.
>
> Figured by hand Divine—there's not a gem
> Wrought by man's art to be compared to them;
> Soft, brilliant, tender, through the wave they glow,
> And make the moonbeam brighter where they flow."

There is one large species common in the Straits of Singapore dreaded by the Malays on account of the violence of this power.

Sometimes these animals are colorless, and as transparent as crystal; others are embellished with the most brilliant hues, and seem as if adorned with the richest enamel. Stevens, one of the first voyagers to the East Indies, describes the jelly-fish he saw in the Gulf of Guinea as "a thing swimming on the water, like a cock's comb, but the color much fairer, which comb standeth upon a thing almost like the swimmer of a fish in color and bigness."

Another curious and widely-distributed class of marine animals are the *Annelides* or *Sea-Worms*, the bodies being composed of rings and joints. Some species are only met with in the high seas, swimming freely, while most of the others are to be found on the sea-shore, burrowing in the sand or mud, or living under stones, or amidst seaweed. A few construct a sheath or case for themselves, in which they ordinarily live, but which are not essential to the existence of the tenant, as they can leave it without inconvenience, and wander at liberty for their food elsewhere. Their bodies are formed of more or less numerous rings, each of which is furnished with feet, which are the chief organs of motion, and are truly wonderful. They are generally in the form of small tubercles, and for the most part are composed of two branches. Their summit or tip is frequently armed with one or more bundles of bristles, which play an important part in the history of the animals. They form an ornamental appendage to the worm, and at the samo time are used as organs of defence and offence. Notwithstanding they live in situations in which they are seldom seen by the human eye, yet in some species these organs have a remarkable degree of brilliancy, shining with a metallic lustre and splendor of the richest kind. The common *Sea-Mouse,* for instance, has a very large bundle of them attached to each foot, which are very fine and of considerable length. Gold, azure, purple, and green play on their surface in a thousand reflections, and these rainbow colors are in perfect harmony with the changing reflections and rings of the body. The wing of the butterfly has not received a more brilliant dress than these worms, concealed at the bottom of the waters, and sometimes buried in black and fetid mud. They are brilliant as gold, and changeable to every hue of the rainbow. The colors they present are not surpassed in beauty by the scale-like feathers of the humming-bird nor by the most brilliant gems. These bristles, however, are as useful as they are ornamental. Surrounded on every side by enemies, usually dwelling in the waters where the worms live, they require powerful weapons of offence for resistance or for securing their prey.

Some species of these worms are armed with a weapon like a harpoon, a lancet, or a knife. Some have an appendage, falchion-shaped, and others a bayonet fixed upon a musket, while others present the appearance of a barbed arrow. These weapons are used to pierce the bodies of their enemies, and they frequently leave them in the wounds they have made. The tubercles of the feet, from which the barbed arrow-shaped bristles spring, are, in reality, quivers full of arrows, stored there for the use of the animals to protect them from violence; or, as Gosse fancifully observes, "You may imagine you behold the armory of some belligerent sea-fairy, with stores of arms enough to accouter a numerous host."

The number of such-like weapons in these worms is immense. "Let me ask the naturalist," says Dr. Johnson, "to count the number which may be required to furnish the garniture of a single individual. There are worms which have five hundred feet on each side: each foot has two branches, and each branch has at least one spine and one brush of bristles, some of them simple, some of them compound. This individual has therefore two thousand spines at least, and if we reckon ten bristles to each brush, it has also twenty thousand of them! Let us look a little further, not merely to the exquisite finish of each bristle, but to the means by which the host is put in motion. There is a set of muscles to push them forth from their port-holes; there is another to replace each and all of them within their proper cases; and the uncounted crowds of these muscles neither twist nor knot together, but play in their courses, regulated by a will that controls them more effectually than any brace; now spurring them to convulsive energy, now stilling them to rest, and then putting them into action with an ease and grace that charm us into admiration, and fix the belief that even these creeping things participate largely in the happiness diffused throughout creation!"

The *Nereids*, which belong to the same class of sea-worms, have a long body, narrowed towards the inferior extremity, and divided into numerous segments, with well-developed appendages, a head, eyes, horns or feelers, and, in general, a large proboscis, armed with a pair of jaws, curved, hooked, and strong, with teeth on the inner margin. The *Pearly Nereis,* which is one of the finest and commonest of the kind, is thus described: "The upper surface is of a warm fawn brown, but the beautiful flashes of rainbow blue that play on it in the changing light, and the exquisite pearly opalescence of the delicate pink beneath, are so conspicuous as to have secured for it the title of 'pearly' *par excellence.*"

Another species of the group of the Nereids, the "*White-Rag Worm,*" a common inhabitant of the shores of Great Britain, varying from six to ten inches in length, is of a beautiful pearly lustre, exactly similar to that of mother-of-pearl. The foot, when magnified, resembles a horse's hoof, and is a very marvelous piece of Nature's mechanism. The animal swims rapidly in the sea. Another species is of a rich greenish color, varied with bluish shades, reflecting a metallic lustre, and varying like the hues of the rainbow.

With the tribe of sea-worms may be also mentioned the *Sea-Leach* or Skate-sucker, so named because the worm lives on fish, and attaches itself chiefly to the skate, from which it is scarcely ever found free. The mouth of this animal is not provided with jaws, so it sucks up the juices of the body of its host by a kind of pumping process.

The *Leaping-Worms,* found on the coasts of Borneo, are curious creatures. Each step in advance to take them causes them to jump in a rapid manner, and in a series of leaps they reach the margin of the water, when it is impossible to capture them. When lying at rest, they are scarcely distinguishable from the mud in which they lie. They are wedge-shape in form, about three or four inches long, with flat pointed tails, and broad heads and prominent eyes. The sailors have nicknamed them "Jumping Johnnies."

CHAPTER VI.

CORAL—THE ROCK BUILDERS.

NO art can imitate the delicate tracery, the rich color, and the singular forms that coral assumes. It has been called by some, the "Queen of the Ocean," and no term could be more appropriate. A celebrated naturalist, on viewing the coral-beds of the Red Sea, exclaimed, "Where is the Paradise of flowers that can rival such variety and beauty?"

Mr. J. Beete Jukes, in giving his own vivid impressions on seeing some coral-beds in the Pacific, says:

"I had hitherto been rather disappointed by the aspect of the coral-reefs, so far as beauty was concerned; and, though very wonderful, I had not seen in them much to admire. One day, however, on the lee side of one of the outer reefs, I had reason to change my opinion. In a small bay of the inner edge of the reef was a sheltered nook, where the extreme slope was well exposed, and where every coral was in full life and luxuriance."

Mr. Jukes describes them as of every shape; some delicate and leaf-like, others with large branching stems, and others, again, exhibiting an assemblage of interlacing twigs of the most delicate and exquisite workmanship. Their colors were unrivaled, vivid greens contrasting with more sober browns and yellows, mingled with rich shades of purple, from pale pink to deep blue. Among the branches, covered with their beautiful drapery of ocean vegetation, floated fish of various colors, radiant with metallic green or crimson, or fantastically banded with yellow and black stripes. Patches of clear white sand were seen here and there for the floor, with dark hollows and recesses. All these, seen through the clear crystal water, the ripple of which gave motion and quick play of light and shadow to the whole, formed a scene of rarest beauty, and left nothing to be desired by the eye, either in elegance of form, or brilliancy and harmony of coloring.

It is only in the ocean, however, that the glorious homes of the rock-builders are to be seen in perfection; for, immediately after drawing the coral from the water, so rapidly does atmospheric exposure affect them, that it would be difficult to recognize the lovely objects which a moment before were glowing in the still waters.

Such are the grand and mysterious operations of Providence in the depths of the ocean! We will now attempt to describe the singular animals to whom the accomplishment of these marvels is due, but must first mention that coral was formerly supposed to be a marine plant. This ancient notion rested not merely on its shrub-like form, but from the circumstance that its branches are covered with a soft coating while in the water, but which dries up immediately on its extraction. An Italian naturalist perceived small objects in the coral-cells, which he thought were flowers; but at length a French physician at Marseilles discovered that there was life in the coral, and that these assumed flowers were in reality minute animals. Thus, by the aid of the microscope, an object which might be said to belong to mineralogy, and by its trunk and branches to botany, was now admitted to a rank in the animal world. This discovery, the result of thirty years' studious research into the nature of coral, was laughed at by many persons at the time and treated as absurd, but Linnæus, the great Swedish naturalist, saw the truth at once, and did not hesitate to place coral at the head of the zoophytes, or animal plants, an appropriate designation, because it indicates at the same time the double nature of the substances.

A common characteristic of these animals is that their mouths are surrounded by radiating tentacles or feelers, appendages by which they attach themselves to surrounding objects, arranged somewhat like the rays of a flower. By this will be understood the term *polypi*, by which these animals are also known, signifying "many" and "foot." Of these the individuals of a few families are separate and perfect in themselves, but the greater number of zoophytes are compound beings, or each zoophyte consists of an indefinite number of individuals, or polyps, connected together.

This polyp is an extraordinary creature, and has a tenacity of life truly remarkable. If one cut off the branch of a tree, or sever the limb of an animal, these parts will wither and decompose by passing into other parts of matter. Cut a tree carelessly, and its natural symmetry is disfigured; or slit it down its centre, it is destroyed. Animals thus treated die, with the exception of the polyp, for it will put forth new limbs, form a new head or tail, and, if divided, become two separate existences.

If a polyp be cut in two, the fore part, which contains the head and mouth and arms, lengthens itself, creeps, and eats on the same day. The tail part forms a new head and mouth; at the wounded end shoot forth arms; if turned inside out the parts at once accommodate themselves to these new conditions. If the body were cut into ten pieces, every portion would become a new perfect living animal. A polyp has been cut lengthways at seven in the morning, and in eight hours afterwards, each part had devoured a worm as long as itself! How astonishing it is to see a creature so apparently frail in structure, possessing the actions, sensations, and powers of higher organized beings! The stomach is without membrane or cell; the outside surface-cells form a kind of double skin, and the inside consists of a wall of cells running crosswise, with a velvet-like surface, being red or brown grains held together by a sort of gluey substance.

These minute builders of the ocean rocks make their habitations, and form the wonderful coral groves and islands, sometimes hundreds of miles in extent.

The various species of these animals appear to be furnished with glands containing gluten, converting the carbonate of lime which is in the ocean, and other earthy matters, into a fixed and hard substance, twisted—as may be observed in coral—in every variety of shape.

If a piece of coral be examined with the microscope, it will be seen to be covered with a multitude of small pits, which are cells of the most beautiful construction, made with the greatest regularity, and in such a manner that the most experienced builder would pronounce faultless. How this is effected, and what peculiar instincts the little toilers of the ocean possess that enable them to construct their dwellings with s ıch mathematical nicety, are among those mysteries of Nature we cannot comprehend ; but it is certain that large masses of solid rock are framed by these animals, ever working to the music of the waves. "Verily," observes Baker, "for my own part, the more I look into Nature's works, the sooner I am inclined to believe of her, even those things that seem incredible." But here we have the *certainty* of Natures operations: we know that islands and continents are constructed for the habitation of man by these minute animals ; that mountains like the Appenines, and regions to which our own country is but trifling in comparison, are the results of their toil. South-west of Malabar, there is a chain of reefs and islets of coral extending four hundred and eighty geographical miles; on the east side of New Holland are unbroken reefs of three hundred and fifty miles long; and between that and New Guinea, a coral formation of seven hundred miles in length.

The process by which these great changes are effected is still going on extensively in the Pacific and Indian Seas, where multitudes of coral islands emerge from the waves, and shoals and reefs, where the rock builders are ever busy, appear at small depths beneath the water.

How truly wonderful it is to know that the Polynesian Archipelago, now one of the great divisions of the globe, has its foundations formed of coral reefs, the spontaneous growth of once living animals! As one generation of the coral-builders dies and leaves its chalky remains, another succeeds, until the mass of coral appears above the ocean, when the formation ceases, for it is only in that element the laborers can live.

"Ye build ! ye build ! but ye enter not in,
Like the tribes whom the desert devoured in their sin ;

From the land of promise ye fade and die,
Ere its verdure gleams on your wearied eye."

One marvel ceases here, and another commences. The vegetation of the sea, cast on its surface, undergoes a chemical change; the rains assist in filling up the little cells of the dead animals; the fowls of the air and the ocean find a resting-place, and assist in clothing the rocks; mosses carpet the surface; seed brought by birds, plants carried by the oceanic current, animaculæ floating in the air live, propagate and die, and are succeeded through the assistance their remains bestow by more advanced animal and vegetable life; and thus generation after generation exist and perish, until at length the coral island becomes a Paradise, filled with the choicest exotics, the most beautiful birds and delicious fruits.

Here is a glowing theme for the imagination to dwell upon! How wonderful to think that the surface of the globe is being changed by these diminutive living agents; that in tropical climates they are encircling islands with belts of coral, enlarging their coasts, forming stupendous reefs, and working out the plans and the will of the great Architect of the Universe!

We feel surprised, when travelers tell us of the vast dimensions of the Pyramids and other great ruins; but how utterly insignificant are the greatest of these when compared to the mountains of stone accumulated by the agency of various minute and tender animals!

How wonderful is the instinct and design of self-preservation in insects so exceedingly minute as the coral workers or ocean rock builders! POPE graphically says:

"Who taught the natives of the field and wood
To shun their poison or to choose their food?
Prescient, the tides and tempests to withstand,
Build on the wave, or arch beneath the sand?"

To protect their dwellings from the violent storms by which the waters of the deep are frequently agitated, they erect a breastwork, which effectually shields them from wind and wave. In the early stages of their operations they work perpendicularly, so that the highest part of the coral wall, on reaching the surface, is on the windward side, and affords protection to the busy laborers in their operations. These breastworks, or breakwaters, will resist more powerful seas than if formed of granite, rising as they do frequently from a depth of a thousand or fifteen hundred feet, and adapted in a way that no human skill or foresight could equal to the utmost powers of the heavy billows that continually lash against them.

Another observation we may make on this subject, is, that in one species a remarkable arrangement is found; the upper openings of the cells in which they live, have a vase-like form, shutting with a lid: when the animal wishes to expand itself, it opens the lid like a trap-door, and protrudes itself; and when it re-contracts itself and retreats, the lid falls and closes the aperture so exactly that the animal is perfectly protected.

Coral differs in quality and color. The common Red Coral which is used for many ornamental purposes, and is so much admired for its fine color, is chiefly obtained from the Mediterranean, in some parts of which extensive "fisheries" are carried on. It is brought up from the depths of the sea by means of a kind of grappling apparatus dragged after a boat, the pieces being broken from the bottom by beams of wood which are sunk by weights, and then entangled among hemp. Great care is necessary to preserve the pieces from being lacerated. Red coral has a shrub-like branching form, and grows to the height of about a foot, with the thickness of a little finger. Much of the coral obtained from the Mediterranean is sent to India, where it is much prized by the natives. Many of the arms and horse-caparisons of the Oriental chiefs are studded with this beautiful ornament.

Red coral is also found in the Red Sea, the Persian Gulf, Messina, the Dardanelles, and a few other places. The French and the Sicilians are the only people who make coral-fishing a regular source of interest. As the precious substance requires eight or ten years to come to any perfection by the labors of its industrious architects, the spots where it is finished are divided each into ten portions, and only one of these is finished in the year, so that each may remain to "grow" during the time necessary to bring it to maturity.

Black Coral is most esteemed, but it is scarce: the red, white, and yellow are chiefly used for ornamental purposes. The Pink Coral is esteemed for its scarcity.

The ingenuity of man continually exerted to imitate nature, and frequently with great success, is practised in the fabrication of false coral, made with powdered marble and fish-glue, and colored with vermilion and red lead.

Coral beads were anciently worn in India, as sacred amulets or charms. The Romans tied little branches round children's necks to keep off the influence of the "evil eye," a superstition which had also many believers in the middle ages among the inhabitants of England, and which still exists in some foreign countries.

Coral was said to preserve houses from the effects of thunder storms, and to be of much finer color when worn by men than by women. Even at the present time there are people so credulous as to believe that coral necklaces become pale when the wearer is about to be ill. There is no doubt that coral loses its color by time and exposure, and this may have given rise to the superstition. The small pointed branches, mounted with a ring at one end for suspension, are extensively manufactured at Naples as "charms;" and Ferdinand I., King of that country, was a devout believer in their efficacy, and used to point the coral towards anyone whom he suspected of having a malicious influence.

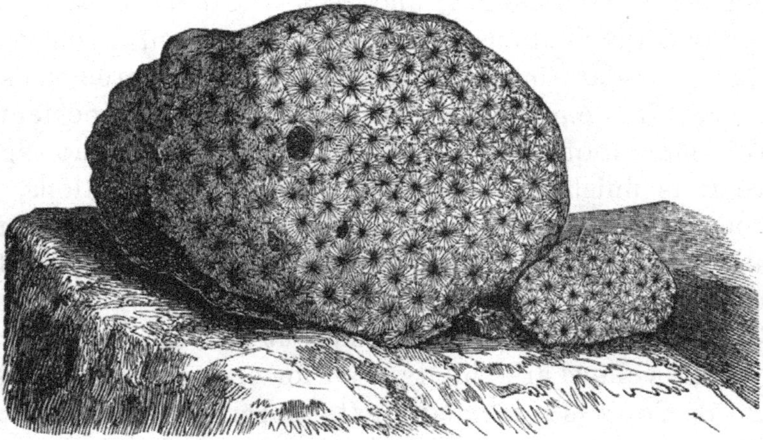

The vast coral reefs are often the source of great dangers to navigators, and numberless instances have occurred of entire or partial destruction of ships, and heavy losses of life in consequence. One case, that happened some years ago in the Indian Seas, nearly proved fatal to the whole crew of a fine large ship called the "Cabalve." The story of this shipwreck, as related in a letter to a friend by one of the surviving officers, is deeply interesting. The vessel was bound for Bombay, and was proceeding on its way at a quick rate, with every feeling of security in those on board, when one morning, between four and five o'clock, the weather being dark and cloudy, an alarm was given of "breakers ahead!" Every effort was instantly made to free the vessel from her dangerous position; but in vain, for she struck on the coral reef, and the shock was so violent that every person was instantly on deck, with horror and amazement depicted upon every countenance at what appeared to be certain destruction. The vessel soon became fixed upon the coral reef, and the sea struck upon her with tremendous violence, staving in the exposed side, washing through the hatchways, and tearing up the decks.

"We were now," observes the officer alluded to, "uncertain of our distance from a place of safety: the surf broke over the vessel in a fearful cascade; the crew despairing and clinging to her sides to avoid its violence, while the ship was breaking up with a rapidity and crashing noise, which, added to the roar of the breakers, drowned the voices of the officers. The masts were cut away to ease the ship, and the cutter cleared and launched in readiness. When the long wished-for dawn at length broke upon us, instead of alleviating it rather added to our distress. We found that the ship had run on the south-east extremity of a coral reef, surrounding on the eastern side those sand-banks or islands in the Indian Ocean, called by the natives Carajos; the nearest of these was about three miles distant, but not the least appearance of verdure could be discovered, or the slightest trace of anything on which we might hope to subsist. In two or three places some rocks in the shape of pyramids appeared above the rest like distant sails, and were repeatedly cheered as such by the crew, until it was perceived that they had no motion, and the delusion vanished. The masts had fallen towards the reef, the ship having fortunately canted in that direction, and the boat was therefore protected in some measure from the surf. Our commander, whom a strong sense of misfortune had entirely deprived of presence of mind, was earnestly requested to get into the boat, but he would not, thinking it unsafe. He maintained his station on the mizzen-topmast that lay along the wreck, the surf which was rushing round the bow and stern continually overwhelming him. I was myself close to him on the same spar, and in this situation we saw many of our shipmates meet an untimely end, being either dashed against the rocks or swept away by the breakers.

"The large cutter full of officers and men now cleared a

passage through the mass of wreck, and being furnished with oars, watched the proper moment and pushed off for the coral reef, which she fortunately gained in safety, but they were all washed out of her in an instant by a tremendous surf; yet out of more than sixty persons whom she contained, only one man was drowned. Our captain, seeing this, wished he had taken advice which was now of no use. Finding I could no longer maintain myself on the same spar, and seeing the captain in a very exhausted state, I entreated him to return to the wreck; but he replied that since we must all inevitably perish, I should not think of him, but seek my own preservation. An enormous breaker now burst on us with tremendous violence, so that I scarcely knew what had occurred to him afterwards, being washed down by successive seas.

"At length, after most desperate efforts, I was thrown on the reef, half drowned and severely cut by the sharp coral, when I silently offered up thanks for my preservation, and crawling up the reef, waved my hand to encourage those who remained behind to make an effort. The captain, however, was not to be seen, and most of the others had returned to the wreck, and were employed in getting the small cutter into the water, which they accomplished, and safely reached the shore. About noon, when we had all left the ship, she was entirely broken up. The whole of the upper works—from the after-part of the forecastle to the break of the poop-deck—had separated, and was driving in towards the reef. Most of the lighter cargo had floated out of her: bales of cloth, cases of wine, puncheons of spirits, barrels of gunpowder, hogsheads of beer, and other articles, lay strewed on the shore, together with a chest of tools. Finding the men beginning to commit the usual excesses, we stove in the heads of the spirit-casks to prevent mischief, and endeavored to direct their attention to the general benefit. The tide was flowing fast and we saw that the reef must soon be covered; we therefore conveyed the boats to a place of safety, and filling them with all the provisions that could be collected, proceeded to the highest sand-bank, as the only place which held out the remotest chance of safety.

"The people now collected together to ascertain who of the crew had perished, when sixteen were missing: the captain, surgeon's assistant and fourteen seamen. We divided our men into parties, each headed by an officer: some were sent to the wreck and along the beach in search of provisions, others to roll up the hogsheads of beer and butts of water that had floated on shore; but the greater number were employed in hauling the two cutters up, which the carpenters were directed to repair."

Such is a graphic account of a fearful shipwreck on a barren coral reef, from one of the survivors among the crew. One can thus form an idea of the dangers to which seamen are exposed by these colossal works of tiny polyps:

> "For often the dauntless mariner knows
> That he must sink beneath,
> Where the diamond on trees of coral grows
> In the emerald halls of death."

prevent mischief, and endeavored to direct their attention
to the general benefit. The tide was flowing fast and we

CHAPTER VII.

PEARLS.

> "Ocean's gems, the purest
> Of Nature's works! What days of weary journeyings
> What sleepless nights, what toils on land and sea,
> Are borne by men to gain thee!"

AMONG the rare and beautiful objects of creation may be mentioned PEARLS, which rank with the most valuable of precious gems, and are highly prized as ornamental appendages by the rich and the noble in all countries.

While admiring these jewels, you may not know, perhaps, at what perils and cost of life they are obtained, for it is necessary to seek for them in the depths of the ocean, and although the divers employed for this purpose are very strong and expert, still in the Indian Sea and the Eastern Archipelago, where the true pearl-oysters are found, sharks are numerous, and it is necessary to take every precaution against those voracious monsters. This occupation was formerly considered so dangerous that only condemned criminals were thus employed, but many thousand persons now obtain a livelihood by these means in the Persian Gulf and at Ceylon. At one time, when the Dutch had possession of this beautiful island, the number of large pearls obtained there was considerable.

These pearl-divers are a hardy race of men, singularly adapted to their hazardous occupation, and very superstitious; for before commencing operations, they consult the "shark-charmer," a wise-acre who pretends to have the power of preserving his dupes from the angry jaws of the great sea-scourge, and makes a good living by it, the office being handed down from father to son as hereditary. The divers have such confidence in their powers, or spells, that they will not descend to the bottom of the deep without knowing that one of the enchanters is present in the expedition. Two of the "charmers" are constantly employed, one going out regularly in the head pilot's boat, while the other performs certain ceremonies on shore, such as consulting the auguries, which, if auspicious, ensure the divers in their perilous submarine occupations by closing the mouths of the sharks at the word of command. The "charmer" is shut up in a room where nobody can see him, from the period of the sailing of the boats until their return. He has before him a brass basin filled with water, containing one male and one female fish made of silver. If any accident should happen from a shark at sea, it is believed that one of these fishes is seen to bite the other. The divers also say that if the conjuror is disatisfied, he has the power of making the sharks attack them, on which account he is sure of receiving liberal presents daily.

The Gulf of Manaar, where the pearls are found (and which separates Ceylon from the continent of India on the north-west), abounds in sharks; and, however the divers may consider their lives "charmed," the risks are lessened by the sea-monsters being alarmed at the unusual number of boats, the noise of the crews, and the constant descending of the baskets for the shells. It is not improbable that the dark skins of the divers are also some protection. It seems that the pearl-divers in the Persian Gulf in former times were so conscious of this advantage of color, that they were accustomed to blacken their limbs in order to baffle their powerful enemy. This is related by one of the earliest of Arabian geographers, who adds, "that the divers filled their ears with cotton steeped in oil, and compressed their nostrils with a piece of tortoise-shell."

The pearl fishery of the Bahrem Islands (in the Persian Gulf) produces a most abundant supply of these ocean gems, the produce of a two months' season realizing nearly five hundred thousand dollars of our money. Persians are chiefly engaged in this pursuit, and the divers belong to that nation.

The method pursued by the Cingalese divers is very simple. They proceed in boats to the place of operation at the season, which lasts about two months, commencing in February and ending in April. Each boat contains about twenty men, half of whom are divers, while the others row the boats and assist their companions in reaching the surface of the water after diving. Five of the divers descend at the time, and when they come up, the other five take their turn; the fatigue and exhaustion of the body is very great in continuing under water, and a minute—in some cases a minute and a half, or nearly two minutes—is about the utmost time these men can sustain their breath. Many divers suffer severely from overtaxing their powers of endurance, and bloodshot eyes and spitting of blood are common to them. It is to be hoped that the modern improvements in diving-bells and suitable apparatus for divers will be much more generally adopted than they have been in a few places, that life may be rendered more secure, and other distressing consequences be obviated.

To facilitate the descent of the diver into the water, a stone weighing about twenty pounds is suspended over the side of the boat, with a loop attached to it, in which he inserts his foot; a bag of network is attached to his toes; his right hand grasps the rope, and after inhaling a full breath, he presses his nostrils with his left hand. He now raises his body as high as possible above the water to give force to his descent, and liberating the stone from its fastenings,

PEARL DIVERS AT WORK.

he sinks rapidly below the surface. As soon as he reaches the bottom, the stone is drawn up, and the diver, throwing himself on his face, collects into his bag as many oysters as he can. This, on a signal, is hauled to the surface, the diver springing to the rope as it is drawn up. The sea, at the oyster-beds, is generally from twenty-four to sixty feet deep. The number of oysters thus collected varies; sometimes several thousand are obtained in one day, and at other times a few hundred only. The oysters are landed from the boats, and are placed underground to putrify, and it is amidst such a mass of corruption that the pearl,

'Purest of Nature's works,"

is obtained.

The pearl-fishers in ancient times used to place the shells in vessels filled with salt, and leave them until all the fish were dissolved, the gems remaining at the bottom. The ordinary operation now is, that as soon as putrification is sufficiently advanced, the oysters are placed in a trough, and sea-water is thrown over them. They are then shaken and washed. Inspectors stand at each end of the trough, to see that the laborers secrete none of the pearls, and others are in the rear to examine the shells thown out. The workmen are not allowed to raise their hands to their mouths while washing the pearls, lest they might attempt to swallow some. Sometimes the pearls, instead of adhering to the shells as is usually the case, are in the bodies of the oysters, which are boiled before being thrown aside as useless. The number of pearls in a shell differs: one may contain a considerable number, while hundreds are without any.

To give an idea of the extent to which the pearl fishery in Ceylon has been carried for several ages, the shore in some parts of the island has been raised to the height of many feet by enormous mounds of shells, millions having been flung into heaps that extend to the distance of many miles.

At the Pearl Islands, near the Isthmus of Panama, the divers use a very simple method of obtaining the oysters. They traverse the bay in canoes that hold eight men, all of whom dive in the water to a depth of from fifty to sixty feet, where they remain sometimes nearly two minutes, during which they collect all the oysters they can in their hands, and rise to deposit them in the canoes, repeating the operation for several hours.

In Sweden the oysters are taken with a pair of long tongs. The fishermen are in small boats, painted white on the bottom, which reflects to a great depth, and enables them to see the oysters and seize them.

The most beautiful and costly pearls are obtained from the East, and are called "Oriental;" the color of those found in Ceylon is generally a bluish silvery white, but they are met with of several other hues. Those from the Persian Gulf are of great purity and richness. The preparation of the pearls for market occupies a considerable number of the inhabitants of Ceylon. After being thoroughly cleaned, they are rounded and polished with a powder made of the pearls themselves, and arranged into classes according to their various sizes and quality. They are then drilled and strung together, the largest being generally sent to India, where they are highly prized, while the smaller ones are forwarded to Europe. The operation of drilling is a very delicate one, and the black people are very expert in it. It is done with a wooden machine in the form of an inverted cone, in the upper flat surface of which are pits to receive the pearls. The holes are made by spindles of various sizes, which revolve in a wooden head by the action of a bow-handle, to which they are attached. During the operation (which is done by one hand, while the other presses on the machine), the pearls are moistened occasionally, and the whole is done with astonishing rapidity.

As to how the pearl is formed within the oyster-shell,

is a subject that has been much debated in ancient and modern times. The illustrious Pliny (who died in the year 79), as one of the most enlightened of the old philosophers, says that "the pearl was produced by the dews of heaven falling into the open shells at the breeding-time. The quality of the pearl varied according to the amount of the dew imbibed, being lustrous if that were pure, dull if it were foul; cloudy weather spoilt the color, lightning stopped the growth, and thunder made the shell-fish unproductive, and to eject hollow husks called bubbles."

The same naturalist also relates a story how the shoals of pearl-oysters had "a king, distinguished by his age and size, exactly as bees have a queen, wonderfully expert in keeping his subject out of harm's way, but if the divers once succeeded in catching him, the rest straying about blindly, fell an easy prey. Although defended by a body-guard of sharks, and dwelling among the rocks of the abyss, they cannot be preserved from ladies' ears."

These are very pretty and fanciful ideas, as were many fictions of the pagans, and the British poet Moore, has alluded to them in one of his sweet melodies :—

> "And precious the tear as that rain from the sky
> Which turns into pearls as it falls in the sea."

Some naturalists have suggested that pearls are the unfructified eggs of the oyster, others that the jewel is a morbid concretion produced by the endeavor of the animal in the shell to fill up cavities; the general opinion, however, seems to prevail thus: most shelly animals which are aquatic are provided with a fluid secretion with which they line their dwellings to render them smooth and polished for their tenderly-formed bodies. This fine even lining is seen in shells of every description. The fluid is laid in extremely thin semi-transparent threads, which gives the interior of the shell the beautiful play of color, so often observed. As for the pearl in the shell, small rounded portions are formed in the lining, which are supposed to be the result of accident, such as grains of sand or other substances getting into the shell, and, irritating the animal inside, causes it, by an instinct of nature, to cover the cause of offence, not having the power to remove it. As the fluid goes on regularly to supply the growth and wear of the shell, the prominences continue to increase, and being more brilliant than the rest of the shell, they become a pearl, a composition of carbonate of lime and a little animal matter.

If a pearl is cut tranversely and observed through a microscope, it will be found to consist of minute layers, resembling the rings which denote the ages of certain trees when cut in a similar manner.

The Chinese, who are never at a loss for expedients, are in the habit of laying a string with five or six small pearls, separated by knots, inside the shells, when the fish are exposing themselves to the sun. These, after some years, are taken out, and found to be very large fine pearls. The same ingenious people also introduce into the shell of a mussel different substances such as mother of pearl, the beautiful white enamel which forms the greater part of the substance of most oyster shells, fixed to wires, which thus become coated with a more brilliant material. Another practice among the Chinese is to serve the purpose of a deception upon the credulous. They place small metal images of their god Buddha in the shells, which are soon covered with a pearly secretion, and become united to the shells. These are sold as miraculous proofs of the truth of their worship. The Chinese are also said to employ a means of procuring pearls artificially by the introduction of shot between the mouth of the animal and its shell.

The pearl-oyster is not the only mollusk which produces pearls: an oyster with a thin transparent shell, which is used in China and elsewhere as a substitute for glass windows, produces small pearls, as also the fresh-water mussel of England, pinna, a genus of the same family with the pearl-mussel, and even in limpets.

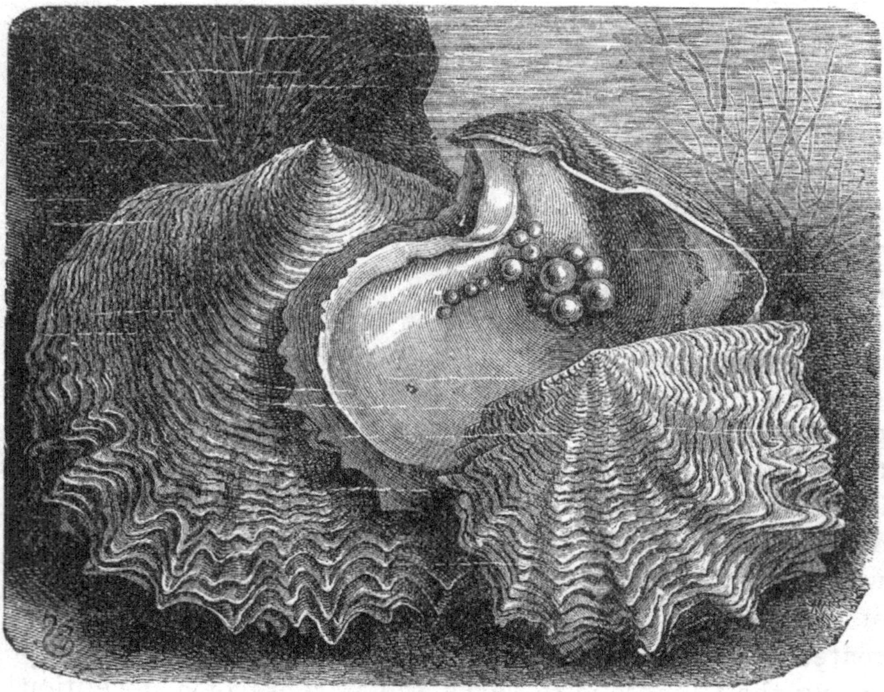

PEARL PRODUCING SHELLS.

The ancients were extravagantly fond of these beautiful jewels: necklaces, bracelets, and earrings were worn in profusion; a string of pearls was estimated by a Roman writer at about forty thousand dollars of our money; the single pearl which Cleopatra dissolved and swallowed was valued at nearly four hundred thousand dollars; and a similar act of folly is reported in later times, in the reign of Queen Elizabeth, when Sir Thomas Gresham, one of London's merchant princes, reduced a pearl to powder worth seventy-five thousand dollars, and drank it in a glass of wine to the health of his sovereign, in consequence of a wager with the Spanish ambassador that he would give a more costly dinner than the other. Quite as absurd was the notion in former times that powdered pearls were unfailing remedies in all stomach complaints.

Pearls are esteemed according to their size, color, form, and lustre: the largest, usually about the dimensions of a small walnut, are called "paragons" and are very rare; those the size of a small cherry are next in rarity, and are called "diadem" or head pearls. They receive names also according to their form, whether quite round, semicircular, or drum-form, or that of an ear-drop, pear, onion, or as they are otherwise irregularly shaped. The small pearls are termed "ounce pearls," on account of their being sold by weight, and the very smallest "seed pearls."

The largest pearl on record is one, pear-shaped, brought from India in 1620, by Gongibus de Calais, and sold to Philip IV. of Spain. It weighed four hundred and eighty grains. The merchant, when asked by the monarch how he could venture to risk all his fortune in one little article, replied with great tact, "because he knew there was a King of Spain to buy it of him." This pearl was said to be in the possession of the princely family of Yousoppoff, in Russia.

Runjeet Sing, the former possessor of the famous Koh-i-Noor diamond, had a string of pearls which was considered nearly equal in value to the "Mountain of Light." They were about three hundred in number, and the size of small marbles, all choice pearls, round and perfect both in shape and color. Two hours before he died he sent for all his jewels, and gave the magnificent string of pearls to a Hindoo temple.

CHAPTER VIII.

SPONGES.

AMONG ancient nations the sponge was used as a soft and elastic lining for the brazen helmets of their soldiers, and many other purposes. It has long been a matter of debate among naturalists whether sponges should be classed among the vegetable or animal kingdoms; they are now generally placed under the order *Zoophyte*, or plant animals.

Aristotle, the greatest of ancient philosophers, who was born three hundred and eighty-four years before Christ, described the sponge as a stationary or rooted animal: but from other statements he made it is certain that he considered its place as between the animal and vegetable. Some modern naturalists have placed sponges among marine vegetables, and their appearance, if one casually looks at them, would seem to justify such an opinion; but the researches of Mr. Ellis, a merchant of London, who made similar branches of natural history a particular pursuit, gave additional interest to this case. In the course of his microscopic investigations, he was astonished at discovering that sponges possessed a system of pores and vessels, in which sea-water passed with all the appearance of the regular circulation of fluids in animal bodies, and a seeming purpose of conveying small minute animals to itself for food.

Afterward, Dr. Grant gave the result of his experiments on the same subject. The account is so interesting that we will give it in his own words. "Having placed a portion of sponge in a watch-glass with some sea-water, I beheld for the first time the splendid spectacle of this living fountain vomiting forth from a circular cavity an impetuous torrent of liquid matter, and hurling along in rapid succession opaque masses, which it strewed everywhere around. The beauty and novelty of such a scene in the animal kingdom long arrested my attention; but after twenty-five minutes of constant observation, I was obliged to withdraw my eye from fatigue, without having seen the torrent for an instant change its direction or diminish the rapidity of its course. In observing another species, I placed two entire portions of this together in a glass of sea-water, with their orifices opposite to each other, at the distance of two inches. They appeared to the naked eye like two living batteries, and soon covered each other with the materials ejected. I placed one of them in a shallow vessel, and just covered its surface and highest orifice with water. On strewing some powdered chalk on the surface of the water, the currents were visible to a great distance; and on placing some pieces of cork or dry paper over the orifices, I could perceive them moving by the force of the currents at the distance of ten feet from the table on which the specimens rested."

So interesting are the sponges, which, although ranked as creatures of very low intelligence, yet are by no means the least curious of those manifestations of the Divine Power

> "That built the palace of the sky,
> Formed the light wings that decorate the fly;
> The Power that wheels the circling planets round,
> Rears every infant floweret on the ground;
> That bounty which the mightest beings share,
> Feeds the least gnat that gilds the evening air."

All of our young readers must be conscious of the useful qualities of the sponge, but many are unacquainted with the manner in which and where they are obtained. The finest qualities of sponge come from the Ottoman Archipelago, and form one of the principal articles of commerce with Turkey. The island of Calymnos is the principal station for the sponge fishery, and more than three hundred boats are employed, averaging each about six tons, and carrying six to eight men, of whom two are rowers. It may be readily seen that this business furnishes occupation for a great number of people. One thousand men are employed in the Grecian Archipelago, alone; and thousands besides with the necessary boats and appliances, are busy in the Gulf of Machia, on the Barbary Coast and elsewhere, so that in many hamlets in these latitudes, from May to September—the best diving time—only old men, women and children are to be seen. The finest qualities are sent in large quantities to our own country, and the common and coarser kinds are forwarded to France, Austria and Constantinople.

The average depth at which the best sponges are found is about one hundred and eighty feet; those of an inferior quality are brought from a lesser depth. The method of diving is much the same as we have described in the coral-fishing. The diver, who goes head-foremost into the water, takes with him a triangular-shaped stone, to which a strong line is attached to assist him in his descent, and direct him like a rudder to any particular spot. On reaching the bottom, the diver tears off a number of sponges, which adhere in masses to rocks and stones, sometimes to large shells, and are either round, flat, or hollow like a funnel; and then, pulling a line, he is drawn up, with the sponges in his arms, by the rowers. An experienced diver will make from eight to ten dives during the day. The proceeds of the fishery are divided into shares, the divers receiving a whole share, and the rowers two-thirds of a share. Formerly the divers used to sell their sponges by weight, to increase which they put sand into them, a practice still continued, though now sold by quantity.

The best quality is brought from the Agean Sea. At daylight, in the summer time, when the weather is pleasant—for it requires smooth water—the boats will leave the shore and proceed to where the water is of suitable depth. The divers then descend as before described. After being busy thus until mid-day, they return to some of those pleasant little nooks which abound in this locality to prepare what they have gathered ready for the market. This is done by pressing out the soft part of the animal. Then they beat and trample the animal until no life is left, after which process the remainder is bleached out by the sun. The skeleton part is thoroughly washed and otherwise treated until it is quite clean, and grows to be a dull yellow color; it is then packed in bags, and shipped to various parts of the world.

The sponge, in its natural state, would not be recognized as that we are accustomed to use daily. In its primitive condition it is covered with a thin dark skin, inside of which there is a liquid like milk, and of the same consistency. If

SPONGE IN ITS NATURAL STATE.

we examine a drop of this liquid by the microscope, it would appear entirely composed of very small transparent grains, nearly of the same size, with some moisture. This jelly matter connects the different parts of the framework of the sponge, and lines the various canals or passages. The pores, or apertures for perspiration, are minute openings on the surface, protected by the framework, and into which the water enters in currents, and after traversing the interior passages, is ejected by means of openings which are larger than the pores, and in many species are elevated above the surface. To examine closely the framework or skeleton of the sponge, it is necessary to macerate it in hot water, which removes the gelatinous matter, and leaves it in a condition to be examined by the microscope. This framework consists principally of two materials, one animal, the other mineral; the first of a thready, horny, elastic nature; the second (the species most commonly used for domestic purposes) of a flinty or chalk material. The thready portion consists of a light pale-colored network, with some few exceptions always solid, and varying considerably in size. The mineral portion has little spines, which, if examined with the microscope, show traces of a central cavity or canal, the extremities of which are closed.

How the growth and increase of the sponge is effected affords matter of the deepest interest, and this, like everything else in nature, shows the unerring wisdom of an all-sustaining Providence.

From the framework or skeleton of the sponge emerge, at certain seasons of the year, a yellow kind of grain, which projects as it increases in size into the cavities of the sponge, and forms the germ or seed of another race; these are egg-like in appearance; and a large portion of its surface becomes covered with little hairs, called eyelashes from their resemblance to such. These hairs act as oars to the little germ, to convey it away as soon as it falls on the water to some other spot to which it may attach itself, The hairs, after accomplishing their purpose, fall off, leaving the germ to gradually develop into the sponge.

This sponge is a natural production, and we have already hinted, has been known from the times of highest antiquity. As is well known, all naturalists are now satisfied of the animal nature of this species of creation, although they were once thought to represent the lowest and most obscure grades of animal existence, and that so close to the confines of the vegetable world, that it was considered difficult in some species to determine whether they were on one side or the other.

According to a generally accepted view, the channels of the sponge perform the two functions of digestion and respiration. The rapid currents of aerated water which traverse them lead into them the substances necessary to the nourishment of these strange creatures, and at the same time carry off all excremental matter. At the same time, the walls of these animals present a large absorbing surface which separates the oxygen with which the water is charged, and disengages the carbonic acid which results from respiration. But science is far from being settled in its views as to the organization and development of these obscure and complex creatures; nor is it more advanced in its knowledge of the duration of life and the quickness of growth in sponges. Nor can it be denied, also, that these beings constitute, in spite of the investigations of modern naturalists, a group still somewhat problematical as to their position in the scale of animal life, and that they still are very imperfectly known as regards their internal organization.

The demand for sponges is increasing annually, and it is only a question of time when the trade must cease. The submarine fields are constantly being cleared, and the destruction is such that the reproduction will cease to be adequate.

In order to prevent this result, it has been suggested that attempts should be made to naturalize the several species of sponges, and that their cultivation and reproduction should be protected. The first thing to consider is what waters have the same or nearly the same temperature as that in which the sponges now dwell. This being comparatively easy, the next and most difficult part would be to transfer the animals in such a shape as would insure reproduction in their new homes. Some such submarine boat as has been used recently in making scientific operations conducted in deep water, might, probably, give the necessary facility for collecting sponges for the purpose. Such a boat can descend to great depths, and its crew can even dwell there for a considerable time, for it is continually fed with fresh air; so that the men could readily select such specimens as were suited for acclimatizing, removing whole blocks of rock along with them.

It might be possible too to collect the very young forms of sponge in the months of April and May, shortly after they have commenced their independent existence, and to transplant them to favorable localities. At the end of a few years, when these true submarine fields would be probably ripe for harvesting, they could be farmed out for methodical collection, which would be effected by means of diving boats.

CHAPTER IX.

SEALS.

"Man bends the ocean monsters to his sway,
No terrors daunt him on his arduous way;
Through frozen waters, or in sunlit waves,
He seeks the SEAL, unnumber'd hardships braves
To gain a prize so rich in useful store."

AT the approach of the Arctic summer, all is bustle and activity among the natives of the Arctic regions. The materials for the summer huts are prepared, and the whole household, consisting of five or six families, move down to the fishing-place, which is generally an island with a low beach, in a southern aspect, for the convenience of launching their boats or drawing the seals which have been taken ashore. They are not confined to any particular spot in the summer, unless abundance of seals are seen; but they generally shift to some other station, which, in the course of former seasons, they may have observed as more suitable.

The Esquimaux have their regular divisions of work. The men are the carpenters; the women are the tailors, shoemakers and cooks, helping their husbands or fathers occasionally in their fishing. It is heavy work for these poor females, but Providence has endowed them with a strength of constitution and powers of endurance far greater than women in more genial climates possess. They have to haul the seals that have been taken by the men, ashore, and convey them to the huts. They also flay and cut up the spoil.

Seal's flesh forms their chief food, and they employ various methods for preserving it for future use. The most common plan is to cut it into thin strips, and dry them over a line in the interior of the huts. The seal-skins, which the Esquimaux have a mode of rendering waterproof, form the chief articles of dress; when tanned, they make excellent shoes.

It may be mentioned here that the Romans believed a seal's skin was a preservative against lightning, and they made tents of it to shelter themselves during thunder-storms. The Emperor Augustus is said by Suetonius never to have traveled without one of these skins, having a great dread of lightning.

The blubber of the seal is most carefully preserved by the Esquimaux, being useful in many ways to their domestic comfort, and more precious to them by far than wine is to others. The oil is the luxury of their meals, and is of a superior quality to that of the common whale; their bread is nothing more than the dried muscular parts of seals or birds. Whatever may be thought of the Esquimaux's partiality for seal-flesh, it is well to remember that our English ancestors considered it a delicacy. The seal and the porpoise are mentioned in the bill of fare of a feast given at the enthronization of George Neville, Archbishop of York, in 1465. The meat is described as tender, but it certainly has a look and smell which would not be agreeable to any but very hungry persons.

The Esquimaux are exceedingly expert in their mode of capturing the seal. This is done either individually or in company, or in winter on the ice. Their *kayaks,* or skin boats are very curious: they are about eighteen feet in length, pointed at the head, and shaped like a weaver's shuttle; they are, at the same time, scarcely a foot and a half wide over the middle, and not more than a foot deep. They are built of a slender skeleton of wood, consisting of a keel and long side-laths, with cross-ribs like hoops, but not quite round. The whole is covered with seal-skin. In the middle of this covering is a round aperture, supported with a strong rim of wood or bone ; the Esquimaux slip into this cavity, their feet resting on a board covered with skin. The lance, harpoon and tackle are arranged before the boatman. He uses his oar or paddle with wonderful dexterity, striking the water on either side alternately, by which means he can proceed at the rate of sixty miles or more in a day. In this frail bark, which only those accustomed to such can manage, the Greenlander fears no storm or the roughest breakers, so long as he retains his oar, which enables him to sit upright ; and if overturned, while the head is downward in the water with one stroke he can recover himself.

> "Train'd with inimitable skill to float,
> Each balanced in his bubble of a boat,
> With dexterous paddle steering through the spray,
> With pois'd harpoon to strike his plunging prey,
> As though the skill, the seaman, oar and dart
> Were one compacted body, and one heart,
> While instinct, motion, pulse, empowered to ride—
> A human nautilus upon the tide."

As the natives are ever on the watch, as soon as they discover a herd of seals—driven usually by stormy weather into some creek or inlet—they endeavor to cut off their retreat, and frighten them under water by shouting, clapping, and throwing stones. As, however, the seals must speedily come to the surface of the water to breathe, they are surrounded and killed with long or short lances.

There are various modes of capturing seals on the ice. As the animals make holes in it for breathing, the Esquimaux seat themselves on stools, watching their appearance at the apertures, and rarely fail to harpoon them, enlarging the holes to withdraw and kill them. Sometimes, on seeing a seal lying on the ice near a hole, the Greenlander slides along

ESQUIMAUX SEAL HUNTERS.

on his stomach towards it, wagging his head, and making a sound like a seal, thus deceiving the poor animal into a belief that it is one of his companions. But the seal is usually wary—that is, the older ones—and takes every opportunity of escaping from its pursuers. When one is seen at sea, a signal is passed to the different boats engaged in the chase, and the animal is surrounded; a careful watch is kept for the moment of its reappearing, and on this taking place, one of the boats having advanced near enough, a dart is hurled with unerring aim. The seal, terrified and wounded, dives in the greatest hurry; but a float being attached to the dart, it is soon forced up again and dispatched. The wounds of the seal are then carefully staunched, to save as much of the blood as possible, and the body is distended by blowing into the cellular part, in order to render the animal buoyant, or, otherwise, it would sink to the bottom as soon as dead.

The chase of the seal, however, is not free from danger, even to the expert fisherman of the Arctic shores. If the animal is not too much exhausted when pursued, it sometimes turns on its adversary, seizes his frail skin boat, and with its sharp teeth pierces a hole, when the *kayak* sinks with its unfortunate owner. Many risks also occur from the lines to which the floats are attached getting foul of the paddle or the arms or neck of the fisherman, when the seal dives suddenly on being wounded. The males are very pugnacious, and have terrible fights among themselves.

Seal-hunting, or fishing, as it is often called, is the great occupation of the Greenlanders, and is also extensively pursued by various nations in other northern parts of the world.

A great many species of seals are met with on the western coast of Greenland; but the most highly prized by the natives is what sailors call the *Sea-Calf*, so named from a supposed resemblance of the voice to that of a calf. These animals live in families, the old male being attended by his progeny for several generations. They are chiefly seen in flocks, amounting sometimes to hundreds. The teeth are very sharp, and the bite is severe. The habits of the seal are filthy, and singularly mischievous. A perpetual tyrant over weaker animals, it is also an object of constant pursuit with others. The white bear—with whom the seal is a great dainty—is constantly on the watch to surprise it when sleeping on the ice; but the cautious animal usually selects a single piece of ice for a nap, from which it may gain a full view of all around, and the proximity of the water may afford a ready means of escape. They are also said to have a great dread of the toothed whales. If a grampus perceives a seal of any species basking on floating ice, it does its best to upset the ice, or beat the seal off with its fins, when the animal becomes any easy prey.

Seals are easily stunned by a blow on the forehead; but from this state they often recover, and are desperate in their revenge. The sea-calf, in particular, is subject to violent fits of anger. After it has been hoisted on board a ship from the boat in which it had been carried, apparently dead from the blows it had received, it has been known to recover unexpectedly, and seizing with its teeth the nearest object within reach, tear away such a portion as it could grasp. Even after death this irritation manifests itself, as the muscular parts of the animal—though stripped of its outer integuments or coverings—still retain the principle of vitality, starting and quivering long after the dismemberment of the body has taken place.

When seals are observed making their escape in the water before the boat reaches the ice, the sailors give a loud, prolonged shout, which, causing them to stop in amazement at a sound so uncommon, sometimes delays their retreat until arrested by the fatal blows of their pursuers.

In the higher latitudes, the *Bearded* or *Great Seals* are mostly found. These are usually of an enormous size, sometimes ten or twelve feet in length, and of proportionate magnitude of body. This seal migrates in families, the elder ones leading the van, while the young follow confusedly behind, playing, tumbling, and frisking along in the highest enjoyment, and frequently in the extravagance of their fun, flinging themselves quite out of the water. The sailors call these antics "seals' weddings."

Though the bearded seal does not yield much oil, yet its fat is esteemed delicious by the northerners. The *Harp Seal*, so named from a large black crescent-shaped mark on each side of the back, belongs also to the ice regions, though sometimes seen on the British coast. It attains the length of eight, and even nine feet.

The seal belongs to the Mammalia, or animals that suckle their young, and constitute the family *Phocidœ.* All the animals of this class are mainly aquatic, but also frequently resort to land, or ice-islands, where they remain for days, and even months, suckling their young, or basking in the sun during the brief summer. The *Fur Seal* seems to possess remarkable powers of agility on land, often escaping when pursued by the men running fast. They cannot walk, but shuffle along, especially over the ice, very quickly. On land the hind feet are never employed, nor the fore feet unnecessarily, but in moving forward it bends the hinder part of the spine underneath it, thus making a kind of arch, and then fixing the latter end, it suddenly straightens out the whole body in front, and in a repetition of this movement consists the peculiar kind of jerking leap for which these animals are remarkable. When the seal ascends an ice-island or rock, the ease with which it accomplishes its purpose is wonderful It then makes especial use of its fore paws, and those which have claws are implanted into them like so many grappling-irons, and, having thus secured a fixed point, they raise their monstrous bodies with the greatest rapidity. The general shape of a seal resembles i its trunk that of a fish and a common quadruped; the head is like that of a dog; the arms, which are destitute of collar-bones, are so hid beneath the skin of the body that only the wrists and hands appear, and they are then so short that they can scarcely be advanced forward at all. But what they lose in extent they gain in power. They are admirably adapted for swimming, and serve also for seizing or holding. The fingers have an intervening membrane, but they can be separated so as to diminish or increase the surface of the paws. In all the species, the fingers can be distinguished through the paw, and in most the nails appear at the termination; but in one group of seals there is this difference, that the membrane or web extends beyond the nails, not joined, but hanging down in the water like broad leathern strips, which the sailors call "flippers." The face is provided with strong whiskers placed on each side of the mouth and at the corner of the eye, communicating with nerves of considerable size, and the slightest impression produces sensation.

The ground color of the hair or skin of the common seal, when the animal is alive and dry, is a pale whitish-gray, with a very slight tinge of yellow. When just out of the water and wet, the color is ash; after death, and as seen in museums, the ground color is pale yellowish-gray, the oil having penetrated the skin and rendered the hair of a more yellow hue. The fur of seals is very smooth, and abundantly lubricated with an oily secretion. There is generally an inner coating of rich fur, through which grow long hairs, forming an outer covering. Another adaptation to aquatic life and a cold climate is the layer of fat under the skin, from which the oil is obtained, and serving, as in the case of the whale, not only for support when food is scarce, but protection from the cold, besides rendering the whole body lighter. The respiration of the seal differs considerably from what has been observed in most animals: the nostrils are habitually closed, instead of being uniformly opened.

Buffon examined a tame seal, and remarked that the period between its several inspirations was very long: the creature opened its nostrils to make a strong expiration, which was immediately followed by an inspiration; after which it closed them, often allowing two minutes to intervene without taking another breath. This power of suspension for a considerable time is of great use, enabling the seals to pursue their prey under water. Seals are often subjected to enormous pressure under water, which must be resisted, at the respective apertures of the body, by an appropriate mechanism. A similar provision is made for the eyes, as well as the nostrils, in more ways, perhaps, than one. At the inner angle of the eye (which is very large and round) there is a third eyelid, which can be drawn over the whole eye. The ears as well as the eyes, can be closed at will, so as to resist pressure.

How very wonderful is the provision thus afforded to the seal, as, in fact, to all created objects, and how the contemplation of such subjects should raise our hearts to the Omnipotent God!

To know and feel His care for all that lives."

Captain Scoresby, who had numerous opportunities of observing the habits of the seal, states that the animal hears well under water, and that music, and particularly a person whistling, draws it to the surface, and induces it to stretch out the neck to the utmost extent, so as to prove a snare, by bringing them within reach of the shooter. Many similar observations of this curious faculty in seals have been related by different writers. One remarks: "In walking along the shore, a few notes of my flute would bring half a score of seals within thirty or forty yards of me; and there they would swim about, with their heads above water, like so many black dogs, evidently delighted with the sounds. For half an hour, or indeed for any length of time I chose, I could fix them to the spot; and when I moved along the water edge, they would follow me with eagerness."

The food of the seal appears to be chiefly fish, although it does not reject other animal food, and it is said to derive part of its nourishment from marine vegetables. It has been found that seals have a remarkable habit of swallowing large stones, for which no probable reason has been yet assigned. The keeper of the celebrated "talking seal" in the Zoological Gardens is reported to have given his pet fifty pounds' weight of fish in a day, but this is by no means a limit of appetite, for double the quantity would no doubt have found a ready reception. This will give you an idea of the vast consumption of fish in its native element. A good-sized Spitzbergen seal in good condition is about ten feet in length and six feet in circumference, weighing about six hundred pounds or upwards. The skin and fat amount to about one-half the total weight. The blubber yields about one-half of its own weight in oil.

It has been supposed that seals can be easily tamed, but such cases are exceptional. Some of the common species, however, have shown great attachment to their owners, and exhibited considerable powers of intelligence. An anecdote is related of a seal that performed very cleverly what it was ordered to do, and would raise itself on its hind legs, take a staff in its paws, and act the sentinel. At the word of command it would lie down on its right side or left, and tumble head over heels. It would give either of its paws when desired, and was equally ready at a *kiss*. Another was kept by Cuvier for a considerable time, and became very tame. When teased it resisted, and when much irritated barked very feebly. It was particularly attached to the old woman who had charge of it, and recognized her at a considerable distance, keeping its eyes upon her as long as she was in sight, and running to her as soon as she approached its enclosure. If free when food was brought, it ran and urgently

SEA-LIONS. ALBATROSS. GOLDEN PENGUIN.

solicited it by the motion of its head, and still more by the expression of its countenance.

Of another species of seal called the *Marbled*, and found on the coast of France, which was kept for several weeks in the Jardin des Plantes at Paris, Cuvier says: "I have never known any wild animal which was more easily tamed, or attached itself more strongly. When it first came it endeavored to escape when I wished to touch it, but in a very few days all its apprehensions vanished; it had discovered my intentions, and rather desired my caresses than feared them. It was in the same enclosure with two small dogs, which amused themselves by frequently mounting on its back, with barking, and even biting it; and although these sports and the vivacity of the attending movements were little in harmony with its own actions and habits, yet it appreciated their motive, and seemed pleased with them. It never offered any other retaliation than slight blows with its paws, the object of which was to encourage rather than repress the liberties taken. If the puppies escaped from the enclosure, the seal endeavored to follow them, notwithstanding the difficulty it experienced in creeping along the ground covered with stones and rubbish. When the weather was cold, the three animals huddled closely and kindly together, that they might contribute to their mutual warmth." The creature did not exhibit any alarm at the presence of man or animals, and did not get out of the way unless when threatened to be trod upon. Though very voracious, it did not show any opposition or anger when robbed of its food. "Often," adds Cuvier, "have I tried him when pressed with hunger, and never opposed my will; and I have seen the dogs, to whom he was much attached, amuse themselves when he was feeding, by snatching the fish from his mouth, without his exhibiting any rage. On the other hand, when their mess was supplied to the seals (for he had a companion), as they were lying in the same trough, a battle was the usual result, and blows with their paws followed, and as usually happened, the more feeble and timid gave way to the stronger."

The seals of the Southern seas are quite different from those of the Northern. The most remarkable of these animals is the *Sea Elephant*, or *Proboscis Seal*, named thus, partly on account of the very peculiar appearance of its short trunk, and also from its being much the largest of its kind, doubling the dimensions of its terrestrial namesake, reaching the enormous length of twenty-five and thirty feet, and being also of a proportionate thickness. Its color is sometimes gray or blue-gray, and more rarely blackish-brown. There is an absence of everything like external ears; it has great whiskers of strong coarse hairs, very long, and twisted somewhat like a screw, with other similar hairs over each eye, supplying the place of eyebrows; the eyes are very large and prominent; strong and powerful swimming paws, having at their margins five small black nails; a very short tail, which is almost hid beneath two flat horizontal fins: these form the distinguishing peculiarities of this strange animal. When the sea-elephant is in a state of repose, its nostrils, shrunk and hanging down, serve only to make the face appear larger: but whenever he rouses himself, when he respires violently, or when about to attack or defend himself, the proboscis becomes lengthened in the form of a tube to the length of about a foot; and then not only is the countenance changed, but the character of the voice is modified in a not less striking manner. Though furnished with large and powerful tusks, the sea-elephant is mild and inoffensive in his habits; but when assailed is a formidable adversary. It has been related that a sailor having killed a young one, and skinned it in the presence of its mother, she came up behind him, and seizing his head in her mouth, so injured his skull, that he died in a day or two afterwards. This is not, however, their usual habit, as has been stated. A young one, petted by an English seaman, became so attached to his master from kind treatment for a few months, that it would come at his call, allow him to mount upon its back, and put his hands into its mouth.

The cry of the female and the young is said to be like the lowing of an ox; but the hoarse, gurgling, singular voice of the male—strengthened by the proboscis—is heard from a great distance, and is wild and frightful. They are found in the Atlantic and Southern Oceans. The great object for which this animal is hunted is for the oil, which is remarkably pure in quality; the skin is used extensively for carriage and horse harness, on account of its thickness and strength.

The *Sea Leopard* is a rare species of seal, in length about nine feet ten inches, which has been found in South Shetland. The *Monk Seal* frequents the southern shores of Europe.

The *Otaries* are a species of seal thus named because their heads are furnished with external ears, of which the others are deprived, and from whom they also differ in other particulars. These include the *Sea-Lion* of the Northern seas, about fifteen feet in length, and found chiefly on rocky coasts and islet rocks, on the ledges of which it climbs, and its roaring is sometimes useful as warning sailors of danger. The old males have a fierce aspect, but it is only when driven to extremities that they fight furiously. The *Sea-Bear*, or *Ursine Seal*, is an inhabitant of the Northern Pacific, and attains a length of about eight feet. The hinder limbs of this animal being better developed, it can stand and walk almost like a land quadruped. It swims with great swiftness, and is fierce and courageous. The skin is much prized for clothing in the regions where it abounds.

POTWAL.

CHAPTER X.

WHALES—THE MONARCHS OF THE OCEAN.

OF all the industrial pursuits which engage the venturous seaman on the wide ocean, those connected with the capture of the WHALE,—

"the mightiest that swims the ocean stream,"

and, it may be said, in point of dimensions the monarch also of creation,—are the most exciting and perilous; requiring the greatest endurance, hardihood and courage, and at the same time yielding, under favorable circumstances, a substantial return for the dangers encountered.

Before relating some of the exciting adventures which occur in the pursuit and capture of the unfortunate whales, a few particulars will be given about the animals themselves.

There are many peculiarities to be observed in these huge monarchs of the ocean. They comprise a class of animated creatures distinct from both fishes and land animals, though partaking of the characters of both. They are classed in the order of warm-blooded Mammalia, that is to say, they breathe as the land Mammalia, and yet are as completely aquatic as true fish, which are cold-blooded. Fish never breathe, and if removed from the water into the air, they immediately die; but whales, if deprived of air, and confined under the water, would be literally drowned. They usually come to the surface to breathe at intervals of eight or ten minutes, but they are capable of remaining under water nearly an hour. The whale has no gills, but a heart with two ventricles or cells, and very elastic lungs in a great bony chest, into which the air is freely admitted, not through the mouth; for, although the animal is of such prodigious dimensions (some species attaining upwards of one hundred feet in length, and a weight of nearly as many tons), yet the throat is so small that it could not dispose of a morsel which is swallowed by an ox. Through what are popularly called "blowers" or *spiracles*, huge nostrils which open on the summit of the head, from eight to twelve inches long, but of small breadth, the whale can send a column of moist vapor forty to fifty feet high; and when this breathing or blowing is performed under the surface of the ocean, a vast quantity of water is also thrown into the air, and the noise made in this operation can, it is said, be heard at the distance of between two and three miles.

Another peculiarity about these wonderful creatures—which belong to the class *Cetacea*, and which comprises not only all the varieties of the whale tribe, but likewise the grampus, the porpoise, the dolphin, the dugong, amd some others of comparatively very small size—is the tail, which is not vertical as in most fishes, but level, by which they are able to reach the surface of the water with greater facility for the purposes of respiration; and such is the strength of this tail that even the largest whales are able, with its assistance, to force themselves entirely out of the water; in the large whales the surface of the tail comprises from eighty to one hundred square feet. In length it is only from five to six feet, but in width it measures from eighteen to twenty-six feet.

Providence has given this immense power to serve as a defense, as well as a means of propulsion, to the huge animal, for the tail is nearly the sole instrument of its protection. With one stroke of it the whale will send a large boat with its crew into the air, and shatter the wood into a thousand pieces. The tail enables the animal to rise in the water by striking a few slight blows with it downwards, when the head is naturally carried in an opposite direction, and when the whale wishes to sink, a few similar strokes with the tail upwards at once serve to bury the head beneath the surface.

Sometimes the animal takes a perpendicular position in the water, with the head downwards, and, rearing the tail on high, beats the waves with fearful violence. On these occasions the sea foams for a wide space around, and the lashing is heard at a great distance, like the roar of a tempest. This performance is called by the sailors "lob-tailing."

The head is of enormous size, being about one-third of the entire bulk of the whale, and the lips, nearly twenty feet long in some species, show a cavity large enough to hold a ship's jolly-boat and crew; but, as I observed before, the throat is very narrow. It is stated to be no more than an inch and a half in diameter even in a large whale, so that only very small animals can pass through it. The basis of the head consists of the crown-bone from each side of which descend the immense jaw-bones, from sixteen to twenty feet in length, extending along the mouth in a curved line until they meet and form a kind of crescent.

In the Arctic seas whales find an abundance of food in the shape of animalculæ, several species of marine worms, jelly-fish, crabs, and especially shrimps, which abound in those regions. John Parry relates that joints of meat hung by his crew over the sides of the ship were in a few days picked to the bone by shrimps.

Some species of whales are entirely destitute of teeth, but Nature has provided them with an apparatus of whalebone, for the purpose of straining out of the water the small animals which form their nourishment. There are several hundreds of these plates on each side of the mouth, the whole quantity in that of a large whale sometimes weighing nearly two tons.

The tongue of the whale is a soft thick mass, not extending beyond the back of the mouth. It was formerly considered a great delicacy of the table, and a right of royalty. The sword-fish, an implacable enemy of the whale, has a similar relish for the tongue, and, it is said, leaves the rest of the carcass untouched. The skin of the whale is naked and smooth, with the exception of a few bristles about the jaws, and is covered with an oily fluid, which renders it very slippery; beneath this is a thick layer, from eight to twenty inches, of a fatty substance, called *blubber*, the most valuable part of the animal, and which yields on boiling nearly its own bulk of thick coarse glutinous oil. It is by this wrapper that Providence enables the whale, a warm-blooded animal, to defy the utmost extremity of cold, and to retain a sufficient proportion of heat even under the icy Polar seas. It also serves to make the specific gravity of the body much lighter than it otherwise would be, so as to resist the pressure of the water at the great depths to which the whale descends. Yet it is this warm covering, so essential to the animal itself, that has excited the cupidity and deadly pursuit of man, causing him to brave the most appalling dangers, trusting to the resources of art in the instruments of destruction where brute force alone could never prevail.

To give an idea of the quantity and the value of the oil obtained from a Greenland whale of sixty feet in length, it has been stated that the weight of the animal being seventy tons would be nearly that of three hundred fat oxen. Of this vast mass the oil of a rich whale comprises about thirty tons, which renders it a valuable capture.

The whale has no external ear, but, when the skin is removed, a small opening is perceived for the admission of sound. This sense may seem imperfect, yet the animal, by a quick perception of all movements made on the water, discovers danger at a great distance. The eyes appear small for such a huge animal, being about the size of those of an ox; but the sense of seeing is very acute. Behind them are the fins; these are about nine feet long and four or five feet broad, and are enclosed by very elastic membranes, also provided with bones, similar in form and number to those of the human hand.

The whale does not attain his full growth under twenty-five years, and is said to reach a very great age. The flesh is red, firm, and coarse, and is eaten raw by the Esquimaux, who also drink the oil with much enjoyment. In the bleak Polar regions, where the means for satisfying hunger are very scanty, the capture of a whale by the natives is an occasion for great rejoicing.

Captain McClure mentions the Esquimaux method of attacking the whale:

"A woman's boat, is *manned* by ladies, having as harpooner a chosen man of the tribe, and a shoal of small fry in the form of *kayaks*, or single men canoes, are in attendance. The harpooner singles out a whale and drives his weapon into its flesh. To the harpoon an inflated seal-skin is attached by means of a walrus-hide thong. The wounded fish is then incessantly harrassed by men in the *kayaks* with harpoons, a number of which, when attached to the whale, baffle its efforts to escape, and wear out its strength, until, in the course of a day, the whale dies from sheer exhaustion and loss of blood.

"The harpooner, after a successful day's sport, is a very great personage, and is invariably decorated with the Esquimaux order of the blue ribbon, that is, he has a blue line drawn down his face over the bridge of his nose."

The whale not only serves for food to the hardy Greenlanders, but is also valuable in many other ways: some membranes of the stomach are used for the upper articles of clothing; the bones are converted into harpoons and spears for striking the seals or darting at sea-birds, and are also employed in the erection of their tents, and some tribes use them in the formation of their boats.

The preceding remarks have applied to the whale tribe generally, but with a more direct allusion to the "Greenland" or "right" whale, as it is called, from its producing the greatest amount of oil. This animal inhabits the seas of the Northern parts of the world, and abounds chiefly in the Arctic regions. The "Southern," or "Cape" whale is a distinct species, the head being smaller in proportion than its Northern relative, and its color a uniform black. It attains the length of from fifty to sixty feet.

The *Northern Rorqual*, which exists in great numbers in the Northern seas, is the largest of the whale tribe, the mightiest giant among giants, attaining the vast length of from one hundred to one hundred and ten feet, with a bodily circumference of from thirty to forty feet. The amazing speed and activity of this immense animal renders it a dangerous object to attack; besides the small quantity of oil it affords does not repay the fisherman for his risk. This whale has no teeth. When struck by a harpoon, it has been known to run off two thousand eight hundred and eighty feet of rope in a *minute*. An old Arctic navigator mentions an instance of a "razor-back," as the great rorqual is called by seamen, dragging a large boat with its crew amongst loose ice, where they all perished.

The *Smaller Rorqual*, measuring from fifteen to twenty-five feet, frequents the rocky bays of Greenland, and is considered a tender morsel by the natives. There is also a "Rorqual" of the Southern seas, an animal of great power and a fast swimmer, very difficult to capture. The most valuable whale in the Southern seas is the *Cachalot* or *Sperm* whale, which supplies the spermaceti and ambergris of commerce. This immense animal, which grows to the length of seventy to eighty feet, is found in almost every part of the warm latitudes. It has some curious peculiarities : the head has in front a very thick, blunt extremity called the snout or nose, and constitutes one-third of the whole length of the animal ; at its junction with the body, the animal has what the whalers call a " bunch of the neck," a large protuberance on the back, immediately behind which is the thickest part of the body, which from this part gradually tapers off to the tail ; and where this commences there is another large prominence called the "hump" after which the body contracts so much as to become finally not thicker than the body of a man. An immense cavity in the head contains cells filled with oil, which is fluid when the animal is alive, and after its death takes a concrete form known as spermaceti. The size of this cavity may be judged from what is said, that in a large whale it sometimes contains a ton, or more than ten barrels of spermaceti. The food of this huge monster consists principally of a species of polypus called "squid" by the sailors, and it is supposed that they are attracted by the shining white of the inner part of the whale's mouth. The sperm whale is generally seen in herds, or "schools" as they are called, consisting of several hundreds. With each herd of females, large males or "schoolmasters" are always associated, who are extremely jealous of intruders, and fight fiercely to maintain their rights. The large whale is generally incautious, and if alone is attacked without much difficulty, and is easily killed, as he frequently after receiving the first plunge of the harpoon appears hardly to feel it, but continues lying like a log of wood before he attempts to escape. Large whales, however, are sometimes very cunning and courageous, and commit fearful havoc with their tails and jaws. When alarmed they are said to perform many unusual actions; one of these consists in moving the tail slowly from side to side on the surface of the water, as if feeling for any object that may be near. It also rolls over and over on the surface, especially when harpooned, and in this way will coil an amazing length of line around it. One of its most surprising feats is leaping out of the water. Darwin remarks that off Terra del Fuego he saw several spermaceti whales performing this stupendous leap, and as they fell into the water sideways the sound reverberated like distant thunder.

The *White Whale* is described as a very beautiful animal, frequenting chiefly the Arctic seas, varying in length from ten to twenty feet. It is usually of a cream color, though some have been seen of a yellowish color, approaching to orange. In the dreary monotony of the icy regions, a lively herd of these animals, by their gambols and the exhibition of their smooth, slippery white bodies, affords a pleasing relief. The shape of this whale is highly symmetrical, resembling a double cone, one end of which is considerably shorter than the other ; the tail is very powerful, and being bent under the body in swimming, is worked with such force as to impel the animal forward with the velocity of an arrow. The food of this whale is said to be cod, haddock, flounders and smaller fish of this description. They are not at all shy, but often follow ships and tumble about amidst the boats in herds of thirty and forty. Fortunately for them, this fearlessness of danger does not often expose them to the deadly harpoon, their comparative little value being their preservative from the whale-fishers. They do not, however, experience the same immunity from the natives of the Greenland coast, where they arrive in great numbers at the close of the year in stormy weather. They are then chiefly captured by nets, which are extended across the narrow sounds between the islands, and when thus entangled they are killed with lances.

Another whale, called the *Deductor*, resembles somewhat the white whale, and appears to be the most sociable of all the Cetacean tribe, herding together in innumerable flocks. This leads, however, to a prodigious slaughter of these poor animals when (although frequenting chiefly the Northern Ocean), they wander away from their usual haunts, and get driven on shore by the fishermen, the main body of the drove following the leading whales as a flock of sheep.

There is an account given of the capture of ninety-eight of these whales, in 1832, on the island of Lewis:

"An immense shoal of whales was, early in the morning, chased to the mouth of the harbor of Stornoway by two fishing-boats, which had met them in the offing. The circumstance was immediately seen from the shore, and a host of boats, about thirty or forty in number, set off to join the others in pursuit, and engage in combat with these giants of the deep. The chase soon became one of bustle and anxiety on the part both of man and whale. The boats were arranged by their crews in the form of a crescent, in the fold of which the whales were collected, and where they had to encounter tremendous showers of stones, splashings of oars, frequent gashings with harpoons and spears, whilst the din created by the shoutings of the boats' crews and the multitude on shore was in itself sufficient to stupefy and stun the bottle-nosed foe into a surrender. On more than one occasion, however, the floating phalanx was broken, and it required the greatest activity and tact before the breach could be repaired and the fugitives regained. The shore was neared by degrees, the boats advancing and retreating by turns, till at length they succeeded in driving the captive monsters on the beach opposite the town and within a few yards of it. The movements of the whales were now violent, but, except when one became unmanageable and enraged when harpooned, or his tail fixed in a noose, they were not dangerous to approach. One young sailor, however, received a stroke from the tail of one of the largest of them, which promised to be fatal. In a few hours the whales were captured, the shore was strewed with the dead carcasses, while the sea presented a troubled and bloody appearance, giving evident proof that it was with no small effort that they were subdued and made the property of man."

The deductor whale has a very prominent head, short and round, with something like a pad over its mouth, which gives it a peculiar appearance. In length it is from sixteen to twenty-four feet, and in circumference ten or eleven feet. Almost the whole body is black, smooth, and shining like oiled silk. When the mouth is shut, the teeth lock into each other like those of a rat-trap. They are generally very fat, and yield a large quantity of good pale oil.

It is impossible not to feel an emotion of pity for the whale—timid and inoffensive, with all its immense power for mischief, apparently unconscious of it until roused by danger—subjected to such cruel treatment by the cupidity of man: the deadly harpoons inflict tremendous wounds, and the blood, rushing in torrents from its sides, crimsons the sea for a wide space around.

The whale has, however, other enemies to contend with besides man. Commodore Wilkes gives an animated account of a sea-fight between a whale and a grampus, or "killer," as this fish is called.

"At a distance from the ship a whale was seen floundering in a most extraordinary manner, lashing the smooth sea into a perfect foam, and endeavoring apparently to extricate himself from some annoyance. As he approached the ship, the struggle continuing and becoming more violent, it was perceived that a fish, apparently about twenty feet long, held him by the jaw, his contortions, spouting, and throes all betokening the agony of the huge monster. The whale now threw himself at full length from the water, with open mouth, his pursuer still hanging to the jaw, the blood issuing from the wound and dyeing the sea to a distance around; but all his flounderings were of no avail, his pertinacious enemy still maintaining his hold and evidently getting the advantage of him. Much alarm seemed to be felt by the other whales around. These 'killers,' as they are called, are of a brownish color on the back, and white on the belly, with a white dorsal fin. They attack a whale in the same manner as dogs bait a bull, and worry him to death. They are armed with strong sharp teeth, and generally seize the whale by the lower jaw. It is said that the only part of the huge monster that they eat is the tongue. The whalers give marvelous accounts of the immense strength of these "killers." They have been known to drag a whale from several boats which were towing it to the ship."

The saw-fish is also a most formidable assailant of the whale. The upper jaw of this fish is prolonged into a projecting flattened snout, the greatest length of which is six feet, forming a saw, armed at each edge with about twenty large bony spines or teeth. An account is given here of a combat that occurred on the west coast of Scotland, between a whale and some saw-fishes, aided by an auxiliary force of "thrashers" (fox sharks). The sea was dyed in blood from the stabs inflicted by the saw-fishes under the water, while the thrashers, watching their opportunity, struck at the unwieldy monster as often as it rose to breathe.

The sword-fish is also said to attack the whale, furnished, also, with a powerful weapon for defensive or aggressive war, in the shape of a bony snout about four or five feet long, not serrated like the saw-fish, but of a much stronger consistency—in fact, the hardest material known.

Beset by powerful enemies, the whale must have a

troublous existence; and if one thing can enlist our sympathies for these animals more than another, it is the well-known attachment they have to each other, and particularly for their young. It is said that when a female whale is wounded, her companions will remain around her until the last moment, or when they are themselves wounded. The whalers strike the young cubs, or "suckers," as they are called, not for their value, for these would hardly produce a barrel of oil, but the men know that the mother will start forth in their defence. She joins her cub at the surface whenever it has occasion to rise for respiration, encourages it to swim off, and seldom deserts it while life remains. She is then dangerous to approach, but affords frequent opportunities of attack. She loses all regard for her own safety in anxiety for the preservation of her young, dashes through the midst of her enemies, and even voluntarily remains with her offspring after various attacks on herself.

"In 1811," says Scoresby, "one of my harpooners struck a sucker with the hope of leading to the capture of the mother. Presently she arose close to the 'fast boat,' and, seizing the young one, dragged about six hundred feet of line out of the boat, with remarkable force and velocity. Again she rose to the surface—darted furiously to and fro, frequently stopped short, or suddenly changed her direction, and gave every possible intimation of extreme agony. For a length of time she continued thus to act, though pursued closely by the boats, and, inspired with courage and resolution by her concern for her young, seemed regardless of the dangers around her. At length one of the boats approached so near that a harpoon was hove at her: it hit, but did not attach itself. A second harpoon was struck, but this also failed to penetrate; so that, in a few minutes, three more harpoons were fastened, and in the course of an hour afterwards she was killed."

GREENLAND WHALE.

CHAPTER XI.

THE WHALE FISHERY AND ITS PERILS.

THE preparation for a cruise among the whales is very exciting; not so much as it used to be, because the supply of oil from other sources, the general use of gas, and other circumstances, have diminished the necessity which formerly prevailed for a means of illumination. Still there is a considerable demand for the valuable products of the whale—the oil, the whalebone, the spermaceti, and the ambergris, which constitute essential articles of commerce.

The Arctic regions have for several centuries been the chief haunts of the whale fishery. There has been, however, of late years a great decrease in the number of whales, and the fishery as a speculation has become more precarious.

Within a period of twenty years, no less than twenty

whale-ships were wrecked or crushed by the ice, and the sufferings of the crews were fearful.

The ships employed in the Northern fishery are constructed expressly for that object, and strengthened so as to encounter exposure in the ice regions. They are generally of from three to four hundred tons, each having a crew of about fifty men—experienced, hardy sailors—accustomed to the dangers of these particular expeditions. Six or seven light swift boats are requisite for each vessel; and another requirement is what is called a "crow's-nest," a kind of watch-tower, placed on the main-topmast to shelter the man on duty, whose office it is to keep a steady look-out with a telescope, for the spout of a whale in the distance, or the approach of drifting ice.

On reaching the Polar seas, the real hard work commences, the men being on watch night and day, and the boats kept ready for instant use whenever a whale is seen. On receiving an indication to that effect from the man in the "crow's-nest," a boat is launched, having a harpooner, a man to steer, one to look after the ropes, together with three or four rowers, and provided with an immense quantity of rope ready for use. The boat is steered rapidly and silently towards the whale, and on arriving within a few yards of it, the harpooner hurls his weapon so that it may enter under one of the monster's fins—a vulnerable part. The harpoon, in its most simple form, is a spear of about five feet in length, with a much flattened point, having sharp cutting edges, and two large flattened barbs. These are attached to a long line at the opposite end of the barbed joint. The gun-harpoon is a short bar of iron with the barbed spear at the end, and a ring with a chain for the attachment of the line. This is fired from a small swivel cannon attached to the whaler's boat; but the difficulty in whale fishing is to secure the capture of the animal, who sinks to a great depth on being struck, alternately rising to breathe, and sinking, so that the only chance of success is to tire it out. This is a critical moment for the crew in the boat, who are exposed to the most violent blows of the whale's head or fins, and still more of the tail, the tremendous power of which has been mentioned. The moment that the wounded whale disappears, a flag is displayed in the boat, at sight of which those who are on watch in the ship give the alarm by stamping on the deck, and those of the crew who are sleeping below, hastily throwing on a few clothes, launch the boats, and proceed to the assistance of their companions.

The greatest care is necessary by the boatman who has charge of the rope, in letting out and guiding the line to which the harpoon is attached. Should it be entangled for a moment, the whale would draw the boat beneath the waves. The time a wounded whale remains under water is generally half an hour, but some stay much longer. The boats take up a position near which it is likely to rise, when each harpooner strikes his weapon into the animal, and long and sharp lances are thrust into its side, until, exhausted with the loss of blood, the whale gives signs of approaching death by discharging blood from the blow-holes or nostrils, sometimes drenching the ice, boats, and men with it. As the huge animal plunges along in agony, its course is marked by a broad line of oil on the sea, issuing from its wounds.

The final capture is generally preceded by an awful and convulsive struggle; the tail lashes the water with fury, and the circles formed on the surface of the violently agitated waves extend to a great distance. When dying, the whale turns over on its side or back, a circumstance announced from the boats by loud cries and striking the flags. No time is lost: the tail is pierced and fastened with ropes to the boats, which drag the carcass to the ships with boisterous cheers.

A curious instance is related of a Dutch whaling crew,

who had, as they thought, secured their capture to the side of the ship, after towing it in triumph from the scene of conflict, missing their prize. The crew were giving vent to their delight, and the security seemed complete, for they were sailing a long distance from the ice-banks. They were having a good dinner to strengthen themselves before proceeding to the nauseous task of cutting up the animal. The feast was prolonged, but at length the men selected for the operation went on deck, with an air of importance, and full confidence. What was their astonishment to find that the whale was no longer alongside! It seems that the ship, driven before the wind, had dragged at the animal, the cord had broken, and the rich prize, which had caused so much peril and fatigue, had sunk to the bottom of the sea!

A dead whale, if left in the water, soon putrifies: it swells to an enormous size, until at least a third of the carcass appears above the surface of the water, and sometimes the body bursts by the force of the air generated within.

After the whale has been secured to the ship's side, the next operation is what is called "flensing," or securing the blubber and whalebone, which occupies about four hours, and is, as may be well imagined, anything but an agreeable occupation. The harpooners, having spikes on their feet to prevent their falling from the slippery surface, begin with a kind of spade and huge knives to make long parallel cuts from end to end, which are divided by cross-cuts into pieces of about half a ton. These are hoisted on deck, and after being reduced into smaller pieces, are put into casks and stowed away in the hold. When the flensing is proceeding and reaches the lips, which contain much oil, the whalebone is exposed and detached by means of bone handspikes and bone knives, and is hoisted upon deck in one mass, where it is split and stowed away. The two jaw-bones, for the quantity of oil they contain, are taken on deck, after which the huge carcass is abandoned to the birds and sharks, which are always waiting for their share, and speedily devour it.

In the early period of the Northern whale fishery, the animals being numerous and easy of capture, settlements were formed on the ice-coasts for boiling the blubber and extracting the oil, which was sent home in casks; but when the whales diminished, and the fishermen were obliged to seek them in the open sea, the capture became more difficult and dangerous, the settlements were abandoned, and the blubber was, for economy's sake, sent home to be boiled. In the different parts to which whale-ships are bound, there are establishments for extracting the oil; those at Hull are on the outskirts of the town. The blubber when conveyed to the boiling house is emptied from the casks into large vats, where it undergoes certain processes for extracting the oil.

The whale fishery in the Southern seas does not present the same amount of dangers which beset the whalers of the ice-regions, and differs in some particulars, being specially for the capture of the sperm whale.

It was well remarked by an old whaling captain that "if the Almighty had gifted the whale with a knowledge of his strength, few indeed would be caught." It is truly so, and there are occasions when the whale, inoffensive in its general habits, displays an amount of power and hostility which forms one of the grandest and most exciting spectacles that could be witnessed. In fact, the dangers which the whalers incur in their hazardous occupation, are most imminent.

As an instance of the spirit of mischief which sometimes animates the ocean monarch, I will relate what happened to the whale-ship, the *Essex*, Captain Pollard, in the Pacific Ocean. A number of sperm whales being signalled by the look-out, three boats were manned and sent in pursuit. The mate's boat was struck by one of them, and he was obliged to return to the ship to repair the damage. While he was thus engaged, a sperm whale, thought to be about eighty-five feet long, broke water about twenty yards from the ship on the weather-bow. He was going at the rate of about three knots an hour, and the ship at nearly the same rate, when he struck the bows of the vessel just forward of her chains. At the shock produced by the collision of two such mighty masses of matter in motion, the ship shook like a leaf. The whale passed under the ship, grazing her keel, and then appeared at about the distance of a ship's length, lashing the sea with fins and tail, as if suffering intense agony. He was evidently hurt by the collision, and greatly enraged. In a few minutes he seemed to recover himself, and started with great speed directly across the vessel's course to windward. Meanwhile the hands on board discovered the ship to be gradually settling down at the bows, and the pumps were instantly rigged. While working at them, one of the men cried out "God have mercy! he comes again!"

The whale had turned about one hundred yards from the ship, and was making for her with double his former speed, his pathway white with foam. Rushing head on, he struck her again at the bow, and the tremendous blow stove her in. The whale dived under again and disappeared, and the ship went down in ten minutes from the first collision.

The crew took to their boats as the vessel was sinking, and after fearful hardships and sufferings, the survivors of this catastrophe reached the low island called Ducies. It was a mere sandbank, nearly barren, and they could only obtain water and some wild-fowl. On this uninhabited island, dreary as it was, three of the men chose to remain, rather than experience again the uncertainties of the sea. The poor fellows were never afterwards heard of. The three boats, with the remainder of the crew, put off for the island of Juan Fernandez, two thousand miles distant. The mate's boat was taken up by the *Indian*, of London, ninety-three days from the time of the catastrophe, with only three survivors. The captain's boat was fallen in with by the *Dauphin*, but with only two men living. Thus, out of a crew of twenty, only five remained to tell the story of the whale's victory.

If the huge monster, in the exercise of his enormous strength, can shatter a large sailing vessel in such a way as to cause its destruction, you may readily imagine what perils are encountered by the hardy crews of the whaling-boats. A singular story is related of a Dutch harpooner, James Vienkes. A wounded whale had disappeared by diving, and the seamen were preparing to deal it a second stab, when the animal, on returning to the surface, struck its head against the boat and dashed it to atoms. Vienkes was hurled into the air, and fell on the monster's back, but contrived to bury his harpoon, which he had not let go, into it, and by means of this and the line he still held in his hand, he secured himself from slipping off. He called the other fishermen to his assistance, but their efforts to approach the whale were in vain. The captain of the ship, seeing no other way of saving his life, called out to him to cut the rope; but the harpooner was unable to do this, as his knife was in his trousers pocket, and he could not let go his hold for an instant. The whale was meanwhile advancing along the surface of the water at a swift rate, and it was fortunate for its rider that it did not dive. The sailors were beginning to despair of their comrade's life, when the harpoon by which he was supporting himself came out of the animal's body. He profited by the circumstance to cast himself into the sea, and struggling against the waves, regained the boats which had been unable to succor him. He was picked up at the moment his strength was exhausted, and his companions, furious at the disaster, pursued the whale, and killed it.

A writer relates: "Being myself in the first boat which approached a whale, I struck my harpoon at arm's length, by which we fortunately evaded a blow which appeared to be aimed at the boat. Another boat then advanced, and another harpoon was struck, but not with the same result, for the stroke was immediately returned by a tremendous blow from the fish's tail. The boat was sunk by the shock, and at the same time whirled round with such velocity that the boat-steerer was precipitated into the water on the side next to the fish, and carried down to a considerable depth by its tail. After a minute or so he arose to the surface, and was taken up along with his companions into another boat."

"In one of my earliest voyages," observes the same writer, "I remarked a circumstance which excited my highest astonishment. One of the harpooners struck a whale: it dived, and all the assisting boats had collected round the fast boat before it rose to the surface. The first boat that approached it advanced incautiously. It rose with unexpected violence beneath the boat, and projected it and all the crew to the height of some yards into the air. It fell on its side, and cast all the men into the water; one was somewhat injured, but the rest escaped."

In the year 1804, the ship *Adonis*, being in company with several others, struck a large whale off the coast of New Zealand, which became furious, and destroyed nine boats belonging to the different vessels, and then escaped. It was captured afterwards, however. Many harpoons of various vessels were found in its body.

This whale was extensively known to the fishermen under the name of "New Zealand Tom."

Sometimes the rope to which the harpoon is attached gets carried off, at a prodigious rate, by a whale in its efforts to escape, and the boat is carried far out to sea, and exposed to fearful perils. The annals of the whale fishery have many thrilling stories of wonderful escapes in such instances. A very remarkable instance occurred in connection with the ship *Independence*, Captain Belair. While cruising in the

Pacific Ocean, a whale was seen, and two boats were sent to capture it. The harpoon was fixed, and the boats were soon out of sight of the ship. An hour or two passed away, when suddenly another whale rose in the water, only a few yards from the vessel. The temptation to effect its capture was too strong for the captain, who ordered the remaining boat to be lowered, and, leaving but one man and two boys to take care of the ship, sprang into the boat with the rest of the crew. The harpoon was plunged into this whale also, and they were carried with the speed of the wind about fifteen miles from the ship. Then the whale plunged perpendicularly into the depths of the ocean. It was not long before they saw him, fathoms deep in the crystal waters, rushing up with open jaws to destroy the boat. By skilfully evading the attack, they escaped twice; but the third time, as the monster rose, he struck the boat in the centre of the keel, scattering the fragments and the crew over the waves, and then, plunging into the deep, disappeared. The captain and the men were now in the water, clinging to the pieces of the demolished boat. They were many miles from the ship, and could not be seen from the deck. The other boats were gone they knew not where. The hours passed slowly away, as they were drifting along at the mercy of the waves, until six o'clock in the evening.

The sun had now disappeared behind the distant waves, and a dreary night was settling down over the ocean. Just then they saw in the distance one of the absent boats returning to the ship. It was, however, far off, apparently beyond the reach of their loudest cries, and their hopes again fell. The boat at length drew nearer, and they redoubled their shouts; and at length they were heard, taken from the water, and carried almost lifeless to the ship.

The utmost care is requisite in "paying out" the rope when the whale is harpooned, so that no impediment occurs. The safety of the boat's crew depends upon the watchfulness of the man entrusted with this important duty. Scoresby, one of the most distinguished whalers that has ever been known on these perilous enterprises, records an instance which had a fatal consequence:

"As soon as the boats came within hailing distance (sent in pursuit of the whales), my anxiety induced me to call out and inquire what had happened. 'We have lost Carr!' This awful intelligence, for which we were altogether unprepared, shocked me exceedingly, and it was some time before I was able to inquire into the particulars of the accident which had deprived us of one of our shipmates. As far as could be collected from the confused accounts of the crew of the boat of which he went out in charge, the circumstances were as follow: The two boats that had long been absent on the outset, separated from their companions, and, allured by the chase of a whale and the fineness of the weather, they proceeded until they were far out of sight of the ship. The whale they pursued led them into a vast shoal of the species. They were, indeed, so numerous that their 'blowing' was incessant, and there could not have been less than one hundred. Fearful of alarming them without striking any, the crews in the boats remained for some time motionless, watching a favorable opportunity for commencing the attack. A whale at length arose so near the boat of which William Carr was harpooner, that he ventured to pull towards it, though it was meeting him, and afforded but an indifferent chance of success. He, however, fatally for himself, succeeded in harpooning it: the boat and fish, passing each other with great rapidity after the stroke, the line was jerked out of its place, and instead of 'running over' the stern, was thrown over the gunwale. Its pressure in this unfavorable position so careened the boat that the side sank below the water, and it began to fill. In this emergency the harpooner, who was a fine active fellow, seized the bight of the line, and attempted to relieve the boat by restoring it to its place; but, by some singular circumstance which could not be accounted for, a turn of the line flew over his arm, in an instant dragged him overboard, and plunged him under water to rise no more! So sudden was the accident, that only one man, who had his eye on him at the time, was aware of what had happened; so that when the boat righted—which it immediately did—though half full of water, they all at once, on looking round at the exclamation of the man who had seen him launched overboard, inquired, 'What has got Carr?' It is scarcely possible to imagine a death more awfully sudden and unexpected."

Some boats of the whale-ship *Aimwell* being in pursuit of these monarchs of the ocean, harpooned one. When struck, the animal only dived for a moment, and then rose again beneath the boat, struck it in the most vicious manner with its tail and fins, broke and upset it, and then disappeared. The crew, seven in number, got on the bottom of the boat; but the unequal action of the lines, which remained entangled with the boat, rolled it over occasionally, and thus plunged the men repeatedly beneath the water. Four of them recovered themselves, and clung to the boat; but the other three were drowned before assistance could arrive.

In the Arctic seas the whalers are exposed to many dangers from the ice. About the year 1856, Captain Deuchars, a most experienced navigator, in command of a fine strong vessel, the *Princess Charlotte*, lost it in Melville Bay. It was a fine morning, and all on board were anticipating a very successful voyage; the steward had just reported breakfast ready, when the captain, seeing the floes of ice closing together ahead of the ship, remained on deck to see her pass safely between them; but they closed too quickly—the vessel was almost through when the points of ice caught her sides abreast of the mizzen-mast, and passing through, held the wreck up for a few minutes, barely allowing time enough for the crew to escape and save their boats.

Poor Captain Deuchars thus lost his breakfast and his ship within *ten minutes.*

A wonderful case of deliverance from apparently certain destruction among the ice is recorded of the *Trafalgar*, an Arctic whale-ship. The account is given by Mr. Gibson, surgeon of the ship:

"Blowing a fresh gale, with rain, the floe to which the vessel was made fast set down under the lee ice, so as to render our situation perilous. Towards midnight we became unexpectedly entangled among heavy pieces of ice and floes, when the ship received some severe blows on her beams. Finding it impossible to get out, we lay to, and in half an hour the ship was close beset. Though I retired to bed when the ship was enclosed, I expected every minute to be called to quit it. Soon after, a large piece of ice pressing on the vessel opposite my bed-cabin, broke two or three of the timbers with a dismal noise. Thinking all was over, I sprang out of bed and found to my great consternation that the ship was under an enormous pressure from numerous large masses of ice surrounding her on all sides, without an opening of water sufficient for about two miles; and no other ship was in sight, although the weather was clear. Most of the crew were providing for shipwreck, and many of the people were supplicating Divine mercy for deliverance. Four days' allowance were cooked with all speed, other provisions were taken on deck, and everything of importance placed in readiness to be thrown on the ice. At noon, the man on the mast-head saw a ship, on which we instantly made signals of distress. At this time a dead silence prevailed throughout the ship, the crew looking on one another in awful suspense. At one time the pressure was so strong that the panels of the captain's state-room were forced out of their framing. About half an hour after this the ship was suddenly thrown upon her larboard side, on which all hands sprang upon deck. I shall never forget the confusion of the poor men, nor their wild looks when they gained the deck—for half of them were below at the time of the shock, and from the smallness of the hatch only one could get up at a time. Some leaped upon the ship's side and were going upon the ice, when the captain cried out to them to behave like men, and to stick to the ship so long as she remained above the water. We all stood on that part of the vessel nearest the ice, with our bags of clothing on our shoulders. For about fifteen minutes we had patiently waited our doom, when, by the interposition of Divine Providence, the wind changed, the ice began to set off from the ship, and in fifteen minutes more she recovered her upright position. The water now rapidly spread among the surrounding ice, and finally the vessel was warped out and floated safely on the waves."

A feaful series of calamities befell a small squadron of six very fine whaling vessels in 1830, during a storm in Baffin's Bay. Masses of ice were driven upon them, by which they were completely beset. The ships were ranged under the shelter of a large floe, having water barely sufficient to float them. Here they formed a line, one behind the other, standing close, stern to stem, and being at the same time so pressed against the ice, that in some places a boat-hook could with difficulty be inserted in the space. The sky darkened, the gale increased, the floes began to overlap each other, and closed upon the ships in an alarming manner. The sailors then attempted to saw out a sort of dock, where they hoped to be relieved from this severe pressure; but soon a huge flow was driven upon them with irresistible violence. The *Eliza Swan*, of Montrose, received the first shock, and was saved only by the ice raising her up. It next struck the *St. Andrew*, of Aberdeen, amidship, breaking about twenty of her timbers, and staving a number of casks; but it then, fortunately, moved along her side, and went off by the stem. It now reached successively the *Baffin*, of Leith; the *Achilles*, of Dundee; the *Ville de Dieppe*, a French ship; and the

Rattler, of Leith, and dashed against them with such tremendous fury, that these four noble vessels, which had braved for years the tempests of the Polar seas, were in a quarter of an hour shattered into fragments. The scene was awful: the grinding noise of the ice, tearing open their sides, and the masts breaking off and falling in every direction, were added to the cries of two hundred sailors, leaping upon the frozen surface with only such portions of their clothes as they could snatch in a single instant. The *Rattler* is said to have become the most complete wreck ever known. She was literally turned inside out, and her stem and stern carried to the distance of a gun-shot from each other; and the *Achilles* had her sides pressed together, and her stern thrust out, and her decks and beams broken into innumerable fragments.

Such are some of the perils which have been related by the hardy travelers of the ocean whose years have been spent in continued struggles, not only with the element,

> "Boundless, endless, and sublime,
> The image of eternity,—"

but with the huge monarch of the waters, whose reign has been disputed by a greater power in creation, who "sees all things for his use."

> "Thou little knowest
> What he can brave, who, born and nurst
> In Danger's paths, has dared her worst!"

CHAPTER XII.

SHARKS—THE PIRATES OF THE OCEAN.

IMAGINE a SHARK seventy feet long, with a tooth four inches and a half in the enamel, or the part visible above the socket, jaws with the bow about thirteen feet, and a mouth capable of stretching more than twenty-six feet around! This was one of the species of *fossil* sharks, an antediluvian animal, which has been discovered in the limestone rocks, the teeth and the vertebræ (small bones or joints composing the spine or back-bone) enabling the geologist to determine the species to which the animal belongs.

A tooth, the size of that mentioned, was shown to the distinguished French naturalist, Lacépède, and, in order to discover the proportions of the animal to which it belonged, he measured first the teeth, and next the stuffed specimens of all the sharks preserved in the Museum of Natural History in Paris, and he found in every instance that the relative proportions they bore to each other was one to two hundred, and he was thus enabled to ascertain the prodigious size and capacity of this formidable antediluvian animal.

Although the sharks of our own time are not of the same monstrous proportions, they are, from their immense strength and voracity, the objects of dread to those who behold them in their native element.

> "The type of horror and remorseless hate,
> Of villainy the worst."

The *White Shark* in particular, one of the largest of the tribe, and frequently weighing as much as a thousand pounds, sometimes measuring from twenty-five to thirty feet in length, abounding in warm latitudes, and attacking everything within his reach, deserves the title given to him of "the pirate of the ocean." When we relate that a lady's work-box has been found in the stomach of one of these sharks, and the papers of a ship that had been thrown overboard: that the baskets, shavings, cordage, ducks, hens, and buffalo-hides, etc., which had been thrown into the sea one morning from Captain Hall's ship, the *Alceste*, were found in the body of a captured monster shortly afterwards; that in another was discovered a tin canister, which, on being opened, was found to be nearly filled with old coins, you will have some idea of his indiscriminate appetite. He will devour even those of his own species. An anecdote is related of a Laplander capturing a shark, and fastening it to his canoe; he soon missed it, however, without an idea of how it had happened. A short time afterwards he took another shark of much larger size, in which, when opened he found the shark he had lost. An officer states that when some midshipmen had caught a shark, they pulled him into their boat, cut open his stomach, and then sent him back into the water. His body was instantly attacked by the sharks nearest to him, and was torn in pieces. The experiment was repeated with the same result.

The tenacity of life in the shark family is something extraordinary. The fish has been known to be active for many hours in the sea after its head has been taken off. Instances have been known of a shark having taken a bait in the depth of the sea, after its liver had been taken out for the purpose of extracting oil, and also when the whole of the entrails had been removed.

But a far worse character attaches itself to the shark, which is, his preference for human flesh: of all other food, it is this which he most prizes, and numbers of persons fall victims to his voracity in the seas he frequents. It is terrible to think of such a fate, for the huge monster is not only capable of snapping off a limb in a moment, or biting a person in two, but has been known to swallow a man alive. It is also stated on good authority that a shark was taken off the island of St. Margaret, which weighed fifteen hundred pounds, and the stomach was found to contain the whole body of a horse, which had probably been thrown overboard from some ship.

The following horrible tragedy is related: "As the ship *Karnak* was leaving the port of Nassau, a pilot fell overboard from her boat, in which he was being towed. The ship was stopped, and the boat instantly left for his rescue, while two life-buoys were thrown from the ship. The boat got close enough to give him the end of an oar, which he took, and cried, 'For God's sake save me!' The men were about to haul him into the boat, when he was carried down by a large shark which came up at the moment, taking the oar with him.

"A few days after the fatal accident, a shark was captured in Nassau harbor, and on being opened, the pilot's right hand and wrist, with a portion of his shirt (by which the hand was identified), a goat's head, with horns nine inches long, and a turtle's head were found in his stomach."

The French name this fearful animal the *Réquin*, or Requiem (the rest or stillness of death), in allusion to the deadly character of his habits: to add to the horror of his appearance, a phosphoric light is emitted from his huge body when near the surface of the water. To get at human flesh, the shark has been known to bound several feet out of the sea, and seize the unwary sailor occupied in the rigging of the vessel, when in full sail, and to leap into fishing-boats, and grapple with the men at their oars.

"No wonder that every man's hand should be raised

against this ferocious monster; and although of such fearful strength and audacity, he is sometimes overcome. The natives on the African coast show great courage and dexterity in attacking him. The mouth of the shark being placed in the lower part of the head, he is obliged, in order to seize his prey, to turn round in the water, and the negroes, taking advantage of this, thrust a knife into his stomach, the part where he is most vulnerable, for the skin on the upper portion of his body is so hard and rough that it forms a kind of armor, defending him from the bites of any animal he may encounter in the deep. This skin is even made use of by carpenters for polishing hard-grained wood, and it is also employed for other purposes where hardness and strength are required.

An amusing instance of punishing a shark for his greediness was related some years ago. The author of the article says:

"Looking over the bulwarks of the schooner, I saw one of these watchful monsters winding lazily backwards and forwards like a long meteor; sometimes rising until his nose disturbed the surface, and a gushing sound, like a deep breath, rose through the breakers, at others, resting motionless on the water, as if listening to our voices and thirsting for our blood. As we were watching the motions of this monster, Bruce, a lively little negro and my cook, suggested the possibility of destroying it. This was, briefly, to heat a fire-brick in the stove, wrap it up hastily in some old greasy cloth as a sort of disguise, and then to heave it overboard. This was the work of a few minutes, and the effect was triumphant. The monster followed after the hissing prey; we saw it dart at the brick like a flash of lightning, and gorge it *instanter*. The shark rose to the surface almost immediately, and his uneasy motions soon betrayed the success of the manœuvre. His agonies became terrible: the waters appeared as if disturbed by a violent squall, and the spray was driven over the taffrail where we stood, while the gleaming body of the fish repeatedly burst through the dark waves, as if writhing with fierce and terrible convulsions. Sometimes, also, we thought we heard a shrill, bellowing cry, as if indicative of anguish and rage, rising through the gurgling waters. His fury was, however, soon exhausted; in a short time the sounds broke away into distance, and the agitation of the sea subsided. The shark had given himself up to the tides, as unable to struggle against the approach of death, and they were carrying his body unresistingly to the beach."

In the South Sea Islands sharks are caught by means of a log of wood, set afloat with a strong rope attached to it, having a noose at the head. The fish, with his natural impetuosity, gets his head entangled, and, floundering about in attemps to escape, becomes tired out, and is then easily dispatched.

Captain Basil Hall gives an interesting account of the capture of one of these huge monsters. He says:

"The sharp, curved dorsal (the back) fin of an enormous shark was seen rising about six inches above the water, and cutting the glazed surface of the sea by as fine a line as if a sickle had been drawn along it. 'Messenger, run to the cook for a piece of pork,' cried the captain, taking the command with as much glee as if an enemy's cruiser had been in sight. 'Where's your hook, quarter-master?' 'Here, sir, here!' cried the fellow, feeling the point, and declaring it was as sharp as any lady's needle; and at the next instant piercing with it a huge junk of pork, weighing four or five pounds. The hook, which is as large as a little finger, has a curvature about as large as a man's hand when half closed, and is six or eight inches in length, while a formidable line, furnished with three or four feet of chain attached to the end of the mizzen-toptail-halyard, is now cast into the ship's wake.

"Sometimes the very instant the bait is cast over the

SHARK FISHING.

stern, the shark flies at it with such eagerness that he actually springs partly out of the water. This, however, is rare. On these occasions he gorges the bait, the hook, and a foot or two of the chain, without any mastication, and darts off with the treacherous prize with such prodigious velocity that it makes the rope crack again as soon as the coil is drawn out. Much dexterity is required in the hand which holds the line at this moment. A bungler is apt to be too precipitate, and jerk away the hook before it has got far enough into the shark's maw. The secret of the sport is to let the monster gulp down the whole bait, and then to give the line a violent pull, by which the barbed point buries itself in the coat of the stomach. When the hook is first fixed, it spins out like the log-line of a ship going twelve knots.

"The suddenness of the jerk with which the poor devil is brought up often turns him quite over. No sailor, however, thinks of hauling a shark on board merely by the rope fastened to the hook. To prevent the line breaking, the hook snapping, or the jaw being torn away, a running bowline is adopted. This noose is slipped down the rope, and passed over the monster's head, and is made to join at the point of junction of the tail with the body; and now the first part of the fun is held to be completed. The vanquished enemy is easily drawn up over the taffrail, and flung on deck, to the delight of the crew."

A sight of this voracious monster in his own element is never to be forgotten. It has been observed that the word "villain" has never been written in more unmistakable characters on any living creature than the shark. His appearance exhibits every character of ferocity. The head is large; the mouth wide and grasping; but the teeth, the most appalling features of the animal, are remarkable for their power of mischief: there are six rows in the upper jaw, and four in the lower. The teeth are triangular, sometimes two inches in breadth, sharp-edged, and notched like a saw, and as they are so planted in the jaw that each tooth is capable of independent action, being furnished with its own muscles, and as the strength of the jaws is enormous, they form a most terrific and formidable apparatus of destruction.

Although no part of the shark is wholesome for food, the flesh being coarse and leathery, yet it is eaten by the natives of Guinea, after being kept a considerable time to render it tender. The fins being gelatinous are used by the Chinese for making a rice soup. The liver yields an abundance of oil which is much esteemed. I have already mentioned the uses to which the skin is applied.

On some of the African coasts there are human beings so depraved and superstitious as to worship this fearful monster, and who believe that a person swallowed by him is sure to go to heaven. Their mode of adoration is thus: The negroes proceed in their boats to offer sacrifices of goats, poultry, and other things. But far more horrible still is the offering of an infant, reared for the purpose until it reaches the age of ten. The poor child is bound to a post on a sandy point at low water; as the tide rises the sharks arrive, and the infant is devoured, the parents fully believing that it will enter Paradise. We may ask ourselves if it is possible to find a more atrocious and dismal proof of human depravity!

The South Sea Islanders had some strange superstitious ideas relative to some of the shark species. Although they would not only kill but eat certain sharks, the large blue kind (*Squalus glaucus*) were deified by them; and rather than attempt to destroy them, they would endeavor to propitiate their favor by prayers and offerings. Temples were erected, in which priests officiated, and offerings were presented to the deified monsters; while fishermen and others, who were much at sea, sought their favor. Many funny legends were formerly in circulation among the people relative to the regard paid by the sharks at sea to priests of their temples, whom they were always said to recognize, and never to injure. But for the sharks, the South Sea Islanders would be in comparatively little danger from casualties in their voyages among the islands; and although, when armed, they have been known to attack a shark in the water, yet, when destitute of a knife or other weapon, they become an easy prey, and are consequently much terrified at such merciless antagonists.

A fearful instance is related of the rapacity of the shark, when a number of chiefs and people—altogether thirty-two—were passing from one island to another in a large double canoe. They were overtaken by a tempest, the violence of which tore their canoes from the horizontal spars by which they were united. It was in vain for them to endeavor to place them upright, or empty out the water, for they could not prevent their incessant overturning. As their only resource, they collected the scattered spars and boards, and constructed a raft on which they hoped to drift to land. The weight of the whole number who were now collected on the raft was so great as to sink it so far below the surface that they sometimes stood above their knees in water. They made very little progress, and soon became exhausted by fatigue and hunger. Destitute of a knife or any other weapon of defence, they fell an easy prey to these monsters. One after another was seized and devoured or carried away by them, and the survivors, who with dreadful anguish beheld their companions thus destroyed, saw the number of assailants apparently increasing as each body was carried away, until only two or three remained. The raft, thus lightened of its load, rose to the surface of the water, and placed them beyond the reach of the voracious jaws of their relentless destroyers. The voyage on which they had set out was only from one of the Society Islands to another, consequently they were not very far from land. The tide and the current now carried them to the shore, where they landed, to tell the melancholy fate of their fellow-voyagers.

The natives of Tahiti use hooks made of wood, and of the most formidable character, for shark fishing. These are a foot in length and an inch in diameter. They are such frightful implements that no fish less voracious than a shark would venture to approach them. In some, the marks of the sharks' teeth are numerous and deep, and show the effect with which they have been used.

One of the most sad and thrilling episodes of shark encounters was published some years since. A small schooner called the *Magpie* was cruising between the island of Cuba and the Havannah, in search of pirates. One evening the sea and the air were so calm that the vessel lay on the bosom of the water like some huge animal asleep, with her head towards the shore. The crew were engaged in telling those marvelous stories which seamen believe, and never fail to narrate to each other in their hours of idleness, for such occasionally visit even the mariner afloat. Lieutenant Smith, the commander, who had been on the look-out for the pirate ship as long as twilight enabled him to do so, laid aside his glass and descended into the cabin. All above, below, and around was now lulled as in slumber, for the laugh and the voice of the story-teller had become silent. Presently the mate of the watch observed a small black cloud resting over the land. The cloud was gradually increasing, and although the mate saw no ground to apprehend danger, he thought it right to communicate the fact to his superior officer. Mr. Smith commanded him to keep a sharp look-out, and he would join him on deck immediately. A moment after, a squall, as strong as it was sudden, burst from the cloud, and just as the lieutenant had ascended to the deck, the schooner was upset, and immediately sank.

Two of the crew were below, and they went down with

her; the others, twenty-two in number, were left struggling with the quiet deep, for the squall had passed, and the sky and the sea were again tranquil. It was now discovered that the boat had drifted from the vessel, and floated. A rush was made towards her, and several of the men attempted to get into her on the same side. The consequence was that she became half full of water, upset, rolled over and over, and at length lay with her keel upwards. Some got across her keel, others supported themselves by holding on to her with their hands, and thus all were for a time safe.

Mr. Smith now reminded the crew that it was impossible for them to remain long in this predicament, and exhorted them to right the boat and bale the water from her. He was immediately attended to; the men on the keel relinquished their seats, the boat was turned over, and two men were ordered into her to bale out the water. This they commenced doing with their hats, and it seemed probable that by perseverance their task would be accomplished. At this moment a man called out that he saw the fin of a shark. Immediately all was confusion; everyone endeavored to save himself, and in so doing rushed into needless danger. Smith begged them to persevere in attempting to clear the boat of water, and directed those not engaged in baling the water to keep splashing with their legs to frighten away the sharks. Again he was attended to; four men were in the boat baling, and the water was rapidly decreasing, when a noise was heard, and more than a dozen sharks darted in amongst them. In the panic which ensued the boat was again upset, and the men were at the mercy of the marine monsters. At first the sharks played about among the men, occasionally rubbing against them; but presently a loud shriek arose from one of them—his leg was bitten from his body! The attack was now general; shrieks arose from one and another. Some were torn from the boat, and several sank into the abyss, either through being bitten or from fear.

In this critical moment Lieutenant Smith was not dismayed. He still gave orders to the crew firmly and coolly, and was still obeyed by them. The boat was again righted, and the baling again commenced, Smith clinging to the stern while he directed and encourged his crew. For a moment he ceased to splash, while he looked into the boat to see what progress his men were making. At this instant a shark bit off both his legs above the knees. With fortitude scarcely to be believed, he endeavored to conceal the fact from his remaining crew, but, in spite of all his endeavors to suppress it, a deep groan escaped him; he loosed his hold of the boat and was about to sink, when two of his men caught hold of him and placed him in the stern-sheets. Although bleeding and in agony, he still exerted himself for his crew. He expressed his sorrow for their situation, gave them advice affectionately yet coolly, and ended with these words: "If any of you survive this fearful night and return to Jamaica, tell the admiral that I was in search of the pirate when this lamentable occurrence took place; tell him that I hope I have always done my duty, and that I——" At this moment some of the men endeavored to get into the boat, which was thus drawn on one side, and Lieutenant Smith rolled overboard, and sank to rise no more. The boat was now again upset. Some of the bleeding seamen placed themselves on the keel, but one by one dropped into the ocean. It was at eight o'clock when the *Magpie* sank, and before nine all on board of her were eaten by the sharks or drowned, with the exception of two, who succeeded in righting the boat and getting into her. They immediately began baling, and worked until they were nearly exhausted. The sharks swam round the boat, and endeavored to upset her, but failing, and perhaps gorged already, at length departed. The men worked at intervals, until the boat was nearly free from water, and then lay down and slept until after daylight. The morning was fine but sultry. The men were hungry, thirsty,

A FRIGHTFUL ENCOUNTER WITH SHARKS.

and fatigued: they looked around them; an unbroken ocean, a cloudless sky, and a burning sun were all that were within their view. They began to think of the only resource remaining for either—to kill his comrade and devour his flesh. They were men of equal strength, and both had knives. Each, however, seemed unwilling to resort to this horrible expedient except in the last extremity. The man at the stern (for they had separated from each other, in mutual apprehension, by nearly the whole length of the keel) knelt down and prayed, and his comrade followed his example.

As the morning went on they suffered intensely from thirst, and aggravated their sufferings by attempting to allay it with salt water. The madness of despair was beginning to develop itself in one of them when a sail appeared in sight, which afterwards proved to be a brig steering towards them. One flung his jacket in the air, while the other hailed again and again, and sometimes both hailed together, although the brig was at such a distance that it was not possible their cries would be heard. She approached nearer and nearer, and so riveted were their minds on the brig that hunger and thirst were forgotten in the excitement of hope. The people on board the ship appeared to notice them, but just as they had reason to think that such was the case, she changed her course and hoisted additional sail. Still they attempted to gain their attention, and attempted to propel the boat with their hands; but all was in vain; the ship was becoming every moment more distant, and their chance of release from such a horrible condition, of course, fainter.

At this moment one of the sailors conceived the bold project of swimming to the brig, which was by this time two miles and a half from them. His comrade remonstrated with him, so wild and hopeless did the undertaking appear to him, especially as the fins of sharks were seen here and there above the water. After a little hesitation, caused by the appeal of his shipmate, and a short prayer, he jumped over. The splash occasioned by his doing so caused the sharks to disappear, and the man in the boat well knew that they were in search of his comrade. Immediately afterwards, three of them passed the boat towards him.

With the greatest anxiety the sailor in the boat watched his messmate: he swam well, kicking and splashing as he went, to frighten the sharks. Once he beheld one of them close to him; but he only swam the faster, and kicked more vigorously. The wind had freshened, the brig was sailing more fleetly, his cries were unheard by the crew, and he began to think he must yield himself a prey to the sharks. At last he saw a man look over the side of the vessel; he held up both his hands, jumped up in the water, and was at length seen. A boat was got out, the brave swimmer was picked up, and was soon joined by his comrade on board the brig. The sharks were defrauded of their prey. The two survivors of the *Magpie* were tried by a court-martial, and as a reward for their perseverance, industry, and obedience to their commander in circumstances of such peculiar peril, promoted to be warrant officers.

To this family of the sharks, belongs the blue species, to which we have alluded, and which visits the coasts of England, during the pilchard and herring fishery, but whose chief residence is the Mediterranean. It is about seven feet long. The whole of the upper part is of a slate-blue color, and the under side nearly pure white.

The Hammer-headed species are distinguished, as the name implies, from each side of the head being extended—hammer-shaped—into a kind of branch, which has the eyes at the outer extremity. Its habits are of the family character, and it never hesitates to attack man whenever an opportunity offers. The Smooth Shark is so named from the smoothness and softer nature of its skin than its other relations; it is about four feet in length, and is a frequent visitor to the British seas. The Dog-Fish is the most common of the minor members of the shark family. The Spinous Shark, so named from its "prickles," which resemble those on the stems of a rose-bush, is not, happily, a frequent visitor to British waters, though of inferior size to most of the family, being from four to eight feet. The Angel-Fish, or Monk-Fish, or Shark-Ray, closes our list of the "ocean pirates." The depressed form, rounded head, with the eyes on the upper surface, and the singularly expansive pectoral fins (which may, under the imaginative form of wings, have originated the designation of "angel") distinguish this strange, and, on the whole, uncouth fish, which partakes something of the character of the ray and the shark. It is not unfrequent on British coasts, and attains a considerable size, some weighing a hundred-weight. It is a fierce and dangerous fish to contend with, and fishermen tell strange stories of its strength and fury.

The Greenland Shark which abounds in the Northern seas, although smaller than his powerful relative, being usually about fourteen feet long and six or eight feet in girth, partakes of his ferocity, and is a fearful enemy to the whale, whom he frequently worries to death, and feasts upon afterwards, scooping out pieces from his body as large as a man's head. The blubber appears to be a peculiarly "dainty dish" to this Arctic monster, and, while the crew of a ship are employed in cutting up a whale, he will come in for his share, and is so greedy for his favorite food that the men consider themselves safe from his gripe. Insensible to pain and tenacious of life as are all the larger sharks, the Arctic member of this ferocious tribe has been proved to be so in a remarkable degree. A few ugly wounds do not spoil his appetite, and even when pierced through the body with a sailor's knife, he does not desert the whale's carcass until his appetite is fully satisfied. Even when the body is cut into parts, the separate portions continue to show signs of life for some time, and it is unsafe to put the hand into his mouth a good while after the head has been separated from the trunk.

The Greenlanders eat the flesh of this fish both fresh and dried, and twist his rough skin into a kind of rope. This shark is known to have seized a canoe covered with seal-skin (which was probably the attraction) in his mouth from beneath, and by closing his jaws, destroyed both the canoe and its inmate.

The largest of this terrible tribe, the Basking Shark, visits the British seas occasionally, though most abundant in the tropics. He has been seen off the coast of Scotland, and taken, from his enormous length, for the "sea-serpent," attaining upwards of fifty feet. One of this size was captured some years ago at Kuraci, at the mouth of the Indus. Happily, however, his voracity is not proportioned to his size, being satisfied chiefly with sea-slugs, small fishes, jelly-fish, etc. Pennant mentions a basking shark twenty-six feet in length, taken off Anglesea, from which one hundred and fifty-six gallons of oil were obtained.

It is said that the pilot-fish is a guide and companion to the shark in his pursuit of prey. Whether this pretty fish, which is only about a foot in length, really does befriend and assist the ocean monster is not quite certain, but some accounts give an air of probability to the belief. One of the first voyagers to the East Indies, alludes to this circumstance in a fanciful manner. Describing the sharks, he says: "These have waiting on them six or seven small fishes, which never depart, with guards (bands), blue and green, round their bodies, like comely serving-men, and they go two or three before them, and some on every side." We have seen three instances in which the shark was led by the pilot. When the former neared the ship the latter swam close to his snout or near one of his breast-fins; sometimes it darted rapidly forwards or sideways, as if looking for something, and constantly went back again to the shark. When we threw overboard a piece of bacon fastened on a great hook, the shark was about twenty paces from the ship. With the quickness of lightning the pilot came up, smelt at the dainty morsel, and instantly swam back again to the shark, swimming many times around his snout and splashing, as if to give him exact information as to the bacon. The shark now began to put himself in motion, the pilot showing him the way, and in a moment he was fast to the hook."

Dr. Bennett, a Naturalist, says: "I have observed that if several sharks swim together, the pilot-fishes are generally absent; whereas, on a solitary shark being seen, it is equally rare to find it unaccompanied by one or more of these reputed guides. The only method by which I could procure this fish was, that when capturing a shark, I was aware that these faithful little fishes would not forsake him until he was taken aboard; therefore, by keeping the shark, when hooked, in the water until he was exhausted, or, as the sailors term it, "drowned," the pilot-fish kept close to the surface of the water over the shark, and. by the aid of a dipping-net, fixed to the end of a long stick, I was enabled to secure it with great facility."

The pilot-fish, like the mackerel in shape, has five conspicuous transverse bands round the body, and the general color is a silvery grayish-blue. It is common in the Mediterranean and abounds in the warmer parts of the ocean.

CHAPTER XIII.

SEA-HORSES, AND NARWHALS.

ALL the shores and borders of the Arctic zone are crowded with amphibious animals, which appear to form an intermediate link between whales and quadrupeds. Among these we will now notice the *Morse* (derived from the Russian *morss*) or *Walrus* (from the Norwegian *hval-ros*, whale-horse), also called by sailors the Sea-Horse. It is a large, shapeless, unwieldy creature, from twelve to fifteen feet in length, and eight to ten feet in circumference; the head small, the limbs short, and of an intermediate character between fins and legs. The eyes are small and brilliant; the nostrils are large, somewhat round, and placed on the upper part of the snout or muzzle. The lips are remarkably thick and covered with bristles. The neck is short. The insides of the paws are protected by a rough horny kind of coating, of a quarter of an inch thick; the fore-paws, or webbed hands, are from two to three feet in length, and, being expansive, can be stretched to a considerable width. The color varies with age; the young are black, they then become brown, and gradually pale, until in old age the walrus is white. The hairs, thick as a crow-quill, together with the long white tusks and fierce-looking eyes, give the animal a most diabolic look as it raises its head above the waves. Previous to the development of the tusks in the young walrus, the front face, when seen at a little distance, bears a striking resemblance to the human countenance; and this appearance seems to have given rise to the fanciful reports of mermen, or mermaids, in the Northern seas. Captain Scoresby mentions that he has seen a sea-horse in such a position and under such circumstances that it was easy to mistake it for a human being. The surgeon of his ship actually reported to him that he had seen a man's head just appearing above the water!

The most remarkable feature of the walrus consists in the two teeth or tusks, which are directed downwards from the upper jaw, and are sometimes nearly two feet in length, diverging at their points, and weighing from five to ten pounds. They are of beautiful white bone, almost equal to ivory, and are much employed in the fabrication of teeth, chessmen, umbrella-handles, whistles, and other small articles. The Greenlanders and other people of the North make hunting weapons from them, and domestic tools. These tusks not only serve the animal in procuring its food—which is said to be shell-fish and marine vegetables—but are formidable weapons against its foes. They also enable the walrus to raise its unwieldly bulk upon the ice, when its access to shore is prevented.

The speed of this animal in the water is very great, and a contrast to its sluggish appearance on the ice. Large numbers of them crowd together on the shore, and present a curious spectacle. The moment the first lands, so as to be dry, it will not stir until another comes, and urges it forward by beating it with its great tusks; this one is served in the same manner by the next, and so on in succession, until the whole are landed, tumbling over one another in the operation.

In the voyages of the early navigators of the Arctic seas,

they found the walrus, hitherto a partially unmolested animal, easy of capture. Stephen Bennet, the captain of the *God-speed*, a vessel of sixty tons, writes: "We saw a huge morse putting his head above water, making such a horrible noise and roaring, that they in the boat thought he would have sunk it." In another place they found a multitude of these monsters of the sea lying like hogs upon a heap." They shot at them in vain until their muskets were spoilt and their powder was spent, when "we would blow their eyes out with a little pease-shot, and then come on the blind side of them, and with our carpenter's axe cleave their heads; but for all that we could do, of about a thousand were killed but fifteen." They filled a hogshead with the loose teeth found on the island. The navigators became more expert in their cruel onslaught upon the poor animals, for in a subsequent voyage the same captain relates that in six hours they slew from seven hundred to eight hundred, not only for the sake of the teeth, but boiling the blubber into oil. They also contrived to get on board two young walruses, male and female; the latter died on the passage, but the other reached England, and was taken to Court, "where the King and many honorable personages beheld it with admiration." It soon, however, fell sick and died.

Captain Cook, who was among the first to give anything like a distinct account of this curious animal, says:

"We got entangled with the edge of the ice, on which lay an innumerable multitude of sea-horses. They were lying in herds, huddled one over the other, like swine, and were roaring and braying very loud, so that in the night, or in foggy weather, they gave us notice of the vicinity of the ice before we could see it. They were seldom in a hurry to get away until after they had been fired at, when they would tumble over each other into the sea in the utmost confusion. Vast numbers of them would follow us, and come close up to the boats, but the flash of a musket in the pan, or even the

STRATAGEM OF A WHITE BEAR.

bare pointing of one, would send them down in an instant. We never found the whole herd asleep, one being always on the watch. This, on the approach of a boat, would rouse the next, and the alarm being gradually communicated, the whole herd would speedily awake."

The walrus is hunted chiefly for its oil and tusks; the natives of the northern shores esteem its flesh highly, and it is greedily eaten along with the lard and even the skin. It has been calculated that about a thousand walruses were captured yearly in the seas about Spitzbergen.

Though generally of a peaceful and harmless nature, yet when attacked by foes, and especially by man, these huge animals will defend and support each other with remarkable courage and fidelity, fearlessly proceeding to the rescue of an unfortunate associate, and striving even to death for its deliverance. Martens relates having killed some sea-horses on the ice; "the rest came all about our boat, and beat holes through the sides of it so that we took in abundance of water, and were at length forced to row away because of their great numbers, for they gathered themselves more and more together, and pursued us, as long as we could perceive them, very furiously."

A similar incident is given, where a boat's crew proceeded to attack two hundred of these animals, but they made almost desperate resistance; some of them with their cubs on their backs; and one of them tore open the planks of the boat in two or three places.

Captain Phipps relates that two officers engaged in an encounter with a walrus, who, on being wounded, plunged into the water, and obtained a reinforcement of its fellows, who made a desperate attack on the boat, wresting an oar from one of the men, and had nearly upset her, when another boat came to her assistance.

The affection of the mother for its young is remarkable. Captain Cook, in his third voyage, says:

"We hoisted out the boats, and sent them in pursuit of the sea-horses that surrounded us. Our people were more successful than they had been before, returning with three large ones and a young one. On the approach of our boats towards the ice, they took all their cubs under their fins, and endeavored to escape with them into the sea. Several, whose young ones were killed or wounded, and were left floating on the surface, rose again, and carried them down, just as our people were going to take them into the boat, and they might be traced bearing them a great distance through the water, which was colored with their blood. We afterwards observed them bringing them up at times above the surface, as if for air, and again diving under it with a dreadful bellowing. The female in particular whose young had been destroyed and taken into the boat, became so enraged that she attacked the cutter, and struck her tusks through the bottom of it.

Another instance is mentioned, where, in the vast sheet of ice which surrounded the ships there were occasionally many pools, and when the weather was clear and warm, animals of various kinds would frequently rise and sport about in them, or crawl from thence upon the ice, to bask in the warmth of the sun. A walrus rose in one of these pools close to the ship, and finding everything quiet, dived down again, and brought up its young, which it held to its breast by pressing it with its flipper. In this manner it moved about the pool, keeping in an erect posture, and always directing the face of its young towards the vessel. On the slightest movement on board the mother released her flipper, and pushed the young one under water; but when everything was quiet, again brought it up as before, and for a length of time continued to play about the pool, to the great amusement of the sailors."

Man is not the only assailant of the sea-horse. On land its especial foe is the great Polar bear, and between these animals there are often terrible battles. On these occasions the tusks of the walrus stand in good service, for they manage, usually, to beat off the grizly enemy, though at the cost of many severe wounds.

An amusing instance is given of the cunning displayed by Bruin in his chase after the walrus:

"One sunshiny day, one of these animals, about ten feet in length, rose in a pool of water not very far from us, and after looking round, drew his greasy carcass upon the ice, where he rolled about for a time, and at length laid himself down to sleep. A bear which had probably been observing his movements crawled carefully upon the ice on the opposite side of the pool, and began to roll about also, but apparently more from design than amusement, as he progressively lessened the distance that intervened between him and his prey. The walrus, suspicious of his advances, drew himself up preparatory to a precipitous retreat into the water, in case of a nearer acquaintance with his playful but treacherous visitor. On this the bear became instantly motionless, as if in the act of sleep, but after a time began to lick his paws and clean himself, encroaching occasionally a little more upon his intended prey. But even this artifice did not succeed: the wary walrus was far too cunning to allow himself to be entrapped, and suddenly plunged into the pool, which the bear no sooner observed than he threw off all disguise, rushed toward the spot, and followed him in an instant into the water, where he was as much disappointed in his meal as we were of the pleasure of witnessing a very interesting encounter."

At sea, the sword-fish is the most nimble and fiercest enemy of the walrus. We should scarcely imagine from the uncouth and heavy appearance of the animal that it would exhibit any striking traits of intelligence; but it seems that when young it is not difficult to domesticate. Lamont mentions having seen one about the size of a sheep on board a

Norwegian vessel, and the most comical *fac-simile* imaginable of an old walrus. It had been taken alive after the harpooning of its mother, and was as playful as a kitten. It was a great favorite with all on board, and the only thing annoyed it was pulling its whiskers.

Another tusky inhabitant of the Arctic seas is the *Narwahl*, or *Monodon*, or what is popularly called the *Sea-Unicorn*, also an animal of the Mammalian order, about sixteen feet long and eight feet in circumference. In appearance the narwahl resembles a small whale, but with the addition of two long, straight, and pointed tusks, like spears, spirally twisted, directed forwards, and differing in length, the left one being about seven feet and a few inches, and the right one seven feet. It frequently happens, however, that only one of these tusks grows, and the other, somehow strangled, remains shut up in the bone like a nut. This will account for the appellation given to the narwahl of the "sea-unicorn." These tusks are of a whiter and harder substance than ivory. The Kings of Denmark possess a magnificent throne in the Castle of Rosenberg made of this material.

In former times, when the origin of the horns of this animal was not well known, they were supposed to possess miraculous powers for healing diseases. The monks, in particular, fostered the delusion, and pretended that every ill under the sun could be removed by their power. The narwahl has no true teeth in either jaw; the mouth is small and the lips are stiff, but it is able to catch and swallow so large a fish as the skate, the breadth of which is nearly three times as much as the width of its own mouth. It seems probable, however, that the horn serves them in this need, the fish being pierced with it, and killed before devoured. It is used, also, in digging sea-plants from the rocks at great depths, in order to drive from their retreats the shrimps and other animals on which the narwahl feeds. The tail is about twenty inches long and four feet broad. It has no dorsal or back fin, but

FIGHT BETWEEN WALRUS AND POLAR BEARS.

in place of it there is an irregular, sharp, fatty ridge, two inches in height, extending between two and three feet along the back, nearly midway between the snout and the tail. The prevailing color of the animal is bluish-gray on the back, variegated with numerous dark spots, with paler and more gray marks on a white ground at the sides. In old sea-horses the color is wholly white, or yellowish-white, with dark-gray spots. They are quiet and inoffensive in their habits, and swim with great rapidity. When respiring on the surface of the water, after blowing repeatedly, they frequently lie motionless for several minutes with the back and head just appearing above water. When harpooned, they dive to a considerable depth, and on returning to the surface for respiration, are readily killed in a few minutes with the lance. Near the coast they are always seen in flocks in the severest winters. The Greenlanders drive them with their sledges to fissures in the ice, where they are dispatched. The blubber, enwrapping the whole body, is from two to four inches in thickness.

When a number of sea-horses are together, they divert themselves in gambols, when, their horns appearing above the water, as if brandished about like weapons, have a singular effect, and the clattering noise they produce, with a kind of gurgling sound of the animals themselves, would lead one to suppose that some hostile proceedings were going on; but it is merely a playful movement of instruments which, if aggressively employed, would be dangerous. The force with which the narwahl urges its speed may be conceived by the circumstance that its tusk has been sometimes found driven through the planks of vessels.

CHAPTER XIV.

NAUTILI THE FLOATING NAVIGATORS OF THE OCEAN.

> "Spread, tiny nautilus, the living sail,
> Dive at thy choice, or brave the freshening gale
> If unreprov'd the ambitious eagle mount
> Sunward, to seek the daylight in its fount,
> Bays, gulf, and ocean's Indian widths shall be
> Till the world perishes a field for thee."—WORDSWORTH.

AMONG the most interesting and poetical illustrations of the wonders of the ocean are the singular floating animals, of which the Nautilus—called by Byron "the ocean Mab," "the Fairy of the Sea"—will be, undoubtedly, familiar to you from the great beauty of its shell, which renders it a favorite ornament in many houses.

Very interesting stories and verses have been written on the sailing and rowing habits of these curious animals; and their appearance, when seen skimming the water, would strongly favor such ideas. The Dutch naturalist, Rumphius, in giving an account of the rarities at Amboyna, the principal of the Molucca islands, says: "When the nautilus floats on the water, he puts out his head and all his tentacles, and spreads them upon the water; but at the bottom he creeps in a reversed position, with his boat above him, and with his head and tentacles (feelers) on the ground, making a tolerably quick progress. He keeps himself chiefly on the ground, creeping also, sometimes, into the nets of the fishermen; but after a storm, as the weather gets calm, they are seen in troops, floating on the water, being driven up by the

THE NAUTILUS.

agitation of the waves. This sailing is not, however, of long continuance, for having taken in all their tentacles, they upset their boat and so return to the bottom.

Until a comparatively recent period, very little was known of the nautilus; for, although shells were plentifully found on the shores of the warm seas it inhabits, the fish itself, living chiefly at the bottom of the sea, creeping like a snail, or lying in wait for runaway crabs or suchlike food, was difficult to obtain. However, a specimen was captured by Mr. Bennett, a naturalist, at the New Hebrides, and the great naturalist, Professor Owen, described the fish in a valuable memoir. The specimen is still preserved in the museum of the Royal College of Surgeons, in London. Little could be known from the shell itself; but here was the tiny navigator of the ocean, that would ride out a storm in which the strongest man-of-war might founder, revealed in all its most curious mechanism: the oars and aërial sails—disappearing, to give place to its real method of propulsion.

The *Paper Nautilus* has eight tentacles, and one pair of these expand at their extremities into broad and thin membranes, which compose a web of several sorts of fibres, interwoven for the wrapping up of some parts, the fibres giving them an elasticity by which they can contract and grasp the parts they contain—whence the fable received through so many ages, of its sails; the membranous arms of the fish are the organs for secreting and repairing the shells.

The functions of the supposed sails of the paper nautilus were determined by an experiment. One of the "sails" was cut off in several living specimens, the right sail being removed in some, the left in others; and the creatures were then kept in a submarine cage, and supplied with food. Some of them survived the operation for four months, when it was found that the shell had grown only on that side on which the membranous arm had been preserved; thus showing the animal to be the builder of its own habitation, and that the expanded arms do not serve the purposes of sails.

The real rower on the ocean is the beautiful little blue and silver shell-fish, the *Glaucus*, also a tenant of the warm seas, who swims with great swiftness by aid of its conical and oar-like appendages.

A wonderful builder is the nautilus, as may be seen by the chambers it fashions for its own accommodation; for the shell is divided into partitions, and as the animal increases in size it forms another and larger apartment proportionate to its growth, leaving the others empty as it proceeds, until, satisfied with its labors, it becomes the occupant of the highest chamber, though still communicating with the chambers it has abandoned, by means of a membranous tube which passes through the centre of each, enabling the nautilus by throwing air or gas into the empty chambers, or by exhausting them of air, to rise or sink into the water at will.

How truly wonderful is the intelligence displayed by the tiny nautilus in its chambered dwelling! "These beautiful arrangements," Dean Buckland once remarked, "are and ever have been subservient to a common object—the construction of hydraulic instruments, of especial importance in the economy of creatures destined to move sometimes at the bottom, and at other times upon or near the surface of the sea. The delicate adjustments whereby the same principle is extended through so many grades and modifications of a single type, show the uniform and constant agency of some controlling intelligence; and in searching for the origin of so much method and regularity amidst so much variety, the mind can only rest when it has passed back through the subordinate series of second causes to the great First Cause, which is found in the will and power of a great Creator."

The *Pearly Nautilus*, thus named from the shell being lined with a layer of the most beautiful pearly gloss, inhabits the Indian and Pacific Oceans. Nothing can exceed the pure loveliness of this "gem of the deep;" the interior being white, like the finest porcelain, and streaked with reddish chestnut. It is highly prized in Eastern countries, where it is made into drinking cups. The Chinese are particularly expert in manufacturing it into various ornaments.

There are other floating navigators of the deep: among others, the *Snail-slime-fishes*, which frequent the Arctic seas, and are found in immense quantities on the coast of Spitzbergen. The shell is the boat of this animal, which it rows through the water by a dip of its raised fins. In this act the open extremity of the shell is its prow, the opposite end occupies the place of a poop, and the margin of the body resembles and performs the office of a keel. A writer says: "I have often seen it with admiration and pleasure. He can move in a retrograde manner. When weary with rowing, or when touched, the little boatman contracts his oary fins, and drawing within the shell, sinks to the bottom, where he rests for a short time. Then again he rises upwards, rowing obliquely until the surface is attained, when his course is held in a straight line over the trackless surge. When taken out of the shell, although without injury and in the water, he immediately dies."

Before quitting the nautilus, we may add, that the shells of this "ocean navigator" abound in the coral seas, and are cast on shore in such profusion, that many tons' weight are collected at New Caledonia and the Fiji Islands, and are conveyed to Sydney. The young shells when polished obtain a high price.

The *Argonaut* differs from the true nautilus, inasmuch as the shell is not divided into chambers, but has one spiral cavity, into which the animal can entirely withdraw itself. From the disproportionate size of the last whorl (a wreath or turning of the spires of *univalves*, or shells of one piece only), it has some resemblance to a canoe, the spire representing the poop. If the waves rise or danger threatens, the argonaut withdraws all its arms into the shell, contracts itself there, and descends to the bottom. The body does not penetrate within the spire of the shell, nor does it adhere to it; at least, there is no muscular attachment, which led to the supposition that it occupied a shell belonging to some other animal. This freebooting stigma does not belong to the argonaut, for experiments have proved that the animal is its own builder, and consequently a rightful tenant of his mansion.

There is a curious and highly interesting floating object to which we may call the reader's attention, the *Sea-Bladder*, called by seamen the "Portuguese man-of-war, and by the French sailors the "galley" or "frigate." This singular zoophyte, or animal plant, for it combines the two natures, is seen floating, sometimes singly, at other times in vast numbers, in the tropical seas, and attracted the attention of naturalists from a very early period. The notion of its sailing properties may have arisen in consequence of the crest which it has the power of erecting along the ridge of his back, which, when caught by the wind, assumes somewhat the appearance of a natural sail, by means of which it seems enabled to glide over the surface of the ocean. This, however, is not the case, as the creature does not move by this means, nor does it appear to possess the power of imparting any special diréction to its course, which is entirely at the mercy of the wind and waves. The body itself, upon which the ridge or crest erects itself, is of a slight half-transparent character, and has somewhat the appearance of an unusually solid soap-bubble, glistening with a more than ordinary amount of various colored hues.

Mr. Bennett describes this body as of delicate crimson tints, as he saw it floating on the waves. There are also veinings of rich purple, and opaline flashes of azure, orange, and green, changing in color at every movement; and its long dependant tentacles or feelers are of the deepest purple.

Dr Collingwood mentions having observed these splendid zoophytes in the Atlantic Ocean, near the equator, sailing by from time to time during the day, and attracting attention by their large size and brilliant color. "They had the appearance of beautiful prismatic shells, standing upright on a rich blue cushion, the cell being radiated from the base or cushion to the circumference, which was fringed with a rich and bright rose color." He captured several specimens, and the largest measured in the bladder eight inches, and the greatest vertical circumference ten inches and a quarter, The long dependant tentacles or feelers are from four to five feet in length, and are capable of being extended much farther when shot off for the capture of prey.

But the glory of these magnificent objects, so developed in their native element, fades, like sea-weeds, as the zoophyte is taken from its watery home, with the exception of the long tentacles, which retain their color (dark purple) until decomposition takes place. "There is no rose without a thorn," is a well-known saying; and this gaily-colored zoophyte has a dangerous stinging property to those who handle it incautiously. An instance is related of a sailor seeing one within reach from a boat, who took it up with his naked hands; the threads or elastic tentacles clung to his arm, causing the man to yell with agony. He was quickly brought on board, and ran about like a maniac, requiring several men to hold him. When secured, and the proper remedies applied, he rolled about for some time groaning with pain; his arm was red, inflamed, and swollen, and remained so for some hours.

Its earliest modern name of "sea-nettle" is derived from that conferred upon this class of marine creatures by Aristotle, in consequence of the burning sting caused by the poisonous tentacles or feelers of several members of this group; a sting which leaves after it a white pimple, like that caused by a nettle.

A remarkable interest is attached to the nautilus from the very remote periods of time to which it can be traced; fossils being found in the most ancient rocks in which shell animals have been discovered, in various parts of the world, living ages before the Flood in temperate and tropical seas. In the London clay, which forms such a large extent of the substratum of the great metropolis, lie buried vast numbers of the pearly shells of the nautilus, which, evidently at a great distance of time, found in that country a congenial climate and home. The largest British specimens of the fossil nautilus occur in the carboniferous limestone, and specimens of these are preserved in the British Museum more than a yard in length, and thick in proportion.

In the museum of the Royal College of Surgeons, in London, is a specimen of the entire animal, soft parts and shell, of the pearly nautilus: a portion of the shell has been removed to show some of the chambers, and the membranous tube or syphon which traverses them. There is also a specimen of the paper nautilus suspended as when floating, with the expanded membranous arms in their natural position spread over the shell which they form and repair.

Resembling somewhat in appearance the nautilus, the shell being chambered and spiral, but differing otherwise in some respects, was the primitive navigator of the ancient seas, the ammonite, of which the shells now only remain, the most beautiful of all our fossils, and found in almost every country in the world, upwards of two hundred species having been described. The name is derived from a fancied resemblance of its shell to the ram's horn ornaments on sculptured heads of Jupiter Ammon. They are of very different sizes, varying to even three or four feet in diameter. The larger ones were formerly taken for petrified snakes, and were found in great numbers at Whitby in Yorkshire.

Sir Walter Scott alludes to this popular superstition in his poem of "Marmion," where the nuns of Whitby exultingly told

> "How of thousand snakes, each one
> Was changed into a coil of stone
> When holy Hilda pray'd."

The visitors to Whitby are still invited to buy a petrified snake, and to add to their natural appearance, the mouth of the ammonite is carved into a head, and eyes are introduced made of colored glass.

The ammonite, with a shell a yard across, would have been an animal large in proportion to its body-chamber, and requiring a certain amount of water to be displaced by its shell, to move at ease along the bottom of the sea in search of its food. The shell of the ammonite, though of the same flat character as that of the nautilus, appears to have been much thinner; but, to compensate for this, there were flutings which are seen in the surface, occasioned by the transverse ribs. The round knobs or bosses studding some of the ammonites were like gems on a diadem, adding strength as well as beauty to their form. The whorls or wreaths of the shell were rounder and more in number than that of the nautilus, and the tubes—the hydraulic instinct by which the chambers were supplied with air, or exhausted, for the ascent or descent of the animal—instead of running through the cells like that of the nautilus, went *round* the chambers of the ammonite.

How strange are the vicissitudes of all created things! While some survive the shocks and rents of time, others are known only as fossil memorials of the primitive world. The nautilus still rides on the crest of the ocean waves, but the ammonite—long, long since removed from the element in which it lived—only remains as a petrifaction to tell of its existence in ages before the Flood.

We also mention the little floating *Pteropoda* or Wing-

GIGANTIC CUTTLE FISH.

shells, the inhabitants of which pass their entire life in the sea far away from any shelter except that afforded by the floating Gulf-weed, and whose organization is peculiary adapted to that sphere of existence. In appearance they strikingly resemble the fry of the ordinary sea-snails, swimming, like them, by the vigorous flapping of a pair of fins. To the naturalist on shore they are almost unknown, but the voyager on the great ocean meets them where there is little else to arrest his attention, and marvels at their delicate forms and almost incredible numbers. They swarm in the tropical, and no less the Arctic seas, where by their myriads, the water is discolored by them for leagues. They are seen swimming on the surface in the heat of the day, as well as in the cool of the evening. In high latitudes they are the principal food of the whale and of many sea-birds.

Another floating inhabitant of the deep is described as the beautiful *Ianthina* or Ocean-Snail, which is quite blind, and has large horny jaws, furnished with sharp, curved, slender teeth. This animal is remarkable for floating shell downwards in the water, and the anterior part of the foot forms a shallow cup, which embraces the smooth anterior rounded part of the float. Thus the fish can raise or lower itself in the water at pleasure. When it wishes to bring its head to the surface of the water, this part of the foot is made to glide over the back of the float. The floats are made of a mucous film containing air; and when cut with scissors, the animal descended to the bottom of the vessel in which it was consigned, and did not make a new one.

The nautili belong to a class called *Cephalopoda*, so named from the singular attachment of the feet to the head —locomotive organs employed as oars or feet when moving along the bottom of the sea, and consisting of a circlet of muscular arms or tentacles, in addition to which many of this class have fins. To this same definition of Linnæus belong the Cuttle-fish, the bony scale on the back of which is employed for making pounce, tooth-powder, for polishing, and other purposes in the arts.

The common cuttle-fish is abundant on the English coasts. Its skin is smooth, whitish, and dotted with red. It attains the length of a foot or more, and is one of the pests of the fishermen, devouring partially the fish which have been caught in their nets. The eggs of the cuttle-fish are frequently cast on shore clustered together. Singularly interesting is the study of these creatures, which are provided with means of escaping danger, in their ink-bags, from which they can at will emit a fluid, darkening the water and thus enabling them to get off. This natural ink of the fish is employed in painting; Cicero tells us that it was anciently used for writing with.

Another property possessed by this class of animals is, that if any of its tentacles or feelers are bitten off, which is often the case—the conger eel having a special relish for the dainty morsel—others supply their place, the power of reproduction being given to them. The whale also regales on the cuttle-fish, and the plaice tribe have the same partiality. The most common species form the bait with which one-half of the cod taken at Newfoundland are caught.

The general description of the cuttle-fish may be thus described: the body oblong, or longer than broad, and depressed, sac-like, with two narrow lateral fins of similar substance with the mantle (the outside skin of shell-fish, which covers a great part of the body, like a cloak). There is an internal shell lodged in a sac on the back part of the mantle, somewhat oval and bladed-shaped, being comparatively thick near the anterior end, where it is terminated by a sharp point, affixed, as it were, to its general outline. The whole shell is light and porous, and is formed of thin plates, with intervening spaces, divided by innumerable partitions, and consists chiefly of carbonate of lime, with a little gelatinous and other animal matter, which is most abundant in the internal harder part of the shell. The eyes are very large, and the head is furnished with eight arms, each of which has four rows of suckers and two long tentacles, expanded and furnished with suckers on one side at the extremity. Cuttle-fish are enabled to leap out of the water by the sudden extension, not of their tails, but of their numerous arms, or other processes from their bodies.

In hot climates some of the species of cuttle-fish grow to a prodigious size, and are furnished with a fearful apparatus of arms with suckers, by which they can rigidly fasten upon and convey their prey to the mouth. In the eight-armed species which inhabit the Indian seas these tentacles are said to be no less than nine fathoms in length.

Extraordinary stories have been related of these animals. Pliny mentions the head of one which was as large as a cask, the arms thirty-six feet long. They are described as first darting from side to side in the pools, and fixing themselves so tenaciously to the surface of the stones that great force was required to remove them. When thrown upon the sand, they progressed rapidly in a sidelong shuffling manner, throwing about their long arms, ejecting their inky fluid in sudden violent jets, and staring about with their shining eyes in a grotesque and hideous manner. As food it was highly prized by the ancients, and is still much esteemed in some parts of the world. It is regularly exposed for sale in the markets at Naples, Smyrna, and in the bazaars of India. In a curious Japanese book there is a picture of a man in a boat engaged in catching cuttle-fishes with a spear; and also a fishmonger's shop in Japan, where a number of enormous cuttle-fishes are represented hanging up for sale.

Columbus describes the mode of fishing with the cuttle-fish pursued in his time by the natives of Santa Marta:

"They had a small fish, the flat head of which was furnished with numerous suckers, by which it attached itself so firmly to any object as to be torn in pieces rather than abandon its hold. Tying a line of great length to the tail of this fish, the Indians permitted it to swim at large. It generally kept near the surface of the water until it perceived its prey, when, darting down swiftly, it attached itself by its suckers to the throat of a fish, or to the under shell of a tortoise, nor did it relinquish its prey until both were drawn up by the fisherman, and taken out of the water."

In this way the Spaniards witnessed the taking of a tortoise of immense size, and Fernando Columbus himself affirms that he saw a shark caught in this manner on the coast of Veragua.

This account, strange as it may seem, has been corroborated by various navigators, and the same mode of fishing is said to be employed on the eastern coast of Africa, at Mozambique, and at Madagascar.

The South Sea Islanders have a curious contrivance for taking the cuttle-fish, which resort to the holes of the coral rocks, and protrude their arms or tentacles for the bait, but remain themselves firm within the retreat. The instrument employed for taking them consists of a straight piece of hard wood, a foot long, round and polished, and not half an inch in diameter. Near one end of this a number of the most beautiful pieces of the cowry or tiger-shell are fastened, one over the other, like the scales of a fish or the plates of a piece of armour, until it is about the size of a turkey's egg, and resemble the cowry. It is suspended in an horizontal position by a strong line, and is lowered by the fisherman from a small canoe until it nearly reaches the bottom. The fisherman then gently jerks the line, causing the shell to move as if it were inhabited by a fish. The cuttle-fish, attracted, it is supposed, by the appearance of the cowry (for no bait is used), darts out one of its arms, which it winds round the shell and fastens among the openings between the plates. The fisherman continues jerking the line, and the fish puts out successively its other arms until it has fastened itself to the shells, when it is drawn up into the canoe and secured.

DEVIL FISH.

In conclusion, we will mention that the cuttle-fish belongs to a period before the Flood, like the nautili; their undigested fossil remains are frequently noticed within the ribs of the *Ichthyosauri* and *Plesiosauri* in the limestone rocks, showing that then, as in the present day, to eat and to be eaten was the general law of nature.

CHAPTER XV.

MODES OF FISHING IN VARIOUS COUNTRIES.

"A thousand names a fisher might rehearse
Of nets intractable in smoother verse."—OPPIAN.

THE space devoted to this subject here must of necessity be brief. It will therefore be understood by the reader that many important and interesting details will have to be omitted. Though, as announced by the heading of this chapter, it is proposed to consider the manner of catching fish; this cannot be done without treating to some extent of the fish themselves, and the implements employed. This at once opens up a subject so extensive and varied, and withal so desirable to know and enjoy, that we have been somewhat embarrassed as to just what it would be desirable to omit in the list of description.

It will be noticed that the American fisheries have not been given the importance here that their magnitude would seem to demand. Of course, this omission has been purposely, and we believe the reader will decide, before he has finished reading this chapter, wisely made. In the first place, it is not proposed to present a compendium of dry, and, to some extent, uninteresting facts; and, secondly, we have deemed it best not to cumber these pages with descriptions of what many of our readers daily see and are therefore familiar with. On the contrary, we have in our illustrations compared primitive modes of fishing in foreign latitudes, with the more modern appliances, and, to some extent, European methods with our own. By this plan our matter must certainly be more picturesque, vivid, and interesting.

The use of nets for entrapping the finny inhabitants of the deep date from the earliest periods. Besides the frequent mention of them in the Holy Scriptures, we find illustrations in the bas-reliefs of Assyria, Greece, and Rome, and in the mural or wall paintings of Egypt. The latter nation delighted in fishing, and, not contented with the abundance afforded by the Nile, they constructed in their grounds spacious sluices or ponds for fish, like the *vivaria* of the Romans, where they fed them for the table, and amused themselves by angling. The fishermen, who composed one of the sub-divisions of the Egyptian castes, generally used the net in preference to the line. The ancients entertained a number of prejudices relative to the wholesomeness or injurious qualities of certain fish. The priests in Egypt were prohibited from eating fish of any kind. For fear of leprosy, the people also were forbidden the use of any fish not covered with scales. Moses adopted the same principles with the Jews: "Whatever hath fin or scales in the water in the seas, them shalt thou eat; whatever hath no fins or scales in the waters, that shall be an abomination to you."

The Greeks and Romans used nets; trawling at sea was also a favorite mode of angling, and harpoons were in general use, by means of which many large fish were secured. Some mosaics discovered at Palestrina represented men engaged in taking fish out of a ready decoy by means of small hand-nets. Arrian, in his "Indian History," mentions a people on the coasts of the Persian Gulf, who had nets capable of covering a quarter of a mile of sea, not made of twine, for hemp and flax were unknown in the land, but of the inner bark of palm trees, being, in fact, papyrus nets.

In the dialogues composed by Elfric to instruct the Saxon

youths in the Latin language, which are yet preserved in the Cottonian manuscripts, a fisherman is asked how he secures his prey, and he answers, "I ascend my ship and cast my net into the river; I also throw in a hook, a bait, and a rod;" which shows that in the earliest periods of the history of that country, nets of various kinds were employed for entrapping fish; indeed, although St. Wilfred is said to have taught the people of Sussex the use of the net (probably an improved kind), such means have been employed in different ways from remotest times. Until late years fishing nets have always been made by hand, and generally the thread has been a more or less thick twine of hemp, or flax, the thickness of the twine and the size of the mesh depending upon the kind of fish for which it was made; recently, however, great improvements have been made in the manufacture of nets, and machinery of the most beautiful minute kind has been invented for the purpose.

A great variety of nets are in use among fishermen, but the principal are the seine, trawl, and drift nets. The first is a very long but not very wide net, one side of which is loaded with pieces of lead, and consequently sinks; the other, or upper, is buoyed with pieces of cork, and is consequently kept on the surface of the water. Seines are sometimes upwards of a thousand feet in length. When stretched out they constitute walls of network in the water, and are made to enclose vast shoals of fish. The trawl is dragged along the bottom of the sea by the fishing-boat; and the drift-net is like the seine, but is not loaded with lead, and is usually employed for mackerel fishing. In the two fishery exhibitions at Arcachon and Boulogne in France, several years ago, a number of curious implements for the capture of the inhabitants of the deep were shown. In one corner were curious tongs for taking eels. Long stretches of netting for the sardine fishery, woven with thread so fine that it might be used for the manufacture of ladies' hose, were festooned over a division of the buildings. At another place was a leech-lifter, and near it were deadly traps for taking crabs and lobsters. From the roofs hung stretches of Scotch-made herring-nets, by far the best of their kind; and with such a wall of meshes floating in the sea as these nets present to the fish, each stretch being about a mile long, and with a fleet of a few hundred boats nightly centered on some well known fishing-ground, the wonder is, not that fishes are scarce and dear, but that a single herring could escape.

In 1864 an attempt was first made to fish by the electric light at Dunkirk, on the coast of France. A magneto-electric machine was afterwards employed. The light was constant at one hundred and eighty feet under water, and it extended over a large surface. As soon as the submarine lantern was immersed, shoals of fish of every description came to sport in the illuminated circle, while the fishermen outside of it spread their nets from the boats. The light illuminating the deep sea, the fish arriving in shoals, attracted by the fictitious sun, the boats at the edge of the lighted circle, the deep silence interrupted only by the grating of the electro-magnetic machine, formed altogether an imposing sight.

Before leaving this part of our subject, we may notice a curious invention stated in Rymer's "Fœdera," for which Charles I. granted a patent in 1632 to a physician, "for a fish-call or looking-glass for fishes in the sea, very useful for fishermen to call all kinds of fishes to their nets."

A singular method of getting fish is that in which other animals are employed for the purpose. Birds are thus trained by the Chinese. Falcons are not more sagacious in the pursuit of their prey in the air than in another element. They are called *alvoau,* and are about the size of a goose, with gray plumage, webbed feet, and have a long and slender bill, crooked at the point. Their faculty of diving, or remaining under water, is not more extraordinary than that of many other fowls that prey upon fish, but the wonderful circumstance is the docility of these birds in employing their natural instinctive powers at the command of the fishermen who possess them, in the same manner as the hound, the spaniel, or the pointer submit their respective sagacity to the huntsman or the fowler. The number of these birds in a boat is proportioned to the size of it. At a certain signal they rush into the water and dive after the fish, and the moment they have seized their prey, they fly with it to their boat, and though there may be a hundred of these vessels together, the birds always return to their own masters; and amidst the crowd of fishing-junks which are sometimes assembled on these occasions, they never fail to distinguish that to which they belong. When the fish are in great plenty, these astonishing purveyors will soon fill a boat with them, and will sometimes be seen flying along with a fish of such size as to make the beholder suspect his organs of vision; and such is their sagacity that when one of them happens to have taken a fish which is too large for a single falcon, the rest immediately lend their assistance. While they are thus laboring for their masters, they are prevented from paying any attention to themselves by a ring which is passed round their necks, and is so contrived as to frustrate every attempt to swallow the least morsel of what they take. They eat thankfully what is afterwards given them in reward. One of the old domestic sports of the Earls of Monteith, in their island home of Talla, was fishing with geese. A line with a baited hook was tied to the leg of a goose, which was made to swim in water of proper depth. A boat well filled escorted this formidable knight-errant. A marauding fish would take hold of the bait, and put his mettle to the test. A combat ensued, in which, by the display of both contending heroes of much strength and agility the goose always came off victorious, and would drag his prisoner to the boat in triumph.

No nation on the earth puts in practice a greater variety of modes for catching fish than the Chinese. One method is to nail on each side of long narrow boats a plank two feet broad, covered with white shining japan, and placed by a gentle inclination so that its lower edge just touches the surface of the water. This device is used at night, with the intent that the reflection of the moon should increase its deceptive influence; and whether the fish which are sporting around are dazzled by the splendor, or merely mistake the lustrous plank for the sparkling water, it is impossible to say, but in their moonlight gambols great numbers either fall on the plank and are secured, or fairly vault into the body of the boat.

In some places the Chinese soldiers have acquired the dextrous art of shooting fish with bows and arrows. To the arrow a long piece of packthread is attached, by means of which, when the fish is pierced, it is drawn to hand. In other places the muddy bottom is so replenished with the finny tribes, that men standing up to the waist in the water strike them with sticks. Besides these various devices, another is in general use, and consists in stretching out a net on four pieces of bamboo suspended by a long pole.

The South Sea islanders are expert fishermen, and their methods for the capture of the finny tribe are numerous, and some very ingenious. They have a singular mode of taking a remarkably timorous fish, which is called the needle, on account of its long, sharp head. A number of rafts are built, each about fifteen or twenty feet long, and six or eight wide. At one edge a kind of fence or screen is raised four or five feet by fixing the poles horizontally one above the other, and fastening them to upright sticks placed at short distances along the raft. The men on the raft go out at a distance from each other, enclosing a large space of water, having the raised part or frame on the outside. They gradually approach each other till the rafts join, and form a connected circle in some shallow. One or two persons then go in a small canoe towards the centre of the enclosed space, with long white sticks, which they strike in the water with a great noise, and by this means drive the fish towards the rafts. On approaching these the fish dart out of the water, and in attempting to spring over the raft, strike against the raised fence on the outer side, and fall on the surface of the horizontal part, when they are gathered into baskets or canoes on the outside. In this manner great numbers ot these and other kinds of fish, that are accustomed to spring out of the water when alarmed or pursued, are taken with facility. Fishing-nets are remarkably well made, and those for casting are used with great dexterity, generally as the islanders walk along the beach. When a shoal of small fish appear, they throw the net with the right hand, and sometimes enclose the greater part of them.

Next to the net the spear is most frequently used. This is darted at the fish, sometimes with one hand, but more frequently with both, and very successfully. When fishing on the reefs, they wear a kind of sandal made of closely-netted cords of the cloth plant, to preserve their feet from the edges of the shells, the spikes of the sea-urchins, etc.

It would be interesting to gaze upon a group of fishermen standing on a coral reef or rock, amidst the roar of the billows, and the dashing surf and foam that broke in magnificent splendor around them. With unwavering glance they have stood, with a little basket in one hand and a pointed spear in the other, striking with unerring aim such fish as the violence of the wave might force within their reach.

The shell, or shell and bone hooks, are curious and useful, answering the purpose of hook and bait, the small ones being made circular, and bent so as to resemble a worm; but the most common one is that used in catching dolphins, albicores, and bonitos. The shank of the hook is made with a piece of the mother-of-pearl shell, five or six inches long and three-quarters of an inch wide, carefully cut and finely polished, so as to resemble the body of a fish. A barb is fastened by a firm bandage of finely twisted flax; to the lower part of this the end of the line is securely fastened. When taken out to sea, the line is attached to a strong bamboo cane about twelve or fifteen feet long. When a shoal of fish is seen, the natives who angle, sit in the stern of the canoe, and hold the rod at such an elevation as to allow the hook to touch the edge of the water, but not to sink. When the fish approach it, the rowers ply their paddles briskly, and the light bark moves rapidly along. The deception of the hook is increased by a number of hairs or bristles being attached to the end of the shell, so as to resemble the tail of a flying-fish. The victims, darting after and grasping their prey, are at once secured. During the season two men will sometimes take twenty or thirty large fish in this way in the course of the forenoon.

The most ingenious method, however, of taking these large fish is by means of a *mast.* A pair of ordinary-sized canoes is usually selected for this purpose, and the lighter and swifter the more suitable they are esteemed. Between the fore part of the canoes a broad, deep, oblong kind of basket is constructed with the stalks of a strong kind of fern, interwoven with tough fibers of a tree: this is to contain the fish that may be taken. To the fore part of the canoes a long curved pole is fastened, branching in opposite directions at the outer end; the foot of this rests in a kind of socket fixed between the two canoes. From each of the projecting branches lines with pearl-shell hooks are suspended, so adjusted as to be kept near the surface of the water. To that part of the pole which is divided into two branches strong ropes are attached; these extend to the stern of the canoe, where they are held by persons watching the seizure of the hook. The tira, or mast, projects a considerable distance beyond the stern of the canoe, and bunches of feathers are fastened to its extremities. This is done to resemble the aquatic birds which follow the course of a small fish. As it is supposed that the bonito follows the birds with as much ardor as it does the fishes, when the fishermen perceive the birds they proceed to the place, and usually find the fish. The undulation of the waves occasions the canoe to rise and sink as they proceed, and this produces a corresponding action in the hook suspended from the mast; and so complete is the deception that if the fish once perceives the pearl-shell hook, it seldom fails to dart after it, and if it misses the first time is almost sure to be caught the second. As soon as the fish is fast, the men in the canoe, by drawing the cord, hoist up the mast and drag in the fish, suspended as it were from a kind of crane. When the fish is removed, the crane is lowered, and as it projects over the canoe the rowers hasten after the shoal with the greatest speed.

These and a variety of other methods of fishing are pursued by daylight, but many fish are taken by night. Sometimes the fishery is carried on by moonlight, occasionally in the dark; but fishing by torchlight is the most picturesque. The torches are bunches of dried reeds firmly tied together. Sometimes the natives pursue their nocturnal sport on the reef, and hunt the hedge-hog-fish. Large parties often go out to the reef and it is a beautiful sight to see a long line of rocks illuminated by the flaring torches. These the fishermen hold in one hand, and stand with the poised spear in the other, ready to strike as soon as the fish appears.

The Indians on the coasts of the Pacific have also a singular mode of taking the Candle-fish, or *Eulachon*, a most valuable acquisition to their domestic comforts. Immense shoals approach the shores in summer. and are caught in moonlight nights, when they come to sport on the surface of the water, which may often be seen glittering with their multitudes. The Indians paddle their canoes noiselessly amongst them,

THE GLOBE FISH.

HALIBUT.

and catch them by means of a monster comb or rake—a piece of pine-wood from six to eight feet long, made round for about two feet of its length at the place of the hand-gripe, the rest flat, thick at the back, but having a sharp edge at the front, where teeth are driven into it, about four inches long and an inch apart. One Indian, sitting in the stern, paddles the canoe; another, standing with his face to the bow, holds the rake firmly in both hands, the teeth pointing sternwards, sweeps it with all his force through the glittering mass, and brings it to the surface teeth upwards, usually with a fish, and sometimes with three or four, impaled on each tooth. This process is carried on with wonderful rapidity. This fish, although not larger than a smelt, enjoys the distinction of being probably the fattest of all animals, comparatively speaking: to boil or fry it is impossible, as it melts entirely into oil. Even in a dried state the Indians use it as a lamp, merely drawing through it a piece of rush pith as a wick, and the fish then burns steadily until consumed. By a peculiar mode of preparation, these fishes are preserved as a winter food, and notwithstanding their great fatness, they are said to be of an agreeable flavor. Drying is accomplished without any cleaning, the fish being fastened on skewers passed through their eyes, and hung in the thick smoke at the top of sheds in which wood fires are kept burning. They are then stowed away for winter.

We will now glance at the White Porpoise fishing in the St. Lawrence River. The animal mentioned is a species of whale, and is chiefly common in those quarters, being valuable for its oil, which gives a brilliant light only surpassed by gas, and its skin, which is manufactured into leather which has no equal for quality. The fish was formerly taken in enclosures made of light and flexible poles fixed in the beach, within which the porpoise pursued the small members of the finny tribe during high tide, and where, its appetite once satisfied, it became heavy and almost asleep from gluttony, and seemed to forget for several hours the dangers that surrounded it as the tide went out. The fishermen, silent, and on the look-out on the cliff, having seen that the waves had retreated, give the signal: two or three light skiffs (either bark or wooden canoes), manned by three or four expert rowers, appear upon the waves, which they scarcely touch with their oars. Standing in the bow of each of these canoes, a man with bare and muscular arm, a steel spear in his hand, intently follows with his eye the track of the fish, indicating the course to be taken, whether to the right or left, and strikes the mortal blows. Often after one of these vigorous strokes, which are enough to kill the largest porpoise, the spearsman may be seen, when he does not strike aright, urging on the pursuit for a new contest of speed between his skiff and the wounded animal: sometimes the blood which reddens the surface of the water indicates the course to be followed, and sometimes the sound of the subdued breathing of the porpoise, which comes to the surface of the water to breathe, throwing up a stream which descends in the form of a curve. The porpoise might break through this fence of flexible poles, eighteen or twenty inches apart, but it is afraid, and it returns by the way it came: a new stroke is given, but it is by a harpoon which has a rope attached to it. The struggle becomes more intense and exciting. The paddle at the stern of the frail skiff is alone put in requisition. It is now the boatman's turn to display his skill. The animal leaps out of the water, stops, dives, and turns about in every way; a white foam rises on each side of the boat, and its progress, hitherto so swift, is suddenly stopped; the animal is fatigued by its wound, wants to breathe, but fear keeps it below the water, and immediately the man in the bow rolls up at his knee the line which he had allowed to run out, and the boat is brought silently forward towards the victim. Again he stands up and with one hand brandishes the spear, while with the other he suddenly pulls the rope, inflicting fresh wounds: the fish once more leaps, but this time is the last, for a vigorous blow aimed at the spine between the head and the neck is fatal.

Another plan is to use nets for entrapping the porpoise. The weight of one of these fishes is about two thousand five hundred pounds: the largest are sometimes four thousand pounds, and these are about twenty-two feet long and fifteen in circumference.

We may remark here that the flesh of the common porpoise was formerly much esteemed in England, and was reckoned fit for the royal table. Among the singular directions for the management of the household of King Henry VIII., we find among the dainty dishes to be "set before the king" a porpoise, "and if too big for a horse-load, an extra allowance to be given to the purveyor." In the time of Queen Elizabeth it was still used by the nobles of England, and was served up with bread-crumbs and vinegar.

A curious mode of fishing the Gar-fish or Sea-Pike, in the Ionian Islands, is mentioned by a tourist. A small triangular raft is formed of three pieces of bamboo, each a foot and a half long; a little thwart is inserted, in which a small mast is fixed; it is then rigged with a sail, etc., in imitation of the boats of the country. The fisherman, taking his station on a projecting rock, with deep water alongside, and an off-shore breeze, commits his little raft to the wind, carrying with it a line of about two hundred feet in length. A float is fixed at about every six feet, and from each float depends a fine hair-line with a baited hook. When the fish bites it draws the bait down violently once, and then seems to resign itself to death. The fisherman waits till ten or twelve are hooked; he then hauls in his raft, relieves it of its freight, and again launches it for another cruise. Fifty or sixty are sometimes caught in this way during half an hour.

The gar-fish is not uncommon on English coasts, and is abundant in the Baltic. It attains a length of two or three feet. The upper parts of the body are of a dark greenish-blue mackerel tint, and a curious circumstance is that its bones are green. It has been noticed that when this fish is taken by the hook, it mounts to the surface often before the fishermen have felt the bite, and there, with its slender body half out of the water, struggles with the most violent contortions to wrench the hook from its hold.

In various chapters of this book we have already mentioned the mode of capturing the large inhabitants of the deep—the whale, the seal, the shark, sea-unicorn, and others. We must not omit another important fish of large dimensions, the *Tunny*, sometimes nine feet in length and upwards of a thousand pounds in weight, and belonging to the Mackerel family. This fish is found in the Mediterranean and the Atlantic Ocean, but chiefly in the former, where this particular fishery is of great importance, and constitutes one of the greatest branches of Sicilian commerce. The fish appear at the latter end of May, at which time the *tonnaire*, as they are called, are prepared for their reception. This is a kind of aquatic castle, formed, at a considerable expense, of strong nets fastened to the bottom of the sea by anchors and heavy-laden weights. The *tonnaires* are fixed in the passages amongst the rocks and islands that are most frequented by the tunny-fish. Care is taken to close with nets the entrance into these passages, except one small opening, which is called the "outer gate." This leads into the next compartment, which we may term the "hall." As soon as the fishes have entered here, the fishermen who stand sentries in their boats during the season shut the outer entrance, which is done by letting down a small piece of net, portcullis-fashion, which effectually prevents the tunnies from returning by the way they came. The inner door of the "hall" is then opened, which leads to another compartment, and by making a noise on the surface of the water the tunnies are soon driven into it. As soon as the whole have been got into this compartment, the inner door of the "hall" is again closed, and the outer entrance is opened to receive more fishes. This last compartment of network is called the "chamber of death." This is composed of stronger nets and heavier anchors than the others.

As soon as a sufficient number of tunny-fish has been collected here, the slaughter begins. The fishermen attack the poor defenceless animals on all sides, who dash the water about in their efforts to escape, but are at length subdued, and yield themselves a prey to their conquerors.

"There is something," says a witness of this fish massacre, "extremely exciting in seeing the wholesale capture of a herd of these great black fish, intermixed, as they generally are, with the forms of many of their large congeners, and occasionally with a sword-fish or a dolphin besides; and no one ever left the spot after one of these enormous hauls without feeling that, however superior the whale fishery may be in enterprise, it cannot yield its votaries half the pleasures or charms of these scenes."

A very questionable kind of pleasure, however, we think it must be to many, to see the agonies and the butchery which must necessarily take place on these occasions.

The *Sturgeon* fishery is carried on to a very considerable extent in the Russian dominions on the coasts of the Caspian and Aral Seas. They are caught in an enclosure formed by large stakes, representing the letter Z repeated several times. These fisheries are open on the side nearest the sea, and closed on the other, by which means the fish, ascending in its season up the rivers, are caught in these narrow angular retreats, and are easily killed. The Hon. Captain Keppell, describing the method of catching sturgeon in the fishery of Karmaizack, says:

"Two persons are in each boat; one (generally a female)

rows, while the other hauls in the fish. The instruments used consist of a mallet and a stick, with a large unbarbed hook at the end. Every fisherman has a certain number of lines. One line contains fifty hooks; these are placed at regular distances from each other; they are without barbs, sunk about a foot under water, and are kept in motion by small pieces of wood attached to them. The sturgeon generally swims in a large shoal near the surface of the water, and upon being caught by one hook, he generally gets entangled with one or two others in his struggles to escape. Immediately on our arrival the boats pulled from shore. Each fisherman proceeded to take up his lines. On coming to a fish he drew it with his hooked stick to the side of the boat, hit it a violent blow on the head with the mallet, and, after disengaging it from the other hooks, hauled it into the boat. On every side the tremendous splashing of the water announced the capture of some huge inhabitant of the deep."

The sturgeon belongs to a numerous species inhabiting both sea and fresh water—those of the former, and the largest kind, being especially plentiful in the Caspian and Black Seas, where they attain a length of from twenty to twenty-five feet, and have been known to weigh nearly three thousand pounds. The flesh has the appearance and consistency of veal, and was highly esteemed by the ancients. Pliny states that it was brought to table with much pomp, and ornamented with flowers, the slaves who carried it being also decorated with garlands and accompanied with music.

In England, when caught in the Thames and within jurisdiction of the city, it is reserved for the Sovereign as a "royal fish." In the *Illustrated London News* for the 15th of April, 1860, is a notice of a fine sturgeon thus taken, and forwarded to the Queen at Windsor by order of the conservators of the river.

The famous *caviare* of the Russians is made from the roe of the sturgeon, freed from its membranes, washed in vinegar, and dried in the open air. It is then salted, put into a bag and pressed, and finally packed in small barrels for sale.

The principal fishery of the *Conger Eel* in England is upon the Cornish coast. They are chiefly caught by what are termed "bulters," which are strong lines, several hundred feet long, with hooks about eight feet apart, baited with sand-launces, pilchards, or mackerel. The bulters are sunk to the ground by a stone fastened to them. Sometimes such a number of these are tied together as to reach to a considerable distance. It is not unusual for a boat with three men to bring on shore from one to two tons as the produce of a night's fishing, the conger being caught most readily at night.

On some of the French coasts the conger fishery is still more abundant than in Cornwall.

The great sea-conger has so great a resemblance to the common eel, the inhabitant of our rivers and ponds, that many persons believed the former was merely an eel of larger growth; but the difference may be readily discerned. The conger, whether large or small, has always the snout and upper jaw projecting beyond the lower one; whilst the fresh-water eel is remarkable for its protuberant lower jaw. The tail is also more lengthened and pointed, the dorsal fin commencing much nearer the head, and the teeth of the upper jaw, although slender, placed so close together as to form a cutting edge. The internal structure of these fishes differs more widely, the conger having a great many more bones than the eel, particularly towards the tail, and in possessing a greater number of vertebræ (the spine or backbone).

The common conger of the Atlantic coasts is a large fish, sometimes exceeding ten feet in length, and weighing upwards of a hundred pounds, but its ordinary dimensions are from five to seven feet. It is entirely a marine species, although frequently found in the mouths of rivers, its object being, it is thought, that of feeding on the fish that ascend or descend the stream. Of these it devours large quantities, not objecting to crabs and shell-fish, which the strength of its jaws permits it to masticate without difficulty. The smaller kinds of fish it swallows entire, and thus fortified by good nourishment, it becomes a formidable adversary when hauled into the boat by a fisherman's line, or found among the rocks, where it is sometimes left by the retiring tide.

This does not seem to be a matter of complaint in *our* time. The conger, however formidable, also finds a dangerous adversary in the spiny lobster of the Mediterranean Sea, which is said to enter into a fierce battle with the conger, and generally becomes the victor, from the superiority of its weapons of defence, the claws, which lacerate and wound the monstrous eel, proving the death summons.

The conger, when properly cooked, has a most delicious flavor, but somehow or other there is a great antipathy to this fish, as being, probably, too much of the serpent form; but travelers in Cornwall find a conger-pie delicious, and those persons who have visited the Channel Islands will not easily forget the delicious soup that is made from this fish. Even as far back as the reign of Queen Elizabeth, there was a singular mode of *curing* congers in Cornwall, which was merely to slit them in half, and without any further preparation to hang them up in a kind of shambles erected for that purpose; such parts of them as were not *gone*, were considered fit for use, and exported to Spain and Portugal.

The *Sand-Eel* fishery, although of a very primitive character, being mostly carried on with spades, shovels, three-pronged forks, rakes, and in fact any implement of a *raking* character at hand, is very exciting and amusing. Large shoals are observed frequently swimming near the shore, and it often happens that, instead of retiring with the ebbing tide, they dig into the sand, and remain there until the water covers them again. Advantage is taken of this, and hundreds of men, women, and children set to work with the readiest implements they can find, and the scene becomes very animated. When dug from the sand, the fish leap about with singular velocity, and the gathering of them affords a fine amusement to the younger parties, who are commonly the most numerous and eager in this pursuit. It is remarkable with what ease and rapidity these slender and delicate-looking fish penetrate the sand, even when it is of a pretty firm texture. They are a favorite meal with many, and are sometimes salted and dried; but their principal use is as bait for the capture of more valuable fishes, there being scarcely any other found to answer the purpose so effectually. This well-known fish scarcely ever exceeds seven or eight inches.

The Mackerel belongs to the same family as the tunny-fish previously described, but is a comparatively small member as regards size, being usually about fourteen inches long and about two pounds in weight. This beautiful fish is readily caught by bait, and particularly when the bait—which is usually a piece from one of its own kind—is moved quickly through the water. The boats engaged for this fishing are often under sail. Besides the line, drift-nets and seines are employed. The size of the mesh is one inch and one-sixth from knot to knot when the twine is wet, or in the square, from one corner to another. A row of corks runs along the head-line, and the lower border is left suspended by its own weight. The number of nets in each boat depends upon its size. A boat may carry eleven score of nets, and as these are fastened in length to each other, they will extend to a distance of a mile and three-quarters. The boats on the various fishing-grounds are shot across the course of the tide twice between evening and morning; for fish avoid the nets during the day, and scarcely less so during very dark nights. This latter circumstance is caused by the light produced in the sea by luminous animals, which then appears most conspicuous; and hence a hazy atmosphere is judged beneficial. The use of lights is employed in some countries. Bloch, in speaking of the mackerel fishery, says, that at St. Croix, on the approach of night, when the sea is smooth, they prepare their torches, and hold them as close to the water as possible. The fish soon show themselves, and rise above the surface, when the nets are immediately shot, and soon taken in with abundant success.

When the shoal of mackerel approaches the land the seine comes into operation. This consists of a single net, which is about seven hundred feet in length, and seventy in depth at the middle. The full size of the mesh from corner to corner is two and three-quarter inches at the sides, which is the same dimension allowed to the drift-net; but for about two hundred feet of the hollow, the size of the mesh is lessened to two and a half inches, to prevent the fish from being hung in the meshes; for if this should happen, the net would not be raised from the bottom, and fish and net would be lost. Shoals of mackerel are rapid in their motion, and exceedingly uncertain, as well as easily alarmed. They rarely stay long at the surface, and when they sink below it is doubtful in what direction they may again appear. The whole proceedings are, therefore, full of excitement, and great haste is employed to enclose them in the circle of the seine.

The mackerel is a favorite article of food, but its flesh soon changes; and a capture that might have proved valuable, may be rendered worthless if the fishes are not at once sent to the market. A principal object of the French fishery is to prepare the mackerel salted for use at home, for which purpose they are immediately stored in bulk on board the boats. In the west of Cornwall, also considerable numbers are salted, chiefly for the use of miners, who seem to prefer salted fish to even the fresh that abound in the finest condition in their markets.

It was formerly supposed that great migrations of mackerel took place, but it is now believed, as in regard to the herring, that they merely leave the deep water and approach the coast for the purpose of spawning. The mackerel is of less importance than the herring fishery. It is a restless, ever-wandering rover, and unlike the herring in its habits in that respect. It is found in large numbers in the Mediterranean.

The Herring fishery affords one of the best illustrations of British enterprize. We must now proceed to the Norfolk coast, for it is there that this most valuable fish is found in the greatest abundance, perhaps more so than in any other part of the world. The name of the fish is derived from the German *heer*, "an army," in reference to the vast shoals in which they arrive. The herrings appear on the Norfolk coast in the last week of September for the purpose of spawning, and are then in the best condition to become the food of man. Having fulfilled this obligation of nature, they return to their former haunts about the commencement of December. A few, however, may be found at other periods of the year, particularly about midsummer; and, although small, they are much esteemed for their delicate flavor. The Yarmouth herring has less oil than the Scotch herring, but is unrivalled in point of quality. It seldom measures more than fourteen inches in length, in girth six inches and a half, and it weighs about nine ounces. The vessels employed by Yarmouth in this fishery are usually decked boats, of from forty to fifty tons burthen, and carrying a crew of ten men. Besides the boats belonging to the town, there are many others called "cobles," which come from Scarborough, Filey, and other northern ports. Each fishing-boat is provided with from sixty to one hundred nets, each net about fifteen yards long upon the rope, fastened by small cords called "seizings." These nets are floated by corks placed at intervals of a few feet from each other; the warp which supports the whole is frequently a mile in length, and is borne up by small buoys. The nets themselves are usually made in four parts or widths, called "lints," one being placed above another, and so forming a wall in the sea, against which the fish are invited to drive their heads.

This fishing is carried on during the night only, it being supposed that the stretching of the nets in the daytime would drive away the shoal. In the dusk of the evening the nets are thrown over the side, and the boat is then steered under an easy sail, or allowed to drift with the tide until daylight, when the nets are hauled in. A single boat has sometimes, in one night, taken twelve or fourteen *lasts* of hearings, each "last" numbering ten thousand fish, or, by the fisherman's calculation, thirteen thousand two hundred; but it often happens that a boat does not obtain more than this quantity during the season. The average catch for each boat is about thirty "lasts" (three hundred thousand); but a boat has been known to bring in the enormous quantity of two hundred and sixty-four thousand herring *at one time.* Like all fisheries, the result is very uncertain. It is a curious and bountiful provision of nature that forces the herring, and other fish usually distributed through the deep, to congregate together, and visit the shores in such immense abundance, at a time when they are in the highest perfection, and when most fitted for human food.

The herring dies as soon as it leaves the water, hence the phrase "as dead as a herring." The fishes are therefore salted as soon as caught, and when the boat has reached land they are brought to shore, and carried to the fish-house in "swills," which are open coarse wicker baskets. Arrived at the fish "office," the herrings, after being sufficiently salted, remain on the floor for twenty-four hours if intended

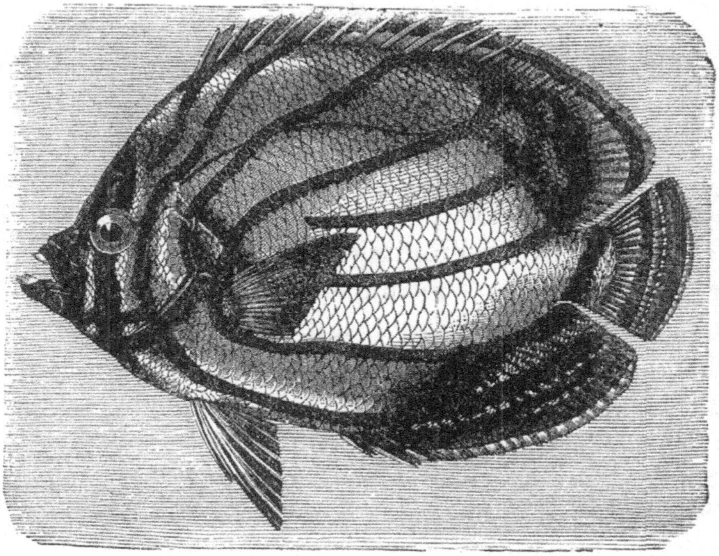

WANDERING CHAETODON.

LUMP FISH.

to be slightly cured, or for ten days if intended for the foreign market; they are then washed in large vats filled with fresh water; "spits," (pieces of wood about four feet long and of the thickness of a man's thumb) are passed through their heads or gills, and they are then hung up in rows to the top of the building. Wood fires are then kindled under them, and are continued day and night, with slight intermissions to allow the fat and oil to drop, until the fish are sufficiently cured, which, if they are intended for the foreign market, is at the end of fourteen days, but if for home consumption, three or four days suffice. The first are called "red" herrings, from the deep color which they acquire, and the others are known as "bloaters." When cured, the herrings are taken down and placed in barrels which contain each about seven hundred fish. From thirty to forty thousand barrels are sent yearly from Yarmouth to the towns on the Mediterranean coasts. The annual supply of herrings at Billingsgate Market is estimated at one hundred and twenty thousand tons, valued at one million two hundred pounds sterling! The greatest enemy to the herring fishermen is the dog-fish, which, in pursuit of the herring, frequently becomes entangled in the nets, and does great damage to them in endeavoring to escape.

The herring fisheries sometimes suffer very considerably from the ravages of this fish, the popular name of some of the smaller species of shark, owing this designation to their habit of following their prey like dogs hunting in packs. These predaceous fishes are seldom abundant when the herrings are in a compact body; but sometimes they commit great destruction when a shoal is first drawn in near land. They have been known to consume as many herrings as would fill a dozen barrels out of one boat's nets in the course of an hour. They are also very destructive to the nets when they get entangled, their hard fins tearing them to pieces. In like manner they make sad havoc with other fishes. Occasionally only a few escape with their heads, the tails of others are snapped off, and pieces bitten out of the belly. A cod-fish sometimes comes up a mere skeleton, stripped to the bone on both sides.

The Dog-fish attains a length of three or four feet, and is found in the Atlantic, the Mediterranean, and the South seas. One of the most abundant species on English coasts is the common dog-fish, which sometimes appears in prodigious numbers, twenty thousand having been taken at Cornwall at one time in a net, and the fishermen of the Orkneys and Hebrides, where they are much used for food, sometimes load their boats to the water's edge with them.

Another voracious enemy of the herring (and the pilchard) is the Hake, a member of the Cod family, with the same predatory instincts. It is sometimes three or four feet in length, coarse in quality, but valuable as a "stock" fish. It is generally taken by lines, like cod and ling, but in the spawning season, when it keeps near the bottom, it is sometimes caught by trawl-nets.

Allied to the herring, but differing in some respects, being nearly equal in size, but rather thicker, and the lines of the back and belly being straighter, the scales also being larger and fewer, is the Pilchard, a fish also of immense importance in the British fisheries, and plentiful on the coasts of Devonshire and Cornwall. These fish congregate in deep waters, within limits extending from the Scilly Isles, as far, sometimes, as the Irish, Welsh and Cornish coasts. A portion strikes the land north of Cape Cornwall, and turns in a north-easterly direction toward St. Ives, constituting its summer fishery. The great bulk passes between the Scilly Islands and the mainland. "To look from Cape Cornwall," says an eye-witness, "or from any of the high lands of St. Just, and see this immense mass of fishes, extending as far as the eye can reach, approaching the shores and reddening the waters, is a sight of great interest and beauty, and such as would repay any exertion to see."

The seine or net used in St. Ives Bay for capturing pilchards is nearly twelve hundred feet long, and nearly sixty feet in depth. More than two hundred and fifty of such nets are kept at St. Ives, each having its own boat to carry it. Every seine or net-boat, when its turn arrives, is attended by one or two tow-boats with stop-nets, and also by a smaller boat called the "follower," used principally for carrying the men to and from the larger boats. When the *huers* or sentinels stationed on the hills perceive a shoal of pilchards, they immediately signal to their respective boats, and by signs give the necessary directions for their capture. They are enabled to do this by observing on the water a reddish hue, like that of sea-weed (very different from their color out of water), and the denser the shoal of fish, the deeper is this hue. As soon as the seine-boat and tow-boat are within reach of the shoal, they start for the same point in opposite directions, and are rowed rapidly round the fish, while the nets which they carry are being shot or cast into the sea. When the seine and the stop-net meet, they are immediately joined, and form a complete circular wall round the pilchards about eighteen hundred feet in circumference, and reaching from the surface to the bottom, the nets being kept in a vertical position by corks strung on their head-ropes and leads on their foot-ropes. This net-work enclosure, with all its contents, is then warped towards the shore into the securest part of the bay, out of the reach of the strong tidal current, and there moored with anchors so placed as to keep it as open or as nearly circular as possible. Within this large net a small one, called the tuck-net, is introduced at low water, so that the fish are raised to the surface, dipped up in baskets into the boats, taken to shore, and carried in barrows to be cured and salted. The St. Ives seine fishery does not differ materially from that in Mount's

Bay, except that in the latter place, owing to the greater depth of water, the nets are about thirty feet deeper, and they are also longer. Besides the method of capturing pilchards with deep nets in shallow water in the day-time, there is a far more common mode in Cornwall of taking them in shallow nets, in deep water, by night. As these drift-nets are always spread in the open sea, where they might be destroyed by vessels sailing over them, their head-ropes are sunk about eighteen feet below the surface, and kept suspended at that depth by cork buoys fixed at regular intervals. By this contrivance, not only are the nets preserved, but larger quantities of fish are taken. These nets, each with a driving-boat attached, are left to go with the wind or tide all the time the net remains in the water.

As soon as the pilchards caught by the seine or drift-nets are landed, some are sold in the neighboring towns and villages, and the rest, when cured and placed in barrels, are exported to the Mediterranean, where, during Lent, they are much sought after.

The method of curing the pilchards is very simple. They are placed in cellars, and women are employed in arranging them in layers, with salt between. After remaining in bulk about five weeks, during which oil and other matters drain from them, they are put into troughs of water, washed quite clean, and then carefully laid in casks, where they are subjected to heavy pressure for about a week. The oil thus expressed flows out from holes at the bottom or crevices in the sides of the untightened casks, and as this reduces their contents, more fish are added, until each cask, when the pressure is removed, weighs at least four hundredweight. The capital employed in the Cornish pilchard fishery amounts to at least two hundred and fifty thousand pounds, and affords employment to about ten thousand persons.

The Sprat was formerly considered by naturalists to be the young of the herring, as well as that of the pilchard; it is now generally admitted to be a distinct species. This fish comes into season in November, and continues so all the winter months, during which the sale, especially in London, is immense. About five hundred boats are annually employed in the sprat fishery. So great is the abundance sometimes, that thousands of tons are sold to farmers for manure. Most fish are caught on dark and foggy nights.

The Whitebait, little fishes from three to six inches in length, the delicious flavor of which the reader may have often enjoyed, are caught by means of bag-nets, sunk four or five feet below the water. They are very abundant in many parts of the British coasts, particularly in the estuary of the Thames in spring and summer, when they arrive in shoals to deposit their spawn. For several months they continue to ascend the river with the flood tide, and descend with the ebb tide, not being able to live in fresh water. It was formerly supposed that this fish was the young of the shad, or sprat, but is now regarded as a distinct species.

The Sardine, a fish of the same genus with the herring and pilchard, smaller than the latter, abounds in the Mediterranean, and is found also in the Atlantic Ocean. The sardines of the west coast of France, which are largely imported into other countries, are generally young sprats, and sometimes young herrings. This "sardine" fishery is a great business in France, and especially at Concarneau, where as many as thirteen thousand men aid in the fishery. This is conducted in a way remarkable for the extravagance it involves. The sprat fisheries on the British coast—indeed, all other net fisheries—are carried on in the most primitive way; but the French have made it a "bait" fishery, and use the roe of the cod, which is brought at a considerable expense from the North seas for the purpose. The fish are gutted, beheaded, sorted into sizes, and washed in sea-water, then dried on nets or willows; they are then placed in a pan, kept over a furnace, and filled with boiling oil. The fish are plunged into the cauldron, two rows deep, arranged on wire gratings. They are afterwards placed to drip, the oil being carefully collected, after which they are packed in the tin boxes with which we are so familiar. It is said that, besides the quantity exported, as many as four millions are annually prepared for the home market.

We need not enter into any particulars about the Cod fishery on the banks of Newfoundland, which presents nothing new or very interesting except in the value attached to every part of this valuable fish. The tongue of the cod, whether fresh or salted, is a great delicacy; the gills are used as baits in fishing; the liver, which is large and good for eating, also furnishes an enormous quantity of oil, now much esteemed for consumptive patients; the swimming-bladder furnishes an isinglass; the head is eaten, and the Norwegians give it, with marine plants, to their cows, to produce a greater quantity of milk; the vertebræ, the ribs, and the bones are given by the Icelanders to their cattle; even the intestines and eggs are eaten. The coast of Iceland abounds in fish, especially of the cod tribe, and this abundance has not only from a very early time supplied the inhabitants with their chief food, but enabled them to procure other necessaries. As the principal fishings begin on the Newfoundland coast, at the Feroe Islands, in Norway, and in Iceland at the same time, it seems evident that the cod is not a migratory fish, but a dweller where it finds its food. The Icelanders fish chiefly from open boats, and sometimes from decked ones. Only the largest boats, with six or twelve oars, are used in the cod fishery, and in these the natives often go out many miles to sea in the depth of winter to fish. They are a most hardy set of mariners. Their mode of capturing the cod is either by small drift-nets, deep-sea or hand lines, and the ordinary long line. The fish caught by the net are different from those taken by the line, being more plump, with smaller heads. The number of Iceland boats employed in the cod fisheries average nearly five thousand, and the number of persons employed exceeds ten thousand.

The modern cod-smack usually carries from nine to eleven men and boys, including the captain. The line is chiefly used for the purpose of taking cod or haddock. Each man has a line of three hundred feet in length, and attached to each of these lines are one hundred "snoods," with hooks already baited with mussels, pieces of herring, or whiting. Each line is laid "clear" in a shallow basket or "skull;" that is, it is so arranged as to run freely as the boat shoots ahead. The three hundred feet line, with one hundred hooks, is called in Scotland a "taes." If there are eight men in a boat, the length of the line will be two thousand four hundred feet, with eight hundred hooks (the lines being tied to each other before setting). On arriving at the fishing-ground, the fishermen heave overboard a cork buoy with a flag-staff affixed to it, about six feet in height. The buoy is kept fixed by a line reaching to the bottom of the water, and having a stone or small anchor fastened to the lower end. To this line, called the "pow-end," is also fastened the fishing-line, which is then "paid" out as fast as the boat sails. Should the wind be unfavorable, the oars are used. When the line is all out the end is dropped, and the boat returns to the buoy. The pow-end is hauled up, with the anchor and fishing-line attached to it. The fishermen then haul in the line with whatever fish may be on it. Eight hundred fish might be taken by eight men in a few hours by this operation. Many a time the fish are eaten off the line by the dog-fish and other enemies, so that a few fragments and a skeleton or two remain to show that fish have been caught. The fishermen of "deck-welled cod-bangers" use both handlines and long lines. The cod-bangers' tackling is, of course, stronger than that used in open boats. The long lines are called "grut lines" or great lines. Every deck-welled cod-banger carries a small boat on deck, for working the great lines in moderate weather. This boat is also provided with a well, in which the fish are kept alive till they arrive at the banger, when they are transferred from the small boat's well to that of the larger vessel.

The Haddock, which has a striking family resemblance to the cod, is taken both by trawl-nets and lines, and being in great esteem by fish-eaters for the excellence of its flavor, we ought to be pleased that the fish is so partial to our own coasts, where it appears in vast shoals at particular seasons. Fishermen sometimes find haddocks and other fishes caught in their lines reduced to mere skin and skeleton by the Hag, one of the species allied to the Lamprey family, resembling an eel or worm, and a perfect anatomist in its way. It is believed to enter by the mouth of the haddock, and thus prey upon it: the fish thus treated is called a "robbed" fish. As many as six hags have been taken out of a single haddock, and they are also said to make their way into fishes through the skin, and are hence sometimes called "borers." It is supposed, however, that the hags are swallowed by fishes, and, in retaliation, work out their insides.

The Coal-fish—a relative of the cod, with a very vulgar name, derived from its black coat, but a fish of really handsome form, and about two or three feet in length—takes a bait with extraordinary eagerness: when a boat falls in with a shoal, they may be kept beside it by being thus attracted till the whole are captured. It is abundant in all Northern seas, and is taken on the British coasts. In many parts of Scotland they are well known to juvenile anglers, who take them in plenty from the end of piers, often with a rude tackle and almost any kind of bait. In the winter-time, while the fry of this fish is in the harbor of Orkney, it is common to see five or six hundred people, of all ages, fishing for them with small angling-rods about six feet long, and a

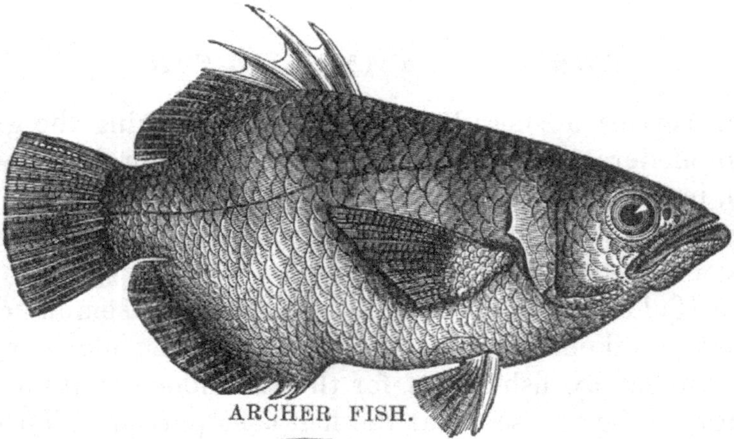

ARCHER FISH.

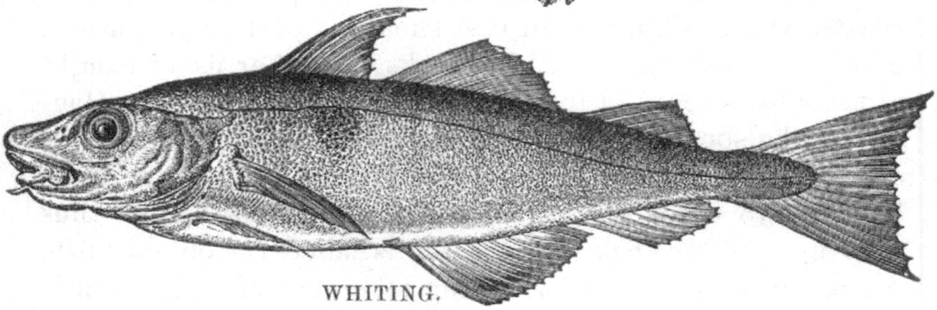

WHITING.

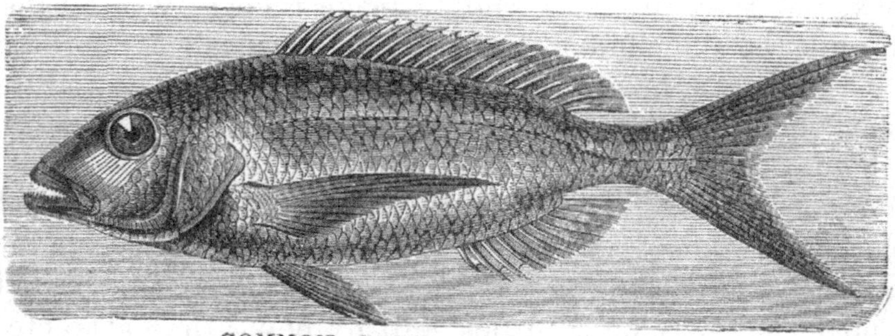

COMMON CARP.

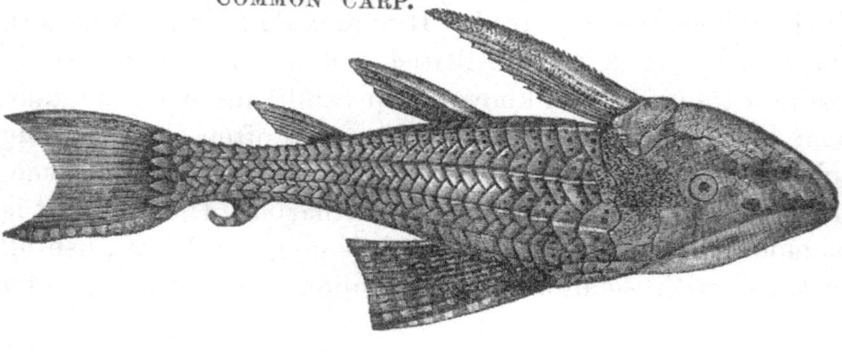

line a little longer; but with this simple apparatus they kill vast numbers. The whole harbor is covered with boats.

Other members of the cod family are caught much in the same manner as their representative, and are very valuable as food, especially the Ling. The sounds (air-bladders), are pickled, and the roes are preserved in brine, and eaten as food, or used as a means of attracting fish by throwing it about the nets, as is often done by French fishermen. The Common Hake, a fish sometimes measuring three feet, is also plentiful on the English and Irish coasts, and very voracious. When enclosed in a net with pilchards—as frequently happens on the Cornish coast—it gorges itself with them: It is to this species, and the common cod when dried and salted for exportation, to which the name of "stock" fish is usually applied. Forty thousand hakes have been landed on the shores of Mount's Bay in Cornwall in a single day, and the quantity captured on the Irish coast is immense. Galway Bay is sometimes called the "Bay of Hakes" from the numbers of that fish taken.

The Turbot, an especial delight of fish epicures in all times, is taken, with other flat fish, by lines and hooks, the fishermen going out in parties of three to a "coble," each man carrying his long line, the united ends of which are a league in length, and draw after them fifteen hundred baited hooks; these lines, as they are to lie across the current, can only be shot twice in twenty-four hours, when the rush of the water slackens as the tide is about to change. The Italians christen the turbot the "sea pheasant," from its flavor. The Romans were particularly fond of this dainty, and frequent allusion to its size occur in their writers; thus:

"Great turbots and late suppers lead
To debt, disgrace, and abject need."

"The border of broadest dish
Lay hid beneath the monster fish."

But the size mentioned by the ancient writers is of a fabulous character. However, it sometimes attains a weight of between seventy and ninety pounds. It is now chiefly obtained by beam-trawling, a triangular purse-shaped net about seventy feet long, usually having a breadth of about forty feet at the mouth, and gradually diminishing to the end of the net, which is about ten feet long, and of nearly uniform breadth. The turbot is of all the flat fishes the most valuable. The Brill belongs to the same tribe, as well as other less important fishes. The turbot is shorter, broader, and deeper than almost any other kind of flat fish. It generally keeps close to the bottom of the sea, and is found chiefly on banks where there is a considerable depth of water. Some of the banks in the German Ocean abound in turbots, as the Doggerbank, and yield great quantities to the London market.

In proportion to the benefits derived from the spoils of the Turtle, the shell of which is so ornamental and useful in the arts, the ingenuity of man has been sharpened by his eagerness to acquire them. The modes by which the people of Celebes take them are by the harpoon and the net, or by falling on the females when they resort to the strand to lay their eggs. The turtle is turned on its back, when, unable to turn again, it lies helpless. It sometimes also falls into the hands of the dwellers on the coast through means of their fishing-stakes, into which it enters like the fish, and from which it can find no outlet. It is then killed and robbed of its upper shield; but, as the shells adhere fast to each other, and would be injured by being torn off, the usual plan is to wait a few days, by which time the soft parts become decomposed, and the shells are removed with little trouble. When the turtles lie floating on the sea either for the purpose of sleep or respiration, the fishermen approach them quietly with a sharp harpoon, carrying a ring at the butt-end, to which a cord is attached. The harpooner strikes, and the wounded animal dives, but is at last secured by the cord. In the South Seas skilful divers watch them when so floating, and getting under the animals, suddenly rise, and so seize them. Mr. Darwin describes a curious method of capturing turtles which he witnessed at Keeling Island in 1836:

"I accompanied," he remarks, "Captain Fitzroy to an island at the head of the lagoon: the channel was exceedingly intricate, winding through fields of delicately-branched corals. We saw several turtles, and two boats were then employed in catching them. The water is so clear and shallow, that although at first a turtle quickly dives out of sight, yet, in a canoe or boat under sail, the pursuers, after no very long chase, come up to it. A man standing ready in the bows at this moment dashes through the water *upon the turtle's back*; then clinging with both hands by the shell of the neck, he is carried away until the animal becomes exhausted, and is secured. It was quite an interesting sight to see the two boats thus doubling about, and the men dashing into the water to secure their prey."

But the most singnlar mode of capturing turtles is that practised on the coasts of China and the Mozambique, by the aid of living fishes trained for the purpose, and thence named "fisher-fishes." This fish is a species of remora (sucking-fish), and the islanders who use it are said to proceed in the following manner:

They have, in their little boat, tubs containing many of these little fishes, the top of whose head is covered with an oval plate, soft and fleshy at its circumference. In the middle of this plate is a very complicated apparatus of bony pieces disposed across in two regular rows, like the laths of Persienne blinds. The number of these plates varies from fifteen to thirty-six, according to the species: they can be moved on their axis by means of particular muscles, and their free edges are furnished with small hooks, which are all raised at once like the points of a wool-card. The tail of each of the trained fishes in the tubs is furnished with a ring for the attachment of a fine but long and strong cord. When the fishermen perceive the basking turtles on the surface of the sea, knowing that the slightest noise would disturb the intended victims, they slip overboard one of their fish tied to the long cord, and pay out line according to their distance from the turtles. As soon as the fish perceives the floating reptile, he makes towards it, and fixes himself so firmly to it that the fishermen pull both fish and and turtle into the boat, where the fish is very easily detached from its prey, and the turtle is secured.

Crabs, which belong to the highest order of Crustaceans, (a hard covering) are taken by traps—baskets which readily permit an entrance, but not their escape, and which are baited with meat or animal garbage of some kind—or pots, or are caught in the holes of the rocks at low tide with a rod and hook. These animals require very careful handling when found on the rocks or the sea-shore. Their fighting propensities are not confined to other prey, but they have fierce encounters among themselves, by means of their formidable claws, with which they lay hold of their adversary's legs, and dexterously amputate them.

The Hermit Crab is one of the most curious of this numerous family. A more daring little burglar could not be found than this animal, appropriating to its own use the shells of whelks and periwinkles, after basely dislodging and killing their lawful owners. It is curious to see this crab busily parading the sea-shore, dragging its old incommodious habitation behind it, unwilling to part with it until another and more convenient one is found. It stops first at one shell, turns it, pssses by, then goes to another, looks at it attentively for a time, and then tries it. Not being found suitable, it resumes the old one, and in this manner frequently changes, until at length it finds one light, roomy, and commodious; into this it enters, and takes up its abode. Frequently two of them will have a severe contest for possession, and a fierce fight ensues. With such very bad instincts and unscrupulous habits, it is not surprising that the

HERRING FISHING.

hermit crab should be a very suspicious animal. On the slightest alarm it retires into its shell, guarding the entrance to it with its largest claw. The structure of the animal renders it equal to most emergencies. The part which in the lobster becomes a fan-like expansion at the end of the tail, is an appendage to the hermit crab for firmly holding on by the shell, and so tenacious is the hold that it may be torn to pieces, but cannot be pulled out. As they increase in size, the hermit crabs are compelled to enter on a fresh career of crime. The ancients were well acquainted with the predaceous habits of this little marauder.

Crabs are inhabitants of almost all seas. The different kinds vary much in the form of the carapace, or back, which in some is round or nearly so: in others longer than broad; in some prolonged in front into a kind of beak, etc.; also in smoothness or roughness, with hairs, excrescences, or spines; in the length of the legs, etc. The King Crab, an inhabitant of tropical seas, is a remarkable species, having a tail which forms a long and powerful dagger-like spine, sometimes exceeding in length the whole body. Some of these crabs exceed two feet in length, and in the Asiatic islands the spine is often used for pointing arrows; in tropical America the shell is used as a ladle. At Labuan and Singapore Dr. Collingwood met with a new species of crab, the "Pill-maker." It is a small creature of its kind, many being the size of large peas. Its habit is to take up particles of sand in its claws, deposit them in a groove beneath the thorax, and afterwards eject them as pellets or pills from its mouth, after having extracted what nutriment they may contain.

The crab (as also the prawn) may be quoted as exercising the virtue of conjugal affection to the highest degree, for the male takes hold of his mate, and never quits her side, swimming with her, crawling about with her; and if she is forcibly taken away, the faithful animal will seize hold of and endeavor to retain her.

A traveler mentions a curious example of instinctive stratagem in a crab on the shores of the Pacific, about six inches in circumference, which covers itself with decaying vegetable rubbish, mud, sand, etc., and thus lies in ambush for its passing prey. It maintains a sluggish character until taken into the hand, or otherwise alarmed, when it becomes very active. The spines upon its body to retain the rubbish, the short but strong claws easily concealed, the eyes placed at the end of long foot-stalks, curving upwards and thus raised above the mass, show the beautiful adapation of its structure to its habits.

Prawns in general form resemble lobsters, cray-fish, and shrimps, but belong to a family remarkable for a long saw-like beak projecting from the carapace or back. There are many species, and some of those inhabiting the warm seas attain a large size. Many of them are semi-transparent, and have very fine colors. The common prawn is from three to four inches in length, is generally taken in the vicinity of rocks at a little distance from the shore, and osier baskets—similar to those employed for catching lobsters—are employed for their capture, and nets.

Shrimps are generally taken by nets in the form of a wide-mouthed bag, stretched by means of a short cross-beam at the end of a pole, and pushed along by the shrimper, wading to the knees in water. Sometimes a net of larger size is dragged along by two boats. The common shrimp is about two inches long, and the short beak readily distinguishes it from the prawn. When alarmed, it buries itself in the sand by a peculiar movement of its fan-like tail.

Dr. Collingwood mentions a new species of shrimp, which he discovered in the warm seas, of a deep violet color (those on the Atlantic coast are of a greenish-gray color, dotted with brown), and with the claw of remarkable construction.

"I placed it," he says, "in a basin of water with a small crab, whose appearance appeared violently to offend it. Whenever the crab came in contact with the shrimp, the latter produced a loud sound, the explanation of which is as follows: the shrimp possessed two claws—one large and stout, and the other long and slender. When irritated, it opened the pincers of the large claw very wide, and then suddenly closed them with a startling jerk. When the claw was in contact with the bottom of the basin, a sound was produced as if the basin were struck; but when the claw was elevated in the water, the sound was like the snap of a finger, and the water was splashed in my face."

The same authority called this animal the "trigger" shrimp, from the action of this claw resembling that of a pistol trigger. If only put upon half-cock, this trigger closed without noise.

How wonderful are the means that the Omnipotent Creator has provided (as in all things) for the protection of the shelly inhabitants! The hard covering accommodates itself to their growth, and at the same time is sufficiently light as not to interfere with the movements and functions of the interesting tenant. All the various tribes of shell-bearing animals are thus defended from the injuries and attacks to which their situation exposes them. Thus, some are protected by multivalve, or more than two formed tubular shells, the tenant protruding its organs at the summit, which is defended by the lid, consisting of more than a single piece; in the univalve, or one shell, the animal protrudes itself at the sides, and has no valve, as in the common barnacle. The bivalves, or animals of two shells, bury themselves in the sand, perforate rocks, or suspend themselves by the byssus, or thready filaments; others, again, as the oyster, fix themselves to any convenient substance.

In the common Periwinkle (a molluscous, or soft-shelled animal), the mouth of its shell is closed by a horny covering; this is called the "patch," which is attached to the foot, or rather neck, by its convex or lower surface: this is the lid.

The Mussel, belonging to the molluscous animals, and the common species of which are very abundant on our own and English coasts, are much used as bait by fishermen. As an article of food it is much consumed in our own country, but especially so in Europe. The French people are remarkably clever in their method of cultivating this shell-fish by artificial means. About four miles from Rochelle there may be seen a wonderful mussel "farm," which has been a source of considerable profit for hundreds of years. The mussels are grown on frames of basket-work carefully made, and are larger and of finer flavor than the natural fish. In the year 1035, an Irish bark loaded with sheep was thrown in a heavy storm on the rocks near Esnande, on the coasts of Saintonge, and the only person on board who was saved was the captain, named Walton, who amply repaid the services which had been rendered him, for having saved some of the sheep from the wreck, he crossed them with the animals of the country, and this produced a fine race, which is still known under the name of the "marsh sheep." He next devised a kind of net, which was stretched a little above the level of the open sea, where it caught large flocks of shore-birds which skim the surface of the water at twilight or after dark. In order to render these nets thoroughly effective, it was necessary to go to the very centre of the immense bed of mud where these birds seek their nourishment. Walton discovered on examining the poles which supported his nets that they were covered with mussel-spawn. He then increased the number of the poles, and, after various attempts, constructed his first artificial mussel-bed. At the level of the lowest tides, he drove into the mud stakes that were strong enough to resist the force of the waves, and placed them in two rows about a yard distant from one another. This double line of poles formed an angle whose base was directed toward the shore and whose apex pointed to the open sea. This palisade was roughly fenced in with long branches, and a narrow opening having been left at the extremity of the angle, wicker-work cases were arranged in such a manner as to stop any fishes that were being carried back by the retreating tide. Walton had thus combined in one a sort of fish preserve, with a bed for the breeding of mussels. The plan soon became very popular, and the beds were extended in every direction.

The little mussels that appear in the spring are called seeds, and are scarcely larger than small beans till toward the end of May; but at this time they rapidly increase, and in July they attain the size of a full-grown bean. They are then fit for transplanting and are placed in bags made of old nets, which are set upon the fences that are not quite so far advanced into the sea. The young mussels spread themselves all round the bags, fixing themselves by means of those silky filaments or threads, called byssus, by which the little animals attach themselves to rocks or other substances. In proportion as they grow or become crowded together within the bags, they are cleared out and distributed over poles lying somewhat nearer the shore, while the full-grown mussels, which are fit for sale, are planted on the beds nearest the shore. It is from this part of the mussel-beds that the fishermen reap their harvests, and every day enormous quantities of freshly gathered mussels are transported in carts or on the backs of horses to La Rochelle, whence they are sent to all parts of France.

As an instance of utility, the common mussel maintains the long bridge across the Torridge River, near its junction with the Taw, at the town of Bideford in North Devon. At this bridge the tide runs so rapidly that it cannot be kept with mortar. The corporation therefore keep boats employed to bring mussels to it, and the interstices of the bridge are kept filled with them. The bridge is supported against the violence of the tide by the strong threads of the byssus which these mussels fix to the stonework.

Closely connected with this subject is that of Oyster farming, which is practiced quite extensively in England and France as well as in our own country. A single visit to the shores of Maryland, Virginia, New Jersey, Long Island and Connecticut would amaze one who had not given this subject much thought, and convince them that it was a

PUNT OR PIROGUE OF THE MARSH.

PILES, WITH BASKET WORK COVERED WITH MUSSELS.

thriving business in more senses than one, and every year becoming more important and extensive.

Farming, as a term descriptive of this calling or industry, may at first seem a misnomer; but the word is significant as used in this sense: Anyone engaging in this undertaking buys or secures a *plot of water*, and proceeds to stake it out in a direct line from the shore; a neighbor secures a plot adjoining, surveys and bounds it in a similar manner. These fenced-in water fields present a novel and picturesque scene to one who beholds it for the first time.

To the oyster farmer the times and the seasons are distinctly worked and rigidly observed. If he sow and cultivate not, neither can he reap. There are comparatively few places on our inland Atlantic shores where oyster culture is not carried on in some one of its various methods; there are places for keeping them alive until wanted, places for breeding them in, and places in which they are fattened. Most oysters cast their spawn in the months of April or May. The spawn is by the fishermen called "spat," and in size and figure each resembles the drop of a candle. As soon as it is cast, or thrown off, these embryo disks adhers to stones, old oyster shells, pieces of wood, or whatever substance comes in their way; a limy secretion issues from the surface of their bodies, and in the course of a day begins to be converted into a shelly substance. It is about two years, however, before oysters acquire their full size, and are ready for the table.

Many curious discussions have arisen as to whether oysters possess the faculty of locomotion. It is well known that, in general, they are firmly attached to stones, to any submarine substance, or to each other, and it is generally believed that they are not endowed with any power of changing their position. It is certain that they are the most inanimate of the mollusca, remaining adhered to the substance under the waves that they have fixed upon, enjoying only the nourishment brought it by the waves, and giving scarcely any sign of life, except the opening and shutting of its valves.

The oyster, particularly when eaten raw, is easy of digestion, and very nutritious. Its best qualities become impaired, however, by cooking, and, though very piquant culinary preparations are made from it, such as sauces, ragouts, etc., these tempting effects are produced by the sacrifice of the best quality of the fish, and should be carefully shunned by the invalid.

The enemies of the oyster are many. The sea-crab seats itself upon the shell, and drills a little hole in his back, and so kills him. On the sea-shore bushels of shells are found quite riddled with holes by this crab. The star-fish was known in ancient times to prey upon the oyster. Oppian says:

'The prickly star-fish creeps on with fell deceit,
To force the oyster from his close retreat.
When gaping lids their widen'd void display,
The watchful star thrusts in a pointed ray
Of all its treasures spoils the rifled case,
And empty shells the sandy hillocks grace.

The drum-fish—in weight about thirty pounds, and about two feet long—swallows oyster and shell; sometimes two or three pounds of shells are found in the stomach of this fish. The star-fishes *hug* the oyster, and wrap their five rays about him, but the embrace is one of death to the poor victim.

It is not surprising that the inhabitants of the ocean should feed partly on shell fish; but it is curious to find animals strictly terrestrial preying upon them. Monkeys are said to descend to the sea to devour what shell-fish they may find on the shore. The ourang-outangs are said to feed in particular on a large species of oyster; and, fearful of inserting their paws between the open valves lest the animal should close and crush them, they first place a tolerably large stone in the shell, and then drag out their victim with safety. Monkeys are no less ingenious. Dampier saw several of them take up oysters from the beach, lay them on a stone, and beat them with another until they had demolished the shells. Even the fox, when pressed by hunger, will eat mussels and other bivalves; and the raccoon when near the shore subsists on them largely, particularly on oysters.

A curious anecdote appeared in "Bell's Weekly Messenger," of 7th January, 1821. A tradesman at Plymouth, having placed some oysters in a cupboard, was surprised on finding in the morning a mouse caught by the tail by the sudden snapping of the shell. At Ashburton, a Mrs. Allridge had placed a dish of oysters in a cellar. A large oyster soon expanded its shell, and at the instant two mice pounced upon the "living luxury," and were at once crushed between the valves. The oyster, with the two mice dangling from its shell, was for some time exhibited as a curiosity. A better natural mouse-trap could not be imagined.

Among birds the mollusks have many enemies. Several of the duck and gull tribes derive a portion of their subsistence from them. The pied oyster-catcher derives its name from this habit. Several kinds of crows likewise feed upon shell-fish. Vultures and aquatic birds detach shell-fish from the rocks.

The consumption of oysters is recorded in earliest history, but their cultivation in the manner just described is a modern invention. This may account in some sense for the excessive and greatly superior production of this country. Though England and France have made lately rapid strides in this direction, their production combined could hardly equal that of our own land. The quantity taken from our waters is far greater than is generally supposed by those not familiar with this important business. The best statistics are necessarily very incomplete, and also uninteresting reading, though much might be said respecting the number of men and boats employed, the packing and pickling establishments with the force employed, the quantity of oysters consumed here and exported, we will merely say in conclusion that the value of the trade in 1877 amounted to over twenty-five million dollars in this country alone.

The Lobsters (which belong to the *Crustacea* or hard-shelled animals), the common species of which is so plentiful on the rocky coasts of our own country, and most parts of Europe, are generally taken in traps, sometimes made of osier twigs, also by nets, sometimes pots, always baited with animal garbage, and in some countries by torchlight, with the aid of a wooden instrument which acts like a forceps or a pair of tongs. They are also taken by the hand, but this requires dexterity, for the claws are powerful weapons of defence: one is always larger than the other, and the pincers of one claw are knobbed on the inner edge, those of the other are serrated. It is more dangerous to be seized by the serrated than by the knobbed claw. A great authority on fish matters says:

"I once heard a clergyman at a lecture describe a lobster as a standing romance of the sea; an animal whose clothing is a shell, which it casts away once a year, in order that it may put on a larger suit; an animal whose flesh is in its tail and legs, and whose hair is on the inside of its breast; whose stomach is in its head, and which is changed every year for a new one, and which new one begins its life by devouring the old. An animal which carries its eggs within its body until they become fruitful, and then carries them outwardly under its tail; an animal which can throw off its legs when they become troublesome, and can in a brief time replace them. Lastly, an animal with very sharp eyes placed in movable horns."

The London market alone requires two millions and a half of crabs and lobsters annually. Large numbers are sent from the Scottish coasts. The west and north-west coasts of

Ireland abound with fine lobsters, and *welled* vessels bring from them supplies for the London market of ten thousand weekly. A large number of lobsters is brought from Norway, as many as thirty thousand arriving from that country in a single day, conveyed in wells on board steam vessels, and kept in wooden reservoirs, some of which may be seen on the Essex side of the Thames. In order that the great mass of lobsters may be kept on their best behavior in these reservoirs, the great claw is rendered paralytic by means of a wooden peg driven into a lower joint: however cruel this may seem, it prevents them from tearing each other to pieces, so pugnacious are the animals. A good-sized lobster, we are informed, will yield about twenty thousand eggs; and these are hatched (being so nearly ripe before they are abandoned by the mother) with great rapidity, it is said in forty-eight hours, and grow quickly, although the young lobster passes through many changes before it is fit to be presented at table. During the early period of growth it casts its shell frequently. This wonderful provision for an increase of size in the lobster is perfectly surprising. It is indeed astonishing to see the complete covering of the animal cast off like a suit of old clothes, when it hides, naked and soft, in a convenient hole, awaiting the growth of its new crust or coat. Lobsters and crabs change their shell about every six weeks during the first year of their age; every two months during the second year; and afterward the changing of the shell becomes less frequent, being reduced to four times a year. Previously to putting off their old shell, they appear sick, languid, and restless. They acquire an entirely new covering in a few days; but during the time they remain defenseless they seek some lonely place, lest they should be attacked and devoured by such of their brethren as are not in the same weak condition. In casting their shells, it is difficult to conceive how the lobsters are able to draw the flesh of their large claws out, leaving the shells of these entire and attached to the shell of the body. The fishermen say that previous to this operation the lobster pines away till the flesh in its claw is no thicker than the quill of a goose, which enables it to draw its parts through the joints and narrow passage near the trunk. The new shell hardens by degrees.

It is supposed that the lobster becomes reproductive at the age of five years. Lobsters are very voracious; they are also full of fighting propensities, and have frequent combats among themselves, in which limbs are often lost; but the limb is soon replaced by the growth of a new one, rather smaller than the old one. In the water lobsters can run nimbly on their legs or small claws, and if alarmed, can spring tail foremost to a surprising distance as swift as a bird can fly. Fishermen can see them pass about thirty feet, and, by the swiftness of their motion, suppose they may go much farther. When frightened, they will spring from a considerable distance to their hold in the rocks, and will force their way into an entrance barely sufficient for their bodies to pass.

Like some of the crabs, lobsters are said to be attached to particular parts of the sea.

> "In shelly armor wrapt, the lobsters seek
> Safe shelter in some bay or winding creek;
> To rocky chasms the dusky natives cleave,
> Tenacious hold, nor will the dwelling leave,
> Nought like their home the constant lobsters prize,
> And foreign shores and seas unknown despise.
> Though cruel hands the banished wretch expel,
> And force the captive from his native cell,
> He will, if freed, return with anxious care,
> Find the known rock, and to his home repair."

In some parts of Europe the fishermen endeavor, by making violent noises, to drive the fish into their nets; but these are so cunning, that when surrounded by the net, the whole shoal will sometimes escape, for if one of them springs over it, the rest will follow like sheep.

The Danish fishermen have a similar mode of taking the horn-fishes, called "green-bone" from the color of their bones. They are timid, and afraid of the nets, and when the shoals approach, the fishermen commence a regular bombardment with stones, and so frighten them into their meshes.

A writer mentions a similar practice in Wales:

"The fishermen," he observes, "commenced their operations at every ebbing of the tide, by stretching a seine across the river, several hundred paces above the coast; and whilst drawing it towards the sea, they incessantly disturbed the water by beating the surface, as well as hurling into it the heaviest stones they could poise. The affrighted fish made at once for the sea, which, however, they could not reach except by passing through the intervening shallows. Here they were pursued by dogs trained for the purpose, and clubbed or speared by the men. I have frequently seen from one to two hundred fine fish, weighing from ten to twenty pounds each, taken in this extraordinary way."

GURNARD.

CHAPTER XVI.

ODDS AND ENDS ABOUT FISHES.

THE description we have quoted of fishes inhabiting the Mediterranean Sea corresponds entirely with the strange and varied character ascribed to them by ancient and modern writers. We will, however, before alluding to any particular species of fishes, give a brief outline of their nature generally. From the earliest ages fishes were most extensively used as articles of diet, and at the present time they form a considerable portion of the food of mankind generally. In some countries they were the only money of commerce, and dried fish were paid as current coin. Mythological honors were rendered to them by the ancients; and in the case of sharks, as mentioned in the chapter on "The Pirates of the Ocean," they are deified on the African coasts. Fish have been perpetuated in coins and sculptures, from which many of the species in ancient use can still be traced.

Fishes people the ocean with their shoals, and serve to keep in check the innumerable creatures of still lower construction, while they themselves are held in check, and afford sustenance to millions which have been placed in our system above them. In *form* they are the most varied beings in creation, and the most inventive fancy could scarcely imagine a shape or appearance to which a resemblance would not be found. They are of hideous or loathsome bulk or the most graceful form, and of gorgeous and resplendent colors; all wondrously adapted to the different modes of obtaining their food, whether by stealth or deceit, strength or swiftness. The general form of a fish is admirably adapted to its native element. In all fishes which require swiftness to secure their prey, the *tail* is the great organ of motion. The absence of any neck gives the advantage of a more extensive and resisting attachment of the head to the body, the greater proportion of which is left free for the play of the muscular masses which move the tail. Besides serving as the rudder or paddle, it is the tail of the fish that enables many of them to make those leaps out of the water to which we have frequently alluded to in these pages. From the enormous whales and sharks to the small stickleback, this power seems to belong to the greater number of fishes.

The *fins* on the upper surface of the fish serve to balance the body; those on the lower surface to turn it, to move it slowly, and to keep it suspended in strong currents; but in all these movements the assistance of the tail is observable.

Some of the fins of fishes are vertical, constituting a kind of keel or rudder. They differ in number, size, and the nature of the rays which support them, being sometimes spiny, and in other cases soft and articulated. Those corresponding to arms or wings are the *pectorals* (the chest), invariably fixed behind the gills.

Paley, in his "Natural Theology," thus sums up the actions of the fins of fishes: "The pectoral, and more particularly the ventral (belonging to the stomach) fins serve to raise and depress the fish: when the fish desires to have a retrograde motion, a stroke forward with the pectoral fin effectually produces it; if the fish desires to turn either way, a single blow with the tail the opposite way sends it round at once; if the tail strike both ways, the motion produced by the double lash is progressive, and enables the fish to dart forwards with an astonishing velocity. The result is not only in some cases the most rapid, but in all cases the most gentle, pliant, easy, animal motion with which we are acquainted. In their mechanical use, the anal fin may be reckoned the keel; the ventral fins, the outriggers; the pectoral fins, the oars [and, we may now add, the caudal fin, the screw-propeller]. And, if there be any similitudes between those parts of a boat and a fish, observe," adds Paley, "it is not the resemblance of imitation, but the likeness which arises from applying similar mechanical means to the same purpose."

Another powerful aid to the buoyancy of fishes is the air or swimming-bladder, which is described as a philosophical apparatus in the body of an animal. It is easy to see at the back-bone of the herring and other fishes a shining pearly-looking membrane, almost enveloped by the roe or milt of the fish. This is the air or swimming-bladder; and it is of this, as found in the sturgeon, the carp, the ling, and many other fishes, when dried and prepared by certain processes, that the substance called isinglass is manufactured.

It is the swimming-bladder that serves the fish for rising or sinking in the waters; but in such fishes as reside at the bottom of the sea or never come to the surface, this bladder is almost always wanting. How truly wonderful is this provision of nature! It would be very worthy of inquiry to know by what method an animal which lives constantly in water is able to supply a repository of air.

The bodies of fishes are nearly the same specific gravity as the water in which they live, owing to the great quantity of fat they contain, so that very little effort is required to keep them at any given height, and their ascent or descent in the water.

The circulation of blood is peculiar. There is but a single heart in fishes, that is, a heart consisting of only two cavities; and these correspond not to the left heart of mammals or birds, but to their right or pulmonic heart.

Respiration is carried on by means of the gills, which take the place of lungs, and consist of a large number of blood vessels, placed near the forward extremity of the animal, and protected by a bony case or covering, often defended by strong spines. The gills are placed in immediate communication with the heart. Water, which is impregnated with atmospheric air, entering at the mouth, is forced out again by the apertures at each side of the neck, and thus maintains almost a constant stream or rush through them, entering and again expelled at intervals. When fishes are taken from the water, the delicate thready structure of the gills immediately collapses; when exposed to the air a kind of suffocation ensues, and death is the result. This is the general principle of respiration in fishes, but in some cases the structure varies.

The smell of fishes in some species is remarkable: they scent their prey at a great distance, and the very perfection of this function is often fatal to them. Some fishes are so allured by scents, that by smearing the hand over with them, and immersing it in water, fishes (not sharks, let us hope) will often flock toward the fingers, and may easily be taken. Fishermen have the habit of making their bait more attractive by steeping it in some strong-smelling ingredient. On the American shores, the fishermen use putrid or damaged fish as bait for mackerel. They are thrown in a box hopper, in which a cylinder studded with knives is made to revolve by a crank. This is called the "bait-mill," and by its aid the contents are reduced to a kind of paste, which is thrown into the sea to attract the fish, which are then caught by lines with hooks, having a piece of polished pewter attached as a lure. In all fishes, nostrils or external openings are very apparent, and in these the nerves of smell are distributed.

Taste in fishes (as in animals who almost invariably swallow their food without mastication) cannot be very acute, since their tongue is in great part bony, and is often furnished with teeth and other hard coverings.

The organ of *touch* is in general as imperfect as that of taste: without prolonged members, and flexible fingers capable of grasping, they can scarcely explore the forms of objects by any other means than by their lips. Certain little fleshy tendrils which some fishes possess may supply the imperfections of touch in the other organs.

The bodies of most fishes are covered with small brilliant plates of a horny nature called *scales*, but in some kinds these are wanting, as in the turbot and others, in place of which are found bony protuberances in some species, and in others a very smooth skin without scales, and covered with a thick gelatinous secretion from the body. The scales consist of a substance chemically resembling the composition of bones and teeth. They usually overlap each other like tiles. Some are very thick forming a kind of armor.

In general, fish have large *eyes*, and in particular the pupil is very broad and open, as might be expected in creatures who require great powers of vision in the deep, where light penetrates but scantily. The eyes have no real eyelids, the skin passing over them mostly in a transparent form, to admit light; and they are sometimes opaque or dense. Some varieties of fish, whose eyes are fixed on the upper surface of their bodies, cannot see what prey they swallow; others have no outward indication of an eye. "No tear moistens, no eyelid shelters or wipes the surface; the eyes of fish are only representations of that beautiful and animated organ which is found in the superior class of animals."

"The *teeth* of fishes," says Professor Owen, "whether we study them in regard to their number, form, substance, structure, situation, or mode of attachment, offer a greater and more striking series of varieties than do those of any other class of animals. In number they range from zero to countless quantities. In the sharks and rays the teeth are supported by the upper or lower jaws, as in most quadrupeds; but many other fishes have teeth growing from the roof of the mouth, from the surface of the tongue, from the bony hoop or arches supporting the gills, and some have them developed from the bone of the nose and the base of the skull." In all fishes the teeth are shed and renewed not once only as in mammals, but frequently during the whole course of their lives.

Fishes have but small occasion for the sense of *hearing*, being condemned to reside in the empire of silence, where all around is mute. In most fishes the auditory parts are buried in the skull, and send no process to the surface.

Singular stories, however, are told of fishes being sensible to the sound of music. Ancient writers—Ælian and Aristotle—mention some fishes, and particularly skates, who are attracted in this manner. Two men embark in a boat, one with a musical instrument and the other with a net, and by this music the fishes become so entranced as to be taken easily. A somewhat similar mode is said to be practised by the boatmen of the Danube, who use bells for the purpose. Carp have been known to distinguish the sound of a bell, and the voice of their keeper when called to be fed.

The *brain* of fishes is remarkably small in proportion to the size of the animal, the quantity of nerves arising out of it, and the size of the cavity which contains it. The space thus left vacant is often filled with oil or fat.

Some fishes are not altogether indifferent to the fate of

their brood. We have already alluded to the attachment of the mammalian order for their young. Some fishes leave the depths of the ocean, and deposit their spawn in the shallows, where the young fry are comparatively safe from the voracity of their numerous enemies. Some build nests for their young, as we will further explain in this chapter.

The *eggs* of fishes are generally deposited on the surface of the water, where they float during the period of their development.

It is in the Northern seas that fishes display their most astonishing fecundity—not so much in the variety of species as in the multitude of individuals of a species; and the ocean nowhere else produces an abundance of fish approaching to the myriads of herring and cod in that quarter.

The uses to which fish are applied are numerous. They afford a valuable manure when they are to be had in plenty. Fishery-salt is also a great fertilizer. Pretty ornaments are made from fish-scales, as brooches, bracelets, &c.; the eyes of fishes are also employed by the makers of shell flowers for imitating buds. Mock pearls are made from an essence obtained by scraping the scales off the bleak (a fresh-water fish) and the whitebait. The natives of the north-west coast of this country make from the entrails of fishes bracelets, fishing-lines, thread, work-bags, head-dresses, and needle-cases; fish-hooks and needles are made of the bones.

We have already alluded to isinglass, which is made from the dense membrane which forms the air-bladder of the sturgeon and other fishes. Oil forms a staple article of commerce. The dog-fish is caught principally for the oil from its liver—a large fish yielding about a barrel-full. The skin of this fish is used to refine liquors, clear coffee, &c.

Our English ancestors were firm believers in the curative properties of certain fish. Pickled herrings were applied to the soles of the feet in fevers; pilchards were in great request for the swellings of the gums and legs; the flesh of the tunny was considered an antidote to poison; the teeth of thornbacks, bruised in a mortar, were used for sore eyes; the gall for complaints of the ear; the bones of the sturgeon were reduced to powder and applied in rheumatic cases; oyster and mussel-shells ground to powder were also employed.

Wonderful is the property of several species of fish of inflicting electric shocks so severe as to produce exhaustion and numbness of the nerves exposed to its action. That God should arm certain fishes, in some sense, with the lightning of the clouds, and enable them thus to employ an element so potent and irresistible as we do gunpowder, to astound, and smite, and stupefy, and kill the inhabitants of the water, is one of those wonders of an Almighty arm which no terrestrial animal is gifted to exhibit.

ELECTRIC EEL.

The Torpedo, popularly named by fishermen "numb-fish" and "cramp-fish," a genus of fishes of the Ray order, is a living electrical machine, which has the power of striking its enemies even at a very considerable distance. Fishermen constantly witness evidences of the singular faculty of this fish. As soon as it enters their net they are made aware of the fact by the shocks which are transmitted through the tackle by which it is suspended. These have been known to be sufficiently violent to compel the men to let go when they are drawing their nets, and thus allow the whole haul to fall back into the sea.

The *Torpenididæ*, as this family is termed, has been divided into a number of genera. They have a short and not very thick tail, cylindrical towards the end, and in outward appearance somewhat resemble a skate, and have nearly the same habits. Two species of the torpedo are occasionally found on the southern coasts of England, the common, or *Marmorata*, which sometimes attains a large size, weighing a hundred pounds; and the *Nobiliana*, which is more rare. They are readily distinguished by the spiracles behind the eyes, which are round and fringed at the edges in the former and perfectly smooth in the latter. These and other species are found more plentifully in the Mediterranean. When the torpedo is disposed to "astonish" any one, she furnishes to a careful observer the following premonitory indications of her intentions: the back—which, unlike that of the cat—is gibbous and raised when she is in good humor, flattens as she waxes angry, till the convex surface, gradually drawn in, becomes at length slightly concave; and at the same time the eyes, remarkably prominent during the repose of the creature, are retracted far back in the orbits. These are the precursory signals that the phials of her wrath are to be poured forth; the shock then instantly follows, and the fish as suddenly swells out again, recovering its usual form, generally to prepare for a new attack. These shocks follow in rapid succession: she sometimes inflicts forty or fifty broadsides in the course of one minute, and they are sufficiently powerful to destroy, as by lightning, small animals exposed to their influence.

Cuvier describes the electric apparatus of this fish to consist of a series of honeycombed-looking cells, filled with a thickish gelatinous fluid, and abundantly supplied with nerves, situated between the gills and the head of the fish.

The electrical organs are two in number. The number of cells varies according to the size of the fish; thus, in each organ of one fish were counted four hundred and seventy, and in another large fish one thousand one hundred and eighty-two. This natural electricity can be drawn from the fish by means of a conductor, and a shock is felt through a circuit formed by several persons joining hands.

The electrical effects produced on the fisherman who seize them were noted from early times; but Redi, the Italian naturalist of the seventeenth century, was the first who studied them scientifically. He caught and landed one of them with every precaution. "I had scarcely touched and pressed it with my hand," says the Italian artist, "than I experienced a tingling sensation, which extended to my arms and shoulders, which was followed by a disagreeable trembling, with a painful and acute sensation in the elbow joint, which made me withdraw my arm immediately."

Réaumur also made some observations upon the Torpedo. "The benumbing influence," he says, "is very different from any similar sensation. All over the arm there is a commotion which it is impossible to describe, but which, so far as comparison can be made, resembles the sensation produced by striking the tender part of the elbow against a hard substance." Redi remarks, besides, that the pain and trembling sensation resulting from the touch diminishes as the death of the Torpedo approaches, and that it ceases altogether when the animal dies.

In the seventeenth century the fishermen affirmed that the sensation was even communicated through the line by which it was caught, and even by the water. Redi does not deny this phenomenon, neither does he confirm it. He states that the action of the animal is never more energetic than when it is strongly pressed by the hand, and makes violent efforts to escape. Neither Redi nor Réaumur, however, could explain the cause of the strange phenomenon. It was reserved for Dr. Walsh, a Fellow of the Royal Society of London, to demonstrate the fact that the power was electrical in its nature. This he did by numerous experiments which he made in the Isle of Ré. The following are some of the experiments:

He placed a living torpedo upon a clean wet towel; from a plate he suspended two pieces of brass wire by means of silken cord, which served to isolate them. Round the torpedo were eight persons, standing on isolating substances. One end of the brass wire was supported by the wet towel, the other end being placed in a basin full of water. The first person had a finger of one hand in this basin, and a finger of the other in a second basin, also full of water. The second person placed a finger of one hand in this second basin, and a finger of the other hand in a third basin. The third person did the same, and so on, until a complete chain was established between the eight persons and nine basins. Into the ninth basin the end of the second brass wire was plunged, while Dr. Walsh applied the other end to the back of the torpedo, thus establishing a complete conducting circle. At the moment when the experimenter touched the torpedo, the eight actors in the experiment felt a sudden shock, similar in all respects to that communicated by a shock of a Leyden jar, only less intense.

Another fish little inferior to the torpedo in its "shocking" properties is the electric eel. Its physical properties enable it to arrest suddenly the pursuit of an enemy or the flight of its prey, to suspend on the instant every movement of its victim, and subdue it by an invisible power. Even the fishermen themselves are suddenly struck and rendered torpid at the moment of seeing it, while nothing external betrays the mysterious power possessed by the animal.

At Calabozo on the Orinoco, the electric eel abound in

great numbers. The natives have a unique and most curious mode of securing these formidable eels; the Indians themselves would describe it as "intoxicating by means of horses." From the neighboring savannahs a score or more of half-wild horses and mules are encircled by the natives and gradually driven to some adjacent marsh or stagnant basin surrounded by luxuriant vegetation and known to abound with the Gymnotis. The animals being forced from all sides are surely forced into the water, when a grand battle commences and a wonderful spectacle may be witnessed. The Indians, armed with long canes and harpoons, place themselves around the basin, some of them mounting the trees, the branches of which hang over the water, and by their cries, and still more by their canes, prevent the animals from landing again. The eels stunned by the noise, defend themselves by repeated discharges of their batteries. To one witnessing such a scene for the first time, it would probably seem that the eels must come off victorious. Such a picture is truly indescribable: groups of Indians surrounding the basin; the horses with bristling manes, terror and grief in their eyes, trying to escape from the storm which had surprised them; the eels yellow and livid, looking like great aquatic serpents swimming on the surface of the water, and chasing their enemies, are objects at once appalling and picturesque. In a few moments, many of the horses, benumbed by the repeated shocks of the eels are drowned. But gradually the eels themselves become bewildered or intoxicated, shun in place of attacking, and are slowly driven toward the banks, where they are easily taken by means of the little harpoon thrown at them. Being landed, the eels are transported to little pools dug in the soil and filled with fresh water; such is the terror inspired that the natives are very reluctant to take them from the harpoons, until life has expired, for fear of receiving the terrible shock.

A remarkable power of stinging is possessed by some of the inhabitants of the deep. What is called the Trygon, or Sting-Ray, is able to inflict severe wounds by its muscular and flexible tail, which it winds around the object of attack, and with the sword or spine at its base, pierces and lacerates. This weapon is armed with rows of serrated teeth at each side, every tooth of which is a small saw. The worst and most dangerous wound, however, is when the elastic tail dashes the apparatus, saws and all, into an unfortunate fisherman's thigh (as has frequently happened, in spite of the ordinary precautions), dragging it out again to make a new lunge before the unhappy victim has had time to escape; and so expert is this fish in this small-sword exercise, and so swiftly does stroke follow stroke, that persons who have seen it in operation report that, but for the spouting of fresh blood, and the larger display of raw surface, they would have declared the weapon motionless all the time. The terrible suffering inflicted by this atrocious caudine weapon—which is borne by four other colossal skates, as well as by the sea-eagle—has caused it to be regarded with as much superstitous reverence by fishermen as was the tail of his music-master, Chiron, by the youthful Achilles.

The Sting-Ray fish attains a colossal size in the Mediterranean. He possesses an enormous pair of fins, which, stretching out from either side of the body, offer a striking resemblance to a pair of wide-spread wings; and he has, moreover, a detached head, terminating in a porrect (extended) process, like a beak, and a large pair of piercing bright eyes, whence the origin of its appellation of "sea-eagle."

The Great Weever or Sting-Bull, and the Little or Viper Weever, possess the same formidable properties as the stinging ray. Both are found on English coasts, the former being about a foot long, and the other about four or five inches. Though of such small dimensions, these fishes are troublesome to encounter. The fins are spiny, and the gill-cover is furnished with a strong and sharp spine, which is directed backward, but is capable of being made erect to meet an enemy. This they use by a sudden bending of the body. The little weever buries itself in the sand, watching for its prey, leaving only its snout exposed, and if trod upon, it immediately uses its weapon with great force.

We have, in the chapter on "The Floating Navigators of the Ocean," alluded to the stinging powers of the Physalis, or "Portuguese man-of-war." It is a common trick with sailors to make a novice pick up one of these beautiful creatures, and then enjoy his discomfiture.

The *Acanthuri* (signifying "a spine" and "tail"), tropical fishes, some of which are remarkable for beauty of form and variety of colors, possess also a power of inflicting dangerous stings or wounds, which has obtained for them the name of "doctors" from our sailors, on account of the severe wounds they inflict on such as handle them unwarily. They have teeth trenchant and notched, and a strong spine at each side of the tail as sharp as a lancet, whence they are also called lancet-fishes. With these weapons they defend themselves with courage and success against the largest of their assailants. Many other fishes possess the same power of inflicting stings and wounds: living a life of constant warfare in the deep, Nature has bestowed upon them means of defence and for procuring their prey.

It is worthy of observation that, with very few exceptions, the immense population of the ocean is carnivorous. The principal circumstance that regulates the choice of diet among fishes seems to be the power of mastery. Of terrestrial creatures, a very large number are peaceful, never, under ordinary circumstances, willingly taking the life of even the most helpless around them; but the sea is a vast slaughter-house, where nearly every inhabitant dies a violent death, and finds a grave in the maw of his fellow. Yet let us not arraign the providence of God, as if it were cruel and unkind: a sudden termination of existence is the most merciful mode, as far as we can conceive, by which the overflow of animal life could be checked.

As James Montgomery says:

> "'Twas wisdom, mercy, goodness that ordain'd
> Life in such infinite profusion—Death
> So sure, so prompt, so multiform to those
> That never sinn'd, that know not guilt, that fear
> No wrath to come, and have no heaven to lose."

A very interesting family of fishes, for the peculiar properties which they possess, are the Sucking-fishes—remarkable for having the ventral fins united under the surface of the body to form the apparatus which distinguishes them. To this family belong the Sea-Owl Snail, and one or two British species, including the Lump-sucker. This animal has a grotesque and clumsy form, but the colors which ornament it are very fine, combining various shades of blue, purple, and orange. It attains a tolerably large size—about nineteen inches—weighing sometimes seven or eight pounds. Its sucker is so powerful that a pail, containing some gallons of water, has been lifted, when one of these fishes contained in it was taken by the tail.

To this family Cuvier also referred the far-famed Remora; noticing, however, the different position of the sucking disc, and other important distinctions, on account of which a very different place is now assigned to it. The use of the sucking apparatus is, however, much the same—that of attaching the animal to fixed substances, so that it may remain and obtain its food, where otherwise it would be swept away by the current.

The remora was the subject of much imaginative terror to the ancients, who believed that it had the power to impede or stop the course of a ship. Oppian says:

THE CRAMP FISH.

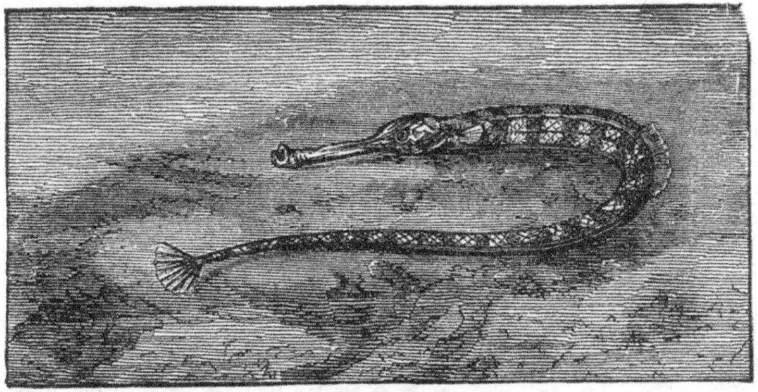

THE PIPE FISH.

THE ARCTIC CHIMAERIDAR.

THE LAMPREY EEL.

"The seamen run confus'd, no labor spared,
Let fly the sheets, and hoist the topmast yard;
The master bids them give her all the sails,
To court the winds and catch the coming gales;
But though the canvas bellies with the blast,
And boisterous winds bend down the cracking mast,
The bark stands firmly rooted on the sea,
And all unmov'd as tower or towering tree."

Pliny writes: "Why should our fleets and armadas at sea make such turrets on the walls and forecastles, when one little fish (see the vanity of man!) is able to arrest and stay, perforce, our goodly and tall ships?"

These are droll fancies; but, tested by the fact, the adhesive powers of this fish are very remarkable, great weights being dragged by it, and retaining its hold with a bull-dog tenacity, even submitting to be torn to pieces before it will relinquish its hold. It is frequently seen among other fishes in the Atlantic, attaching itself to some one or other by its sucker, and often, also, to the rudder or bottom of a ship.

The length of the Mediterranean remora is about eighteen inches, and the length of the head is nearly one-fifth of the proportion of the whole fish. Feeding principally on the small animals diffused throughout the waters of the ocean, it probably receives a sufficiency of food even when attached to a moving object, as a ship or large fish, merely by opening its mouth, which has a very large gape.

Belonging to a distinct family, but employing its mouth as a powerful sucker, is the Sea-Lamprey, a species resembling eels in the rounded shape of the body and a certain similarity of habits. The mouth is circular, armed with hard tooth-like processes, and provided with a flexible lip. So great is the power of suction which it possesses, that a stone has been raised by it out of the water, weighing ten or twelve pounds, and yet the fish measures but from two to three feet.

The historical renown of the lamprey is very great. It

was the favorite dish of the Romans, who kept the fishes in ponds at a great expense. The best lampreys were procured from Sicily as presents to the reigning emperors and high officials. A hundred pieces of gold were sometimes paid for them.

A horrible story is told of Pollio, a friend of Augustus Cæsar, who, on the supposition that lampreys fed on human flesh were more delicate, ordered his slaves, when accused of the slightest fault, to be thrown into his fish-pond. This cruelty was discovered when one of his servants broke a glass in the presence of the Emperor, who had been invited to a feast. The master ordered the slave to be seized, but he threw himself at the feet of the Emperor, and begged him to interfere, and not suffer him to be devoured by the lampreys. On examining into the matter, the Emperor, astonished at the barbarity of his favorite, caused the fish-ponds to be filled up.

Respecting this fish, there is another use to which the mouth or sucker is applied. The whole of its interior arch is studded with rows of teeth, each one of which, on a broad base, is furnished with one or two apparently reversed points, and these teeth which are most remote and concealed are larger than others, and more effectually crowded with these points. For simply biting they are useless, but when the breadth of the mouth is brought into contact with the surface of a fish on which the lamprey has laid hold, by producing a vacuum these roughly-pointed teeth are brought forward so as to be able to act on it by a circular motion; and the limited space of the captive prey is thus rasped into a pulp and swallowed, until a hole is made which may, perhaps, penetrate to the bones, and from the torture of which the most strenuous exertion of the victim cannot deliver it. This is frequent on the mackerel and on other fishes, as the gurnard, coal-fish, cod and haddock.

The "Mail-Cheeked or Gurnard group of fishes offer

some very interesting subjects for notice, including a considerable number of species, all characterized by sharp projecting cheeks, and heads cased in armor of bony plates, among which we may mention the Flying Gurnard, the Sea-Scorpion, and the Father-Lasher.

The name "gurnard" is derived from the growling, grunting noise which these fishes make, by means of the throat and gills, when taken out of the water, and which has obtained for one species the name of "piper." The Romans used to call the latter "lyres," rather, perhaps, on account of their fancied resemblance to an ancient lyre, than to the very unmusical sound they emit. Many of the gurnards are distinguished by beauty of color.

The New Zealand Gurnard, about eighteen inches in length, is a splendid fish: the upper part is brownish-red, the fins are very large and of an emerald green, broadly bordered with azure blue, and having an oval patch of velvety black beautifully relieved with snow-white spots.

The Sea-Scorpion differs from its land namesake, the possessor of one solitary but dangerous tail-sting, the head of the fish being surrounded with goads and prickles, which render it a formidable enemy to contend with, by swelling out its cheeks and gill-covers to a large size, realizing Ovid's description of it,—

> "Scorpœna's poison'd head, beset with spines;"

excepting that the stings, beyond inflicting a sharp pain, are not venomous. Some of these animals are remarkable for their ugliness, and others exhibit very fine colors. They abound in the warm seas, and are often taken on the Atlantic shores, sometimes exceeding a foot in length.

The Marine Sticklebacks, which are thus named from the spines which arm their back, ventral fins, and other parts, are inhabitants of the seas in cold and temperate regions, and are curious little animals, a kind of Liliputian warriors armed at all points for warfare, protected at the sides by shell-like plates, and with spears that play terrible havoc among the crustacea and small animals on which they feed. They are objects of peculiar interest from the beauty of their colors, which they change in a remarkable manner. They are excessively pugnacious and predatory in their habits, the larger species eating the smaller, and destroying the eggs and fry of fishes to a prodigious extent. An observer relates of the fifteen-spined stickleback, about six inches in length,—sometimes called the "sea-adder,"—"that it keeps near rocks and stones clothed with sea-weeds, among which it takes refuge upon any alarm. Though less active than its brethern of the fresh water, it is scarcely less rapacious. On one occasion I noticed a specimen engaged in taking its prey from a clump of sea-weed, in doing which it assumed every posture between the horizontal and perpendicular, with the head downwards and upwards, thrusting its projecting snout into the crevices of the stems, and seizing its prey with a spring. Having taken this fish with a net, and transferred it to a vessel of water, in company with an eel three inches long, the latter was attacked and devoured head foremost; not, indeed, altogether, for the eel was too large a morsel, so that the tail remained hanging out of the mouth, and it was obliged to disgorge the eel partly digested."

A writer relates some interesting observations on the fighting propensities of these animals when confined in a tub of water:

"A few at first are turned in, and swim about in a shoal, apparently examining their new habitation. Suddenly one will take possession of a corner of a tub, or, as it will sometimes happen, of the bottom, and will instantly commence an attack on its companions; and if any of these venture to oppose its rule, a regular and most furious battle ensues. The two combatants swim round and round each

THE STICKLEBACKS—NEST-BUILDING FISHES.

other with the greatest rapidity, biting and endeavoring to pierce each other with their spines, which on these occasions are projected. I have witnessed a battle of this sort which lasted several minutes before the other would give way; and when one does submit, imagination can hardly conceive the vindictive fury of the conqueror, who, in the most persevering and unrelenting way, chases its rival from one part of the tub to another until fairly exhausted with fatigue. They also use their spines with such fatal effect that, incredible as it may appear, I have seen one, during a battle, absolutely rip an opponent quite open, so that it sank to the bottom and died. I have known three or four parts of the tub taken possession of by as many other little tyrants, who guard their territories with the strictest vigilance, and the slightest invasion invariably brings on a battle."

It is pleasing to add for the honor of the sex that the females take no part in these ferocious proceedings; a redeeming feature in the belligerents, however, is the care which they take in building their nests and watching over the welfare of the females and their eggs. The reader may not have heard of nest-building fishes, and, indeed, although the ancients were acquainted with this instinct in some fishes, it was not until 1838 that modern naturalists proved this by the discovery of a stickleback nest. These animals collect small pieces of straw or stick, with which the bottom of the nest is laid among water-plants, and these they cement together by a transpiration from their own bodies, which forms a thread through and round them in every conceivable direction. The thread is whitish, fine, and silken. The sides of the nest are made after the bottom.

Not many fishes are yet known as nest-builders. The Goramy, a native of the China seas, forms at the breeding-season a nest by interlacing the stems and leaves of aquatic grapes. Both male and female watch these nests for a month or more with great vigilance, violently driving away every other fish until the spawn is hatched. The Gobies or Sea-Gudgeons, have similar instincts. Many, however, are known not to construct nests. Salmon and others exhibit an approach to the nest-building habit, in making a place for their eggs in the sand or gravel.

We must now notice the Flying Gurnard, remarkably distinguished from the others of the family to which it is allied by the great size of its pectoral fins, which are long enough, and their webs sufficiently broad, to sustain the fish in the air during its long flying leaps out of the water. These fins, however, are very different in appearance from those of the flying-fish (*Exocetus*, "fishes out of the water"), which belongs to another family. The flying gurnard is an inhabitant of the warm seas; one species is common in the Mediteranean, and is sometimes fifteen inches in length. Its flight is said not to extend more than about forty yards, but it sometimes rises high enough to fall on the decks of large ships. At particular times, and especially on the approach of rough weather in the night, numbers of them may be seen by the phosporic light which they emit, making their passages in apparent streams of fire.

Flying-fishes have the power of raising themselves out of the water, and continuing suspended in the air until their fins become dry, by which means they escape some of their marine enemies, such as the dolphin and many others.

> "So fishes rising from the main,
> Can soar with moisten'd wings on high;
> The moisture dried, they sink again,
> And dip their wings again to fly."

But they run the gauntlet of the long-winged sea birds, which seize them in the air; and between themselves and their swimming and flying enemies, they furnish one of the most singular sights in the warm seas of the tropics. One species of the Exocetus sometimes visits the English coasts, and are said to leap more than two hundred yards in distance, and upwards of twenty feet in height. Although these fishes are called "flying," their action has more resemblance to a long and vigorous leap than the flight of birds. Birds have an elegant, fearless, and independent motion; while that of the fish is hurried, stiff, and awkward, more like a creature requiring support for a short period.

THE FLYING FISH.

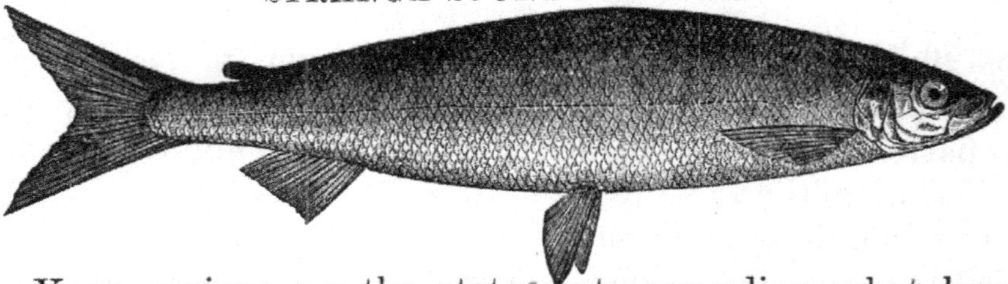

Very curious are the statements regarding what have been called "musical" fish, but how far such a title is warranted is doubtful. It is known that many fishes, notwithstanding their being characterized as mute, are remarkable for giving utterance to a peculiar sound called "drumming." This is very perceptible in the famous *Maigre* of the Mediterranean, the *Umbrina* of the Romans, a fish which swims in groups, and often utters a low bellowing sound beneath the water, which is heard from a depth of one hundred and twenty feet, and is rendered stronger by placing the ear upon the gunwale of the boat.

Lieutenant White of the U. S. Navy, in his "Voyage to the China Seas," published in 1824, relates that being at the mouth of the Cambodia, his crew and himself were extremely astonished by hearing certain unaccountable sounds from beneath and around the vessel. These were various, like the bass notes of an organ, the sound of bells, the croaking of frogs, and a pervading twang which the imagination might have attributed to the vibrations of some enormous harp. For a time the mysterious music swelled upon them, and finally formed a universal chorus all around; but as the vessel ascended the river, the sounds diminished in strength, and soon altogether ceased.

Humboldt was witness to a similar fact in the South Sea, but without suspecting the cause. Towards seven in the evening the whole crew were astounded by an extraordinary noise, which resembled that of drums which were beating in the air. It was at first attributed to the breakers. Speedily it was heard in the vessel, and especially towards the poop. It was like a boiling, the noise of the air which escapes from fluid in ebullition. The sailors began to fear there was some leak in the vessel. It was heard unceasingly in all parts of the vessel, and finally, about nine o'clock, it ceased altogether.

It would form a curious matter of research to ascertain by what organs these sounds are produced at so great a depth, and without communication with the exterior air. The illustrious naturalist further remarks that such of the *Sciænidæ* (the Maigre family) as are the most remarkable for the faculty in question, having the swimming-bladder very large and thick, furnished with extremely strong muscles, and are, in several species, provided with more or less complicated prolongations, which penetrate between the intervals of the ribs. But what renders the phenomenon more unaccountable is that these swimming-bladders have no communication with the intestinal canal, nor, in general, with any part of the exterior.

The interpreter belonging to Lieutenant White's ship stated that the marine music which had so much surprised the crew was produced by fishes of a flattened oval form, and which possess the faculty of adhering to various bodies by their mouths. This fish might have been the Pogonia, which produces much more sound than any of the other Maigre tribe to which it belongs, on which account it is sometimes called the "drum-fish." Schœff reports of them that they will assemble round the keel of a vessel at anchor, and serenade the crew. Some of the species attain a large size—one hundred pounds or more—and are excellent for the table.

Sir James Emerson Tennant, in his account of Ceylon, states: "In the evening, when the moon had risen, I took a boat and accompanied the fishermen to the spot where musical sounds were said to be heard issuing from the bottom of a lake, and which the natives supposed to proceed from some fish peculiar to the locality. I distinctly heard the sounds in question. They came up from the water like the gentle thrills of a musical chord, or the faint vibrations of a wine-glass when its rim is rubbed by a wet finger. It was not one sustained note, but a multitude of tiny sounds, each clear and distinct in itself, the sweetest treble mingling with the deepest bass. They came evidently and sensibly from the depths of the lake, and appeared to be produced by mollusca, and not by fish."

Sounds somewhat similar are heard under water at some places on the western coast of India, especially in the harbor of Bombay.

Among the foremost of queer fish is the Sea-Devil, a most inharmonious name, but which seems to have been given to it on account of its hideous, strange, and uncouth appearance. A species of this extraordinary fish of the Skate family frequents Kingston harbor in Jamaica, where they are seen floating on the surface, or swimming just beneath the water. An interesting account is given by Lieutenant Lamont of the escape of a devil-fish and the capture of another at Port Royal. The lieutenant had been called to the beach by seeing a multitude assembled to look at one of these fishes floating past. His curiosity turned to surprise when he saw, flapping on the water, about twenty yards yards from the shore, a large dark-colored mass, whose shape and size he could not immediately determine, but which seemed prodigiously big beyond anything he could conceive, since it so much exceeded all he had ever seen or heard of fishes. The boats were started off to pursue it, and it was harpooned, but no sooner was the monster struck than it made off with amazing velocity, towing the boat of the harpooner after him. A succession of boats now came up. These strung themselves on to the harpooner one after another, striking each a harpoon as the boats came up. They consecutively formed a long line, but such was the force of the fish that all the boats were drawn out ten miles to sea. Night was drawing on. To bring the chase to a close, another harpoon was struck into the monster, when it made one convulsive effort to get away, and broke loose, carrying away eight or ten harpoons and pikes, leaving every one astonished at the success of its escape.

Another devil-fish was not so fortunate, and Lieutenant Lamont gives the history of its capture within the harbor, which the animal traversed up and down, dragging with such velocity the boat from which it had been struck, that the other boats following could not overtake the fish. Its struggles were tremendous, plunging into the midst of the boats that at length surrounded it, darting from the surface to the bottom of the water, and then rising swiftly, dashing the foam about on every side and rolling round and round to extricate itself from the poles and lines. Unable to get away, it swam off, towing all the boats after it, and then laid itself at the bottom of the water. From this position the stretch and strain of all the boats' crews could not move it. Slackening their efforts gradually, the monster rose again to the surface, when a shower of musket-balls and pikes riddled it through. Until this capture was effected, it was believed that a sea-devil was beyond the might of human art and strengh. The dimensions of this fish were not more than half that of the common size, being only fifteen feet in width. A man, however, entered the mouth with ease, the space being two feet and a half. The weight of the fish was so great, that, with difficulty, forty men with two lines attached to it dragged it along the ground.

A devil-fish taken at Barbadoes required seven yoke of oxen to draw it.

In the account of the fish taken in Delaware Bay (remarks the Hon. Richard Hill in an interesting article on the subject of the devil-fish), it was stated that drawing a boat after it with the celerity of a whale when harpooned, it caused a wave to be raised on each side the trough of the sea, several feet higher than the boat; that during the scuffle the vast fins of the fish lashed the sea with such vehemence that the spray rose to the height of thirty feet, and rained dropping water around to the distance of fifty feet, and yet the measurement of this fish was only half of those generally seen, being only eighteen feet in breadth. Three pairs of oxen, one horse, and twenty-two men, all pulling together, with the surges of the Atlantic to help, could barely convey the monster to the dry beach.

The monstrous skate said by Père Labat to have been observed by the natives of Guadaloupe, and described as fourteen feet broad, and ten feet from the head to the commencement of the tail, with the tail fifteen feet more, altogether twenty-five feet long, was no doubt a kindred species of the devil-fish; and the monster spoken of by the early voyagers as suffocating the pearl-divers in the water, and known by the name of *Manta*, was a similar animal.

Surprising stories are related of these fishes. Le Vaillant

speaks of three that he saw in the Atlantic—one so large that it seemed fifty or sixty feet wide; they all three carried each on his horns a white fish about half a yard long, which appeared to be stationed there on duty as sentinels, to keep watch for the safety of the "devils," and to guide their movements: that these sentinels passed over their backs when they rose too high, and repassed under them until they descended deeper, disappearing and being seen no more for a time, but reappearing and resuming their post as sentinels when the fish again ascended to the surface.

Among other "queer" fish, is the Fishing Frog, or Angler, belonging to the "Wristed" family (so named from the prolongation of the wrist-bones, forming a kind of arm, supporting the pectoral fin on a kind of hand), and one of the most extraordinary and repulsive-looking animals that inhabit the deep.

Let the reader imagine a gigantic tadpole blown out to the size of a porpoise, with an immense head, and a mouth extending on either side far beyond the width of the body, opening to view a capacious den, shagged throughout with hooked and mobile teeth, a triple tier in the upper, and an equal number in the lower jaw, the palate, tongue, fauces, pharynx, and far down the throat, glistening with a like display of ivory fangs; unfishy orbs, resembling those of the "star-gazer" (the "priest-fish," so named from the whites of its eyes, looking constantly heavenward), planted high in the forehead; a scaleless skin, which is reeking cold, and clammy; its surface, from near the tail to the corners of the mouth, as crawling with long wriggling carunculated (fleshy) appendages, like so many worms in agony; the flesh "boggy" to the touch, save where it is padded out with an enormously extended liver, or just over the branchial (apertures for the passage of water from the gills) cavity; a pantry constantly replenished with provisions; add to all these a large pair of Caliban-hand-like fins, planted close under the throat; a fierce, malevolent aspect, and an ungainly mode of wallowing, rather than swimming, through the brine, and it will be apparent, even from this very imperfect sketch, that such a fish scarecrow could not fail to arrest attention, even had their been no other claim to regard than his portentous ugliness.

Of its boldness and voracity many anecdotes are related. A fisherman had hooked a cod-fish, and whilst drawing it up he felt a heavier weight attach itself to his line. This appeared to be a frog-fish of a large size, which he compelled to quit its hold by a heavy blow on the head, leaving its prey still attached to the hook. In another instance one of these fishes had seized a conger eel which had taken the hook; but after the latter had been engulfed in the enormous jaws, and perhaps in the stomach, it struggled through the gill-aperture of its captor, and in that situation both were drawn up together.

An incident is related of its swallowing a large ball of cork employed as a buoy to a butler or deep-sea line.

It has also been stated that when this fish is captured in a net, its rapacious appetite is not in the least diminished, but it generally devours some of its fellow-prisoners.

The sea-frog, as it can live longer out of water than most other fish, is said to pass some of its time on shore. The naturalist, Rondolet, tells a curious story of one being found on land, holding a fox fast by the leg. The cunning quadruped, outreached for once by a fish, had put his foot into the mouth of the sea-frog, who, instantly closing upon it, held it fast as in a trap till next morning, when Rondolet surprised them in this strange position.

The name of "angler" given to this singular fish is derived from its habit of crouching close to the ground, and stirring up with its fins the sand or mud. In the obscurity thus produced the animal moves its appendages, tentacles or feelers, in various directions, by way of attracting as a bait, and the small fishes approaching to examine or seize them are soon conveyed to the capacious jaws of the angler. Nature has added to this provision for obtaining food, inasmuch as a filament shooting up close to the upper lip of the fish carries upon its extremity a little membrane or flag, of brilliant metallic lustre, which, it is supposed, the angler uses as a means of alluring its prey; and the relative position of the flag, the eye, and the mouth, favor such a purpose. The upper part of the body is brown, inclining to dusky, and the lower parts are white. The sea-frog is common in the Northern Ocean and the Mediterranean; it is also taken sometimes on the British coasts.

In the chapter on the "Monarchs of the Ocean." we have alluded to the Saw-fish and the Sword-fish as formidable enemies to the whale; but is not merely on their fellow-inhabitants of the deep that these powerful fishes exercise their aggressive propensities. Some singular instances are related of their attacking even the "wooden walls" that glide tranquilly through their watery domain.

Captain Wilson, of the Halifax packet, states:

"Being in the Gulf of Paria, in the ship's cutter, I fell in with a Spanish canoe, manned by two men, then in great distress, who requested me to save their lines and canoe, with which request I immediately complied, and going alongside for that purpose, I discovered that they had got a large saw-fish entangled in their turtle-net, which was towing them out to sea, and but for my assistance they must have lost either their canoe or their net, or perhaps both, which were their only means of subsistence. Having only two boys with me at the time in the boat, I desired them to cut the fish away, which they refused to do. I then took the bight of the net from them, and with the joint endeavors of themselves and my boat's crew, we succeeded in hauling up the net, and to our astonishment, after great exertions, we raised the "saw" of the fish about eight feet above the surface of the sea. It was a fortunate circumstance that the fish came up with the belly towards the boat, or it would have cut the boat in two.

"I had abandoned all idea of taking the fish, until, by great good luck, it made towards the land, when I made another attempt, and having about three hundred feet of rope in the boat, we succeeded in making a running bowline-knot round the saw of the fish, and this we fortunately made fast on shore. When the fish found itself secured, it plunged so violently that I could not prevail on anyone to go near it; the appearance it presented was truly awful. I immediately went alongside the Lima packet, Captain Singleton, and got the assistance of all his ship's crew. By the time they arrived, the fish was less violent. We hauled upon the net again, in which it was still entangled, and got another three hundred feet of line made fast to the saw, and attempted to haul it toward the shore; but although mustering *thirty hands*, we could not move it an inch. By this time the negroes belonging to Mr. Danglad's estate came flocking to our assistance, making together about one hundred in number, with the Spaniards. We then hauled on both ropes for nearly the day, before the fish became exhausted. On endeavoring to raise the fish it became most desperate, sweeping with its sword from side to side, so that we were compelled to get strong ropes to prevent it from cutting us to pieces. After that, one of the Spaniards got on its back, and at great risk cut through the joint of the tail, when animation was completely suspended. It was then measured, and found to be twenty-two feet long and eight feet broad, and weighed nearly five tons."

An East Indiaman was attacked by a sword-fish with such prodigious force as to drive its "snout" completely through the bottom of the ship, and must have been destroyed by the leak had not the animal been killed by the violence of its own exertions, and the sword remaining imbedded in the wood. A fragment of this vessel, with the sword fixed firmly in it, is preserved as a curiosity in the British Museum.

Several instances of a similar character have occurred, and one formed the subject of an action in the courts of law so recently as 1868, brought against an insurance company for damages sustained by a vessel from the attack of one of these fishes. It seems the *Dreadnought*, a first-class mercantile ship, left a foreign port in perfect repair, and on the afternoon of the third day a "monstrous creature" was seen sporting among the waves, and lines and hooks were thrown overboard to capture it. All efforts to this effect, however, failed: the fish got away, and in the night-time the vessel was reported to be dangerously leaking. The captain was compelled to return to the harbor he had left, and the damage was attributed to a sword-fish, twelve feet long, which had assailed the ship below water-line, perforated her planks and timbers, and thus imperilled her existence on the ocean.

Professor Owen, the distinguished naturalist, was called to give evidence on this trial as to the probability of such an occurrence, and he related several instances of the prodigious strength of the "sword." It strikes with the accumulated force of fifteen double-handed hammers; its velocity is equal to that of a swivel shot, and it is as dangerous in its effects as a heavy artillery projectile would be.

Oppian describes the sword-fish when attacked:

> "He summons to his instant aid
> The oft-tried prowess of his trusty blade;
> Selects some boat, and runs his puissant sword
> Full many an inch within the fatal board."

In remarking upon beautiful fishes, it would be quite out of the limits of a small publication like the present to attempt more than a bare mention of a few species of the ocean inhabitants which possess, in a special degree, the attributes to which this term may be applied. Among the most prominent of beautiful fishes is the Dolphin, which, however, belongs to an extensive family, including the porpoise, grampus, &c., and animals which, on account of their large size, are commonly called whales.

SKATES.

There are, however, many other fish that change color before they die. We have seen species of the cat-fish change from a warm and glowing smalt during the last pangs to a dull leaden hue, losing at the same time the delicate pinky tinge of the sides and abdomen. The common sucking-fish, from a brown, bright, shining, blackish color, changes even in the water to a leaden hue, and as it dies assumes a tan-color, which grows paler by degrees and turns to a dingy white.

When swimming near the surface of the water, and glittering beneath the light of a cloudless sky, the dolphins appear clothed in the richest gold, and to have the starry lustre of the topaz and sapphire. Two species have been named, from the variety and vividness of their tints, the "sea-peacock" and the "blue-fish."

The true dolphin has the snout prolonged into a rather slender beak, whence the French have applied to it the name of "the goose of the sea." It was very differently regarded and designated by the ancients, who looked upon it as a sacred fish, and dedicated it to Appollo, who was worshipped at Delphi with dolphins for his symbols. The name is given to one of the fairest provinces of France—Dauphiny, from which the heir-apparent of the throne formerly derived his title of "Dauphin."

Wondrously beautiful, indeed, are these gay inhabitants of the seas, especially when seen playing and springing from the water, when they assume the curved shape that is not natural to them, but which old painters and sculptors have always given them.

> "Upon the swelling waves the dolphins show
> Their bending backs, then swiftly darting go,
> And in a thousand wreaths their bodies throw."

They are, however, very voracious animals, and are said to prey not only on other fishes, but their own species. The flying-fish in particular comes in for a share of their pursuit. Captain Basil Hall gives a vivid discription of their operations:

"Shortly after observing a cluster of flying-fish rise out of the water, we discovered two or three dolphins ranging past the ship in all their beauty, and watched with some anxiety to see one of those aquatic chases of which our friends, the Indiamen, had been telling such wonderful stories. We had not long to wait, for the ship, in her progress through the water, soon put up another shoal of these little things, which, as the others had done, took their flight directly to windward. A large dolphin, which had been keeping company with us abresst of the weather gangway, at the depth of two or three fathoms, and, as usual, glistening most beautifully in the sun, no sooner detected our poor dear little friends take wing than he turned his head towards them and, darting to the surface, leaped from the water with a velocity little short, as it seemed, of a cannon-ball. But, although the impetus with which he shot himself into the air gave him an initial velocity greatly exceeding that of the flying-fish, the start which his fated prey had got enabled them to keep ahead of him for a considerable time.

"The length of the dolphin's first spring could not be less than ten yards, and after he fell we could see him gliding like lightning through the water for a moment, when he again arose and shot forwards with considerably greater velocity than at first, and, of course, to a still greater distance. In this manner the merciless pursuer seemed to stride along with fearful rapidity, while his brilliant coat sparkled and flashed in the sun quite splendidly. As he fell headlong on the water at the end of each huge leap, a series of circles were sent far over the still surface, which lay as smooth as a mirror.

"The group of wretched flying-fish, thus hotly pursued, at length dropped into the sea; but we were rejoiced to observe that they merely touched the top of the swell, and scarcely sank in it; at least, they instantly set off again in a fresh and more vigorous flight. It was particularly interesting to observe that the direction they now took was quite different from the one in which they had set out, implying but too obviously that they had detected their fierce enemy, who was following them with giant steps on the waves, and now gaining rapidly upon them. His terrific pace was, indeed, two or three times as swift as theirs, poor little things!

"The greedy dolphin, however, was fully as quick-sighted as the flying-fish which were trying to elude him, for whenever they varied their flight in the smallest degree, he lost not the tenth part of a second in shaping a new course, so as to cut off the chase; whilst they, in a manner really not unlike that of the hare, doubled more than once on their pursuer. But it was soon too plainly to be seen that the strength and confidence of the flying-fish were fast ebbing. Their flights became shorter and shorter, and their course more fluttering and uncertain, while the enormous leaps of the dolphin appeared to grow more vigorous at each bound. Eventually, indeed, we could see, or fancied that we could see, that this skilful sea-sportsman arranged all his springs with such an assurance of success that he contrived to fall at the end of each just under the very spot on which the exhausted flying-fish were about to drop. Sometimes this catastrophe took place at too great a distance for us to see from the deck exactly what happened; but on our mounting high into the rigging, we may be said to have been in at the death, for then we could discover that the unfortunate little creatures, one after another, either popped right into the dolphin's jaws as they lighted on the water, or were snapped up instantly afterwards.

"It was impossible not to take an active part with our pretty little friends of the weaker side, and accordingly we very speedily had our revenge. The middies and the sailors, delighted with the chance, rigged out a dozen or twenty lines from the jibboom-end and spritsail-yard-arms, with hooks baited merely with bits of tin, the glitter of which resembles so much that of the body and wings of the flying-fish that many a proud dolphin, making sure of a delicious morsel, leaped in rapture at the glittering prize."

The dolphin, however, in turn becomes the prey of other fishes, and especially of the Fox-Shark, or Sea-Fox as it is sometimes called, a genus of sharks containing only one known species, belonging to the Mediterranean Sea, and the Atlantic, and occasionally seen on English coasts. This powerful fish attains a length of thirteen feet, including the tail-fin, which is remarkably long, nearly half the dimensions of the animal, and which, as a weapon of offence, is very formidable. The furious lashing of this appendage has obtained for this fish the popular name of "thresher." A whole herd of dolphins will take flight at the first splash of this tail, and even the grampus, the largest of the dolphin family, and, it is said, a formidable adversary of the whale, comes off badly in an encounter with the fox-shark.

The numerous and interesting Mackerel family include many species remarkable for rich coloring. The common Mackerel itself, which is described in the chapter on "Methods of Fishing," is a very beautiful fish, with its brilliant blue and green tints, besides its elegant form. The Dory, or John Dory as it is popularly called, is said to derive its name from the golden tint that prevails over it when taken from the water; *jaune* in French being "yellow," and *doré*, "golden." Along the shores of the Mediterranean, where this fish abounds, it is called among other names "St. Peter's Fish," from a legend that the apostle obtained from it the coin to pay the tribute money, and that the impression of his two fingers marks the species to the present day; a distinction, however, which is claimed also for the haddock. The dory is very common on some parts of the Atlantic coasts. The prevailing color of the body (which is oval) is an olive-brown tinged with yellow, reflecting in different lights blue, gold, and white. When the fish is taken, the varying tints of these beautiful colors pass in rapid succession over the body. Though flat in form, the fish swims erect, and both surfaces being thus equally exposed to the light, are alike of a coppery hue.

The Boar-fish, a relative of the dory, is of inferior pretensions as regards shape and color, the mouth having some resemblance to the snout of a hog, which doubtless originated the name. The eyes are very large and prominent, and the body of a pale carmine color, with orange bands on the back.

But the glory of the Mackerel family, at least for splendor of appearance, is the Opah, or King-fish, an inhabitant of the seas of high northern latitudes, and occasionally found on the British coasts, sometimes five feet long and one hundred and fifty pounds in weight. The colors are, indeed, magnificent. The whole back is of a steel blue, which, on the flanks, becomes rich green, reflecting in different lights purple and gold, and a lovely rose-color on the abdomen. Numerous oval spots, some milk-white, others of a beautiful silvery lustre, adorn this groundwork, while small ones ornament the head. The gill-covers are very brilliant, and the iris of its large eye is of a beautiful golden color: all the fins are vermilion.

Among marine members of the perch family, we may mention the Red Mullet as very beautiful in its delicate rose-color, striped with yellow; which colors, however, soon fade after death.

> "On fish a different fate attends, nor reach they long the shore
> Ere fade their hues like rainbow tints, and soon their beauty's o'er."

It was one of these mullets which was so celebrated among the Romans for the excellency of its flesh, its great beauty, and the extravagant prices it brought. In the days of

Horace this fish was valued in proportion to its size, not because the larger were better, but (as happens in the fashionable world frequently in our own time) because they were procured with greater difficulty. Enormous sums were paid for these fishes. Juvenal tells us,

> "The lavish
> Six thousand pieces for a mullet gave,
> A sesterce for each pound."

amounting altogether to a sum of nearly two hundred and fifty dollars of our money, while, according to Pliny, a consul named Asinius Celer gave a sum equal to nearly four hundred dollars of our currency for a single fish of this kind; an infatuation we can only feel paralleled by the "tulip mania" of former days. Neither did the extravagance of these people end even here, for Senaca informs us they were so exceedingly fastidious about the freshness of this fish that, according to the luxurious habits of those days, rich epicures kept aquariums in their dining-rooms, so that the fish could be taken out alive under the table: one reason besides the freshness of the fish, being, that the guests might see them change their colors when they were dying. In these feasts they revelled over the expiring mullet, while the bright red color of health passed through various shades of purple, violet, blue, and white, as life gradually ebbed and convulsions put an end to the revolting spectacle. They also put these devoted fishes into crystal vessels filled with water, over a slow fire upon their tables, a refinement of cruelty which required an "imperial" Humane Society to see after.

The Basse or Sea-Perch is an elegant fish, with chaste and pleasing colors, the upper parts gray with bluish tints shading into silvery white; tolerably common on our coasts during the summer. The armed *Enoplessus*, another member of the Perch family, very abundant in the New Holland seas, is remarkable for its chaste coloring, the ground-shade being of a silvery gray, relieved by eight narrow black bands, which either entirely or in part surround the body. The fins have a yellowish tint. It is about eight or ten inches in length. The Two-banded *Diploprion*, an inhabitant of the coast of Java, also claims the same relationship; the colors are a fine reddish-yellow, relieved by two crossing bands of black; length of the fish about six inches. Another genus is the Mediteranean *Apogon*, about the same length as the last-named fish, but of far more brilliant colors. The prevailing color is of a crimson-red, paler on the lower parts, with three deep black markings. The whole surface of the body is covered over with small black spots or dots.

To the same extensive family belong the Lettered Seranus, a beautifully-marked fish, found on the coasts of the Mediterranean. The general ground-tint of the skin is a reddish-orange, sometimes inclining to olive, and shading to a pale tint on the lower parts. The back is banded, as in the Perch, with dull brown bands, but the most showy marks are the narrow irregular lines of rich blue which run on the nose below the eyes and on the cheeks, which assume the form of written characters (hence the name "lettered"). The ground color of the fins is gray, spotted sometimes with reddish-orange, and sometimes with purple. The Spined Seranus, belonging also to the same warm seas, is of a brilliant red or scarlet, which on the sides assumes a golden tint, and on the belly becomes pale or almost silvery. Upon the sides of the head are three bands of golden yellow, and on the forehead are bands of bronzed green: the fins are tinted with red and yellow. This fish in length is generally from five to seven inches.

The Beautiful *Plectropoma*, also of the Perch family, merits its name from the lovely colors it exhibits. This fish inhabits the tropical seas, and some species are unusually lovely. The ground-tint of the body is olive, crossed by six bands of olive black. A line of blue surrounds the orbit; the fins are tinted with olive and yellow, the pectorals sometimes with a delicate rose-color. This fish is about four or five inches in length. A formidable rival in point of beauty, however, is the the One-spotted *Mesoprion*, of the same family, a native of our seas, and as remarkable for the elegance of its form (length about fourteen inches) as the richness and lustre of the coloring. The back, upper part of the head, and cheeks are of a rich steel blue, the lower part of the cheeks and sides of a rich rose-color, and the belly silvery; the whole body is striped with seven or eight bands on a rose-colored ground, and the others are gamboge yellow. The coloring is subject to a considerable variety in tint, from golden orange to silvery. The Golden-tailed *Mesoprion* is of similar richness.

What is called the "Scaly-finned" family of fishes is a large one, containing about one hundred and fifty species, most of which, however, frequent the Indian and Polynesian seas, and are conspicuous for their splendid coloring. It has been observed that if the "feathered tribes of the warm regions are bedecked with the most brilliant and gorgeous hues, the neighboring oceans contain myriads of the finny race which in this respect excel them. Upon the first of three groups of this family especially, nature has most profusely lavished these splendid ornaments. The purple of the iris, the richness of the rose, the azure blue of the sky, the darkest velvet black, and many other hues are seen commingled with metallic lustre over the pearly surface of the resplendent group, which, habitually frequenting the rocky shores at no great depth of water, are seen to sport in the sunbeams as if to exhibit to advantage their gorgeous dress.

In the chapter on "Submarine Scenery," we have described the *Chœtodon* (signifying I contain a tooth), one of the most beautiful of this family of fishes. Another animal ranged

FATHER LASHER

with the "scaly-fins," is the Archer, a fish about six or eight inches in length, which, when it perceives a fly or other winged insect hovering over the surface or settled on a twig, propels against it with considerable force a drop of liquid from its mouth, so as to drive it into the water; in attacking an insect at rest, it usually approaches cautiously, and very deliberately takes its aim. It is said to be an amusement with the Chinese in Java to keep this fish in confinement in a large vessel of water, in order that they may witness its dexterity. They fasten a fly or other insect to the side of the vessel, when the fish aims at it with such precision that it rarely misses its mark. This Japanese fish is called the *Chelmon rostratus*. Another genus—the *Toxotus jaculata*—shoots its watery deluge to the height of three or four feet, and strikes with unerring aim the insect attacked.

The family of "Riband-shaped Fish" includes the most singular and extraordinary fishes in creation. The form of the body when compared to fishes better known is much like that of the eel, the length being in the same proportion as the breadth; but then it is so much compressed that these creatures have obtained the popular name of "riband-fish," "lath" or "deal-fish.' The body, indeed, is often not thicker, except in its middle, than is a sword; and being covered with the richest silver, and of great length, the undulating motion of these fishes in the sea must be resplendent and beautiful beyound measure. But these wonders of the mighty deep are almost hidden from the eye of man. These metoric fishes appear to live in the greatest depths, and it is only at long intervals, and after a succession of tempests, that a solitary individual is cast on the shore, with its delicate body torn and mutilated by the elements on the rocks.

The family of the "Wrasses," or "Old Wives of the Sea" —as they are commonly called—include some very beautiful species, and are distinguished by their elegant, regular, and oval form. The Rainbow is remarkable for the beauty of its coloring, as the name would imply: it is the ornament of the markets on the coast of the Mediterranean, for the various colors of the fish do not yield in their brilliancy and beauty to the most lovely fishes of tropical seas. The summit of the head and back is of a rich brown, mixed with blue and red; beneath this brilliant tint there is a broad band, with a denticulated margin of orange red; below this band, and at the origin of the gill-ray, the middle portion of the side is colored by a deep blue band. This marking extends to near the tail in a band of ultramarine blue. An ultramarine streak of the lovliest hue arises at the angle of the mouth, crosses the cheek, and is prolonged in fainter hues along the inferior border of the deep blue marking of the side. The dorsal fin is of an olive-color, mixed with red, having the margin light blue.

The Parrot-fish belongs to this numerous family, deriving its name partly from a fancied resemblance in their jaws to a parrot's bill. These fishes are remarkable for their brilliant colors, some of them being of wonderful splendor. One species, found in the Mediterranean, is supposed to be the famous *Scarus* of the ancients, of whose ruminating powers extraordinary accounts have been related.

To the family of the pipe-fishes belong the *Hippocampus*, or Sea-horse, which is, perhaps, more remarkable for the singularity of its form—the upper parts having some resemblance to the head and neck of a horse in miniture—than for any ornament or color, although these are not wanting. The singularity of this fish is in the shape and disposition of the plates on the tail, which are such as to admit of its being easily curved inwards, and by the aid of which the animal twists itself around the stems of marine plants, waiting in that position with its head free, ready to dart at any passing object which it desires to make its prey.

For beauty of coloring, irrespective of shape and other repulsive peculiarities, we may mention the *Chimœra*, or Rabbit-fish, an animal little known, as it frequents the deep recesses of the ocean, and is only an occasional visitant of our coasts. In Norway, however, it is more common, and receives the name of "gold and silver fish," from the resplendent colors which form the ground of the body, set off by dark spots. It is also called by the Norwegians the "sea-rat," from the form of the tail, and "king-fish," from a thready filament, terminating in a tuft, which is found on the head of the male. The colors are very beautiful: the upper parts dark brown, varied with yellowish-brown and silvery; the lower parts bright silver; the eyes large, green, and brilliantly lustrous, so much so, that the Mediterranean fishermen called this fish the "cat." The form of the fish does not correspond with the vivid colors we have mentioned, the repulsive shape of the head, and the rat-like tail, giving it an appearance somewhat allied to sea-monsters.

In concluding these brief notices of a few out of the multitude of beautiful fishes which give a charm and lovliness to the element in which they live, we would have the reader remember that these works of a beneficent Creator are intended to raise our thoughts in reverent admiration to that Holy Being, who made all things for our comfort and delight:

"The inhabitant of the waters, generally speaking, knows no attachments, has no language, no affections; feelings of conjugality or paternity are not acknowledged by him; ignorant of the art of constructing an asylum, in danger he seeks shelter beneath the rocks or in the darkness of the deep; his life is silent and monotonous. The cravings of voracity alone influence his instinct sufficiently to teach him some kind of obedience in his movements to external signs. Although so small a share of enjoyment and intelligence is their lot, fish are, nevertheless, adorned by the hand of Nature with every kind of beauty: variety in their forms, elegance in their proportions, diversity and vivacity in their colors—nothing is wanted to attract the attention of man, and indeed it seems as if that attention was the principal object Nature wished to excite. The splendor of every metal, the blaze of every gem, glitter on their surface; iridescent colors, breaking and reflecting in bands, in spots, in angles, or in undulating lines, always regular, symmetrical, graduating or contrasting, but always with admirable effect and harmony, flashing over their sides: for whom else have they received such gifts, they who at most can barely perceive each other in the twilight of the deep; and if they could see distinctly, what species of pleasure could they receive from such combinations?"

SWORD FISH.

CHAPTER XVII.

SHELLS.

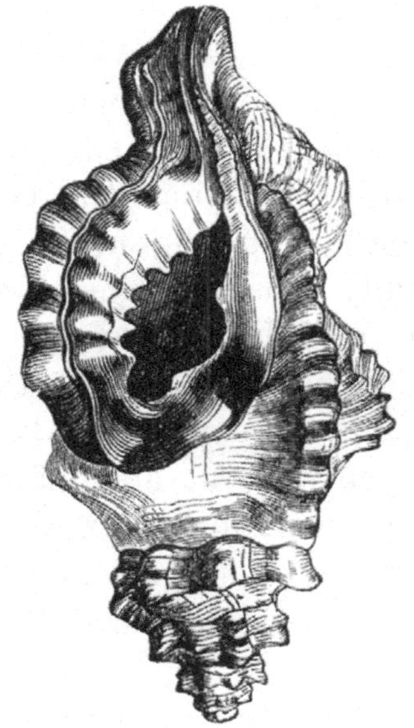

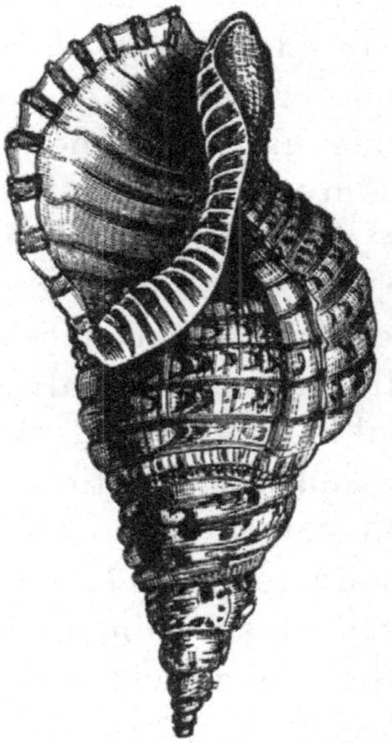

AMONG the many wonderful productions of nature are the sea shells. How beautifully is the wisdom of God manifested in shaping out and moulding them, and especially in the particular angle which the spiral of each species of shell effects, a valve connected by necessary relation with the material of each, and with its stability, and the conditions of its buoyancy.

This is shown in many ways; for in the structure of SHELLS there is a general adaptation of the wants of the animal to which they belong. Thus, there are light shells for the floaters and swimmers, strength for the limpets and periwinkles, and other adjustments as needed for others. What can be more wonderful than the apparatus essential to what are commonly called bivalves, or molluscous animals protected by two shells? The hinge which connects them shows a singular contrivance for the necessities of the animal. It is formed entirely of the inner layer of shell, and consists of either a simple cardinal (a hinge) process, or of serrated projections, or teeth as they are called, with corresponding cavities or sockets into which they are inserted. To this hinge is superadded a ligament, the external substance by which the shells are united, which binds the two parts together, and keeps those composing the hinge in their places. This ligament is highly elastic, being composed of a number of fibres, parallel to each other and perpendicular to the valves which they connect. When the animal is undisturbed, the elastic ligament keeps the valves open, and the functions are carried on without any effort. When danger is apprehended, or circumstances require it, the adductor muscle or muscles contract, overcome the resistance of the hinge, and shut the valves close until they may be opened with safety.

Conchology, is the science which teaches the arrangement of shells into classes, species, etc. Formerly, these beautiful productions of Nature were looked upon as merely pleasing toys and objects of curiosity, but gradually this innocent trifling came to be viewed in its true light, by some collectors worthy of better employment, who put off childish things and went deeper into the subject. In anticipation of this, shell-collectors began to look upon their treasures as an assemblage of gems, and, indeed, the enormous prices given for fine and scarce shells, joined with the surpassing beauty of the objects themselves, almost justified the view which the possessor took of his cabinet of treasures. But after all, these were mere trinkets, and the study of shells and their inhabitants at length became a science of the utmost importance, not only to naturalists generaly, but to the geologist, to whom it is of the greatest value in indicating the difference of strata and their comparative ages.

In Southern Europe some very beautiful shells are found, especially in the Italian seas. Tarento is singularly rich in shells. The Indian seas, more than any other part of the world, abound with the greatest variety of shell-fish, which exhibit a remarkable contrast, comparatively speaking, to the few species found under the parallel latitudes of Africa and America. It is also a singular fact that nearly three-fourths of these shells belong to the animals entirely carnivorous, who, to support life, must be continually carrying on a destructive warfare against the weaker animals of their own class.

Many beautiful shells are brought from the coasts of Chili and Panama in tropical America. From the western coast of Africa are obtained many attractive shells, such as the blood-spotted Harp, the sharp-ribbed Cockle, etc. The small Cowry, well-known as a substitute for coin among the barbarous nations of Western Africa, is the same species as that so abundant in the Indian seas.

Passing to Australia, there are found on the coasts many of the most beautiful and rare rolled shells known: the Snow-spotted kind being most valued. They have two dark bands on a flesh-colored ground, the surface being entirely covered with white dots.

Many deep-sea shells are so firm in their structures, that they are brought to the beaches, especially of the tropical seas, in an entire state, and are eagerly sought after by collectors. Independent of their shape, color and lustre, many of them are valuable, inasmuch as they inhabit the seas at such depths as not to be known in the living state.

The number of shells is far, very far beyond human calculation. An examination of the rocks on the English seashore during the summer will prove this in a slight degree. These are so covered with shells that scarcely a pin's point could be introduced between them. Many apparent grains of chalk are in reality microscopic shells and fragments of marine coral, of which upwards of a thousand have been obtained from one pound of chalk.

The most level and lowest parts of the earth, when penetrated to a very great depth, exhibit nothing but horizontal strata, composed of various substances, and containing, almost all of them, innumerable marine productions. Similar strata, with the same kind of productions, compose the hills even to a great height. Sometimes the shells are so numerous as to compose the entire body of the stratum. They are almost in such a perfect state of preservation, that even the smallest of them retain their most delicate parts, their sharpest ridges, and their finest and most tender processes. They are found in elevations, far above the level of every part of the ocean, and in places to which the sea could not be conveyed by any existing cause. The summits of the Pyrenees and the Andes, at the height of thirteen or fourteen thousand feet above the level of the sea, present them to our notice.

The sea-banks and coasts are covered with broken shells, of which lime is the ingredient. This generally exists in the state of carbonate, the same as in chalk, common limestone, and marble. Many of the more tender shells and shelly matters are broken by the agitation of the waters, and form a variety of sand which is truly a product of the sea, and forms a valuable manure on land. Great deposits of this article are found on the coasts of Devonshire and Cornwall, and in many other parts of the British coast.

A species of shell, the *Cerithium telescopium*, is so abundant near Calcutta as to be used for burning into lime. Great heaps of it are first exposed to the sun, to kill the animals, and then burnt. In some places they are so plentiful as to be used in road making. Mobile, Ala., is built on a shell-bank.

It was formerly believed that shells were not only devoid of vessels, but completely without organs, being composed of the transpiration of particles, chiefly carbonate of lime, cemented together by a kind of animal glue. It is now known that shells possess a more or less distinct organic structure, which in some cases resembles that of the external skin of the higher animals, while in others it approaches to that of the true skin.

In the limited space to which our remarks on the subject of this chapter is necessarily confined, we cannot give more than a brief outline of this exceedingly interesting department of science. We may briefly observe that what are called the *Testacea* (a shell), comprise animals surrounded with a shelly covering, and may be generally described as of three kinds: those that possess a single shell, of whatever form or character, and hence called univalves; those which have two shells, the bivalves, or *Conchæ;* and others having more than two shells, or multivalves. Of these, the univalves are the most numerous and exhibit the greatest variety of forms, being for the most part regularly or irregularly spiral. Among the most common may be mentioned the *Helix*, or snail genus; the *Paletta*, or limpet; and the *Turbo*, or wreath genus, of which the periwinkle is a species.

The shell of the Clam, or Bear's Paw, is described as, perhaps, the most ornamental of bivalves, in regard to form, texture and color. It comes from the South Seas, and is much used for decorative purposes.

Among the most curious shells is the *Murex*, or Purple-shell, so highly valued by the ancients for the exquisite dye it is capable of producing; the *Volute*, or Mitre-shell, including the fine polished spiral shells, without lips or perforation, which are often exhibited on chimney-pieces as ornaments, sometimes embellished with dots and with colored bands. The *Strombus* comprise larger shells, spiral like the volute, but with a large expanding lip, spreading into a groove on the left side, and often still farther projecting into lobes or claws, the back frequently covered with large excrescences, in some species called Cormorant's foot.

And now for a few observations on the use and value of shells. Even as mere objects of attraction they tend to raise the thoughts to that great and glorious Being,

> "Our God, omnific, sole original,
> Wise, wonder-working wielder of the whole:
> Infinite, inconceivable, immense."

who has shaped and adapted them to the wants of numberless creatures, of which science at the most can have but a feeble comprehension. Beautiful, since more exquisite samples of elegance of form and brilliancy of color cannot be found through the wide range of natural objects, whether organized or unorganized; surprising, when we consider that all these durable relics were constructed by soft and fragile animals, among the most perishable of living creatures. Still more surprising is an assemblage of shells, when we reflect upon the endless variation of pattern and sculpture which it displays; for there are known to naturalists more than fifteen thousand perfectly distinct kinds of shells. Every one of these kinds has a rule of its own, a law which every individual of each kind, through all its generations, implicitly obeys.

The formation of the shell itself is but an example of a process at work equally in the animal and vegetable kingdoms. A shell, whether simple or complicated in the contour or color, is the aggregate result of the function opera-

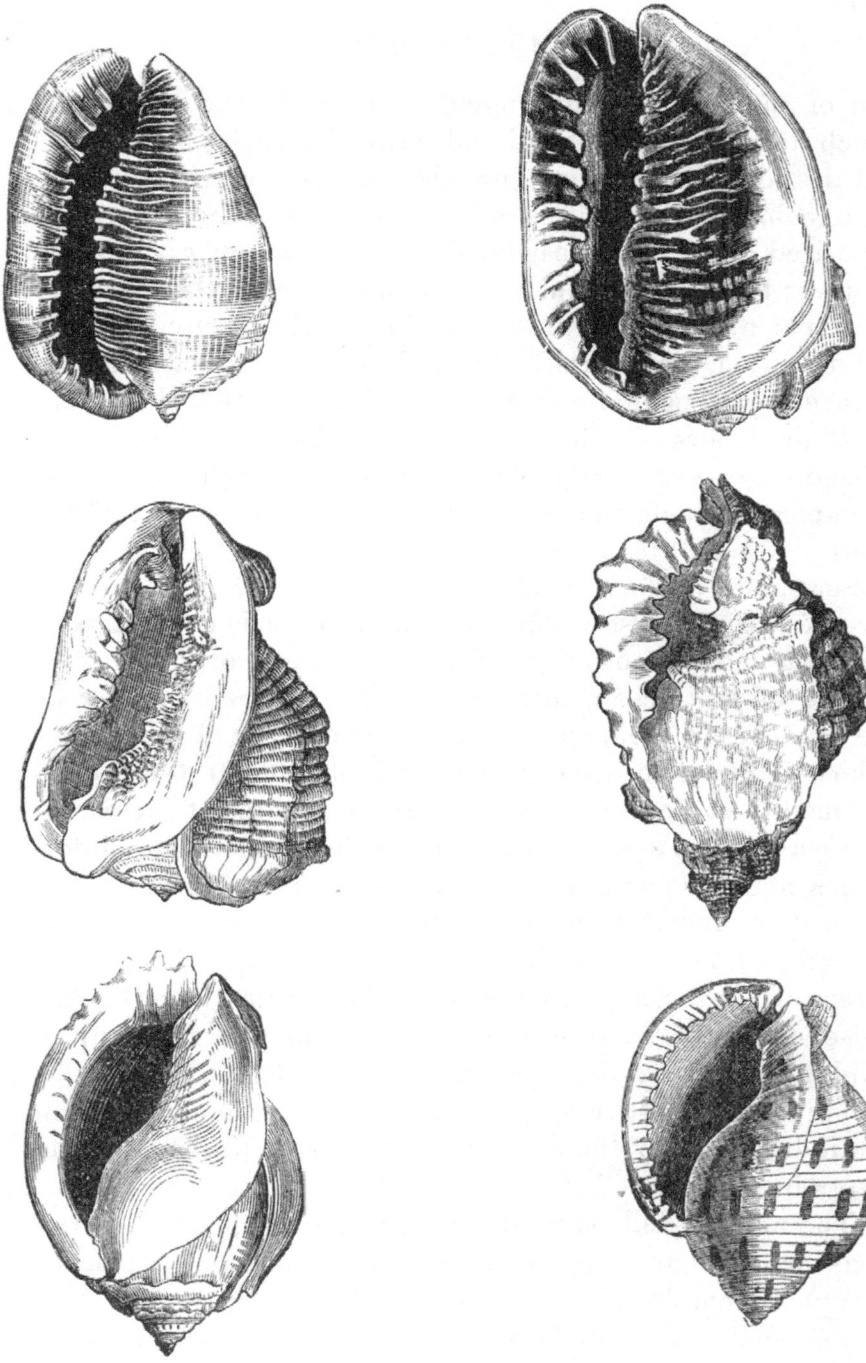

OCEAN SHELLS.

tion of numberless minute membranous cells, the largest of which does not exceed one hundredth of an inch in diameter, and in the majority of instances is less than one thousandth of an inch. In the cavities of these microscopic chambers is deposited a crystalline carbonate of lime, which gives compactness to the beautiful dwelling-house, or rather coat-of-mail, that protects the tender mollusk. How astonishing is the reflection, that myriads of exactly similar and exceedingly minute organs should so work in combination that the result of their labors should present an edifice rivalling, nay, exceeding in complexity, yet order of detail and perfection of elaborate finish, the finest palaces ever constructed by man!

Sea-shells perform also an important part in the economy of the universe. Maury remarks on this subject, that shell-fish and various other tribes that dwell far down in the depths of the ocean, although regarded as being so low in the scale of creation, spread over certain parts of the waters "those benign mantles of warmth which temper the winds, and modify more or less all the marine climates of the earth. The sea-breezes and the sea-shells perform their appointed offices, acting so as to give rise to a reciprocating motion in the waters, and thus imparting to the ocean forces also for its circulation. Sea-shells and sea-insects are the conservators of the ocean. As the salts are emptied into the sea, these creatures secrete them again, and pile them up in solid masses, to serve as the bases of islands and continents, to be in the course of ages upheaved into dry land, and then again dissolved by the dews and rains, and washed by the rivers into the seas."

The use of shells is multifarious: in China, some descriptions are prepared as medicines; as articles of ornament they were employed in the earliest times. Several perforated shells found in Aquitaine, in France, show that they must have been worn as decorations or charms by primitive races. The custom of using shells as necklaces is common not only among savages, but among civilized people at the present day. Nacreous or pearl-like shells are employed for making buttons and other articles; colored and pearl ones form the ornaments of papier-machè work, card-cases, etc. Various small shells are made into flowers and decorations for head-dresses; very beautiful cameos are carved upon some description of shells for brooches, bracelets, ear-rings, and other attractive objects. The Fountain-shell of the West Indies is one of the largest known univalve shells, weighing sometimes four or five pounds. Immense quantities are imported from the Bahamas for the manufacture of cameos. The secret of cameo-cutting consists simply in knowing that the inner stratum of porcellanous shells is differently colored from the exterior. Some shells are manufactured into spoons, handles for knives, cups, lamps, etc. The purest kind of lime is made from calcined shells, and their use as a manure has already been mentioned.

Mother-of-pearl is the beautiful white enamel, or pearly lining, which forms the greater part of most oyster-shells, but especially the larger ones found in the seas of the Pacific and Indian Oceans.

In the cathedral and some of the churches in Panama the upper portions are studded with pearl shells, which give them a strange and not unpleasing appearance.

It has been stated that in many of the houses in the capital, the outer side of the verandah or corridor is composed of coarse and dark-colored mother-of-pearl shells, of little value, set in a wooden framework of small squares, forming windows, which move on slides. Although the light admitted through this sort of window is much inferior to what glass would give, it has the advantage of being strong.

The use of spiral shells as trumpets or horns is traced back to the Romans, and they are thus employed by the Africans, the natives of the Eastern Archipelago and New

Zealand, and also in Japan. The fine Trumpet-shell is found in most warm climates, in the African, the American, and Asiatic seas, also on the coasts of the islands of the South Pacific.

An eminent writer, in speaking of the Tahitians, observes, "The sound of the trumpet or shell used in war to stimulate in action by the priests of the temple, and also by the herald, and others on board their fleets, was more horrific than that of the drum. The largest shells were usually selected for this purpose, and were sometimes above a foot in length, and seven or eight inches in diameter at the mouth. In order to facilitate the blowing of this trumpet they made a perforation, about an inch in diameter, near the apex of the shell. Into this they inserted a bamboo cane about three feet in length which was secured by binding it to the shell with fine braid; the aperture was rendered air-tight by cementing the outside of it with a resinous gum from the bread-fruit tree. These shells were blown when a procession walked to the temple, or their warriors marched to battle, at the inauguration of the king, during the worship at the temple, or when a *tabu* or restriction was imposed in the name of the gods. The sound is extremely loud, but the most monotonous and dismal that it is possible to imagine."

This is the shell generally represented by painters in the hands of the "Tritons" or sea-monsters.

In Ceylon shells of a certain kind are used to contain the sacred oil for anointing the priests. On the western coasts of South America there is a species of limpet which attains the diameter of a foot, and the shell of which is employed by the natives as a basin.

Another general application of shells is as weights to nets and barbs for harpoons and hooks.

To shell-fish, as articles of food, we have already alluded with regard to the lobster, crab, oyster, mussel, etc. The scallops are now almost as much eaten as oysters, but require cooking first.

The giant clam of the Indian Ocean, the shell of which often weighs upwards of five hundred pounds, contains an animal sometimes weighing twenty pounds, which has been found to be very good eating. The rock-limpet is much used by fishermen for bait. In the north of Ireland they are eaten. The whelk is also employed for bait, and many tons' weight of these, cockles, and winkles, are consumed by shell-fish amateurs.

The mention of cockles remindes me of a statement in Drake's "Voyage round the World, the quaint style of which is amusing:

"Our stay being longer than we purposed (in Patagonia) our diet began to wax short, and small mussels were good meat, yea, the sea-weeds were dainty dishes. By reason whereof we were driven to seek corners very narrowly for some refreshing, but the best we could find was shells instead of meat. We found the nests, but the birds were gone—that is, the shells of the cockles on the sea-shore, where the giants had banqueted, but could never chance with the cockles themselves in the sea. The shells were so extraordinary that it would be incredible to the most part; for a pair of shells did weigh four pounds, and what the meat of two such shells might be may be easily conjectured."

The shells called Porcelain-shells by the French and Germans are almost entirely composed of lime, are richly enamelled, and are often very beautiful. They are most abundant and attain their largest size in the seas of warm climates. Only a few species are found on the British coasts. The Cowry-shell, to which we have alluded as a substitute for money, is not of great beauty, being yellow or white, often with a yellow ring about an inch long, and nearly as broad as long. In Bengal three thousand two hunderd cowries are reckoned equal to a rupee, so that a cowry is equal to one-thirty-sixth of a farthing. Yet cowries to the value of two hundred thousand rupees are said to have been imported annually into Bengal. Many tons of cowries are annually imported into England to be used in trade with Western Africa. Of the cowries a very remarkable fact has been stated, that when the animals find their shells too small for the increased dimensions of their body, they quit them and proceed to the formation of new ones of larger size, and, consequently, more adapted to their wants. As soon as the cowry has abandoned its covering, the hinder part of its body begins to furnish anew the shelly matter which is afterwards condensed on its surface. This secretion is continued until at length the shell appears of the consistence of paper; and the mouth or opening of the shell, which at this period is very wide, soon afterwards contracts to its proper form and dimensions. The edges are thickened, and form into those beautiful folds or teeth which are so remarkable on each side of the opening of these shells. The porcelain and cowry-shells belong to a family which includes also the shells called Poached Eggs, and the Weaver's Shuttle, remarkable for its prolongation at both ends.

A well-known shell, distributed over the whole world, is the Fusus (a spindle), so named from its shape. In Scotland it is called the "roaring buckie," from the continuous sound, as of waves breaking on the shore, heard when the empty shell is applied to the ear. Wadsworth alludes to this "voice" of a shell in some sweet lines:

"I have seen
A curious child, who dwelt upon a tract
Of inland ground, applying to his ear
The convolutions of a smooth-tipp'd shell,
To which, in silence hush'd, his very soul
Listen'd intensely, and his countenance soon
Brighten'd with joy; for murmurings from within

> Were heard—sonorous cadence, whereby,
> To his belief, the monitor express'd
> Mysterious union with his native sea."

In the cottages of Zetland, this shell, generally about six inches long, is used for a lamp, being suspended horizontally by a cord, its cavity containing the oil, and the wick passing through the canal.

The shell of the Haliotis (the sea, the ear), is very ornamental, and valued, on account of its pearly lining, for adorning *papier maché* articles. These shells, which are very numerous, and some of splendid appearance, come from the tropical seas, and are commonly called, from their shape, "ear-shells," or "sea-ears." One species, however, is found on the Southern European coasts, and on those of the Channel Islands. From the warm regions are obtained the beautiful Harp-shells, the delicate and brilliant colors of which render them highly prized; also the Fountain-shells to which has already been alluded as used for cameos, and are much esteemed as garden ornaments for their solid and delicately-tinted substance. One of these shells sometimes weighs four or five pounds.

A shell called the Razor, a common species of which is often picked up on the English coasts—some straight, about an inch long and eight inches broad; and another curved like a sword—attain a large size in tropical seas, and are of great beauty. They are found in the sands of all seas, except in the cold regions, the solen, the name of the inhabitant of this shell, burrowing in the sands, and ascending from its holes by means of the foot, which can be lengthened or contracted at will.

What are called Top-shells, from their spiral and very generally top-shape, are frequently found on the English coasts, and many of them are very ornamental, but not equal in this respect to the tropical specimens.

From Australia we obtain a large number of the richly decorated Pheasant-shells, formerly of great rarity, and expensive, but now comparatively cheap.

The Wentletrap-shells, the common kinds of which are found on our own coasts and those of Europe, are very pretty: they are spiral, with many whorls or wreaths, deeply divided, and crossed by remarkably elevated ribs. The *true* shells of this species come from the warm seas, and are generally very beautiful. One kind, called the Precious Wentletrap, is of such rarity and richness, that it is said to have been sold to shell collectors at the price of two hundred guineas, but it may now be had for a few shillings. It is nearly two inches in length, snow-white or pale flesh-colored, with eight separated wreaths. Trough-shells, several small species of which are very abundant on British seashores, are triangular, broader than long, and the valves equal. Some of them have a very attractive appearance.

PENGUINS.

CHAPTER XVIII.

SEA BIRDS.

IN the chapter on "Superstitions Connected with the Ocean," we have alluded to a few marine birds which are considered by seamen as good or evil portents in their passage over the ocean. We will now briefly describe some of the more prominent sea-birds which perform their part in the economy of nature, and derive their chief sustenance from the finny inhabitants of the ocean. They constitute a very extensive family all over the world, ever on the alert to indulge in their fishing propensities, and voracious in their appetites; so that the poor fishes, what with numberless foes in their own element, with sea-birds continually on the watch to prey upon them, together with all the ingenious arts practiced by man to ensnare them, cannot lead the happy and peaceful life which some fanciful writers have imagined them to enjoy.

Many, many miles out at sea the oceanic birds are seen pursuing their predatory instincts, ever restless and untiring, while, nearer shore, thousands in summer seek precipitous coasts and headlands as breeding stations.

In winter, others, scarcely less numerous, flock from their more northern homes, and fill our bays and marine inlets.

A writer describes an interesting spectacle which met his gaze after mounting a rock at Saldanha Bay, near the Cape of Good Hope.

"All of a sudden there rose from the whole surface of the island an impenetrable cloud, which formed, at the distance of forty feet above our heads, an immense canopy, or rather sky, composed of birds of every species and all colors: cormorants, sea-gulls, sea-swallows, pelicans, and, I believe the whole winged tribe of that part of Africa, were here assembled. All their voices, mingled together and modified according to their different kinds, formed such a horrid noise that I was obliged every moment to cover my head to give a little relief to my ears. The alarm that we spread was so much the more general among the innumerable legions of birds as we principally distured the females, which were then sitting. They had nests, eggs, and young to defend. They were like furious harpies let loose against us, and their cries rendered us almost deaf. They often flew so near us that they flapped their wings in our faces, and, though we fired our pieces repeatedly, we were not able to frighten them; it seemed almost impossible to disperse the cloud."

Many of the precipitous rocks and islands of our own country present greatly exciting spectacles at the breeding season. Myriads of ocean birds,

> "Ranged in figures, wedge their way,
> Intelligent of season, and set forth
> Their airy caravan. High over seas
> Flying, and over lands, with mutual wing
> Easing their flight. The air
> Floats as they pass, fanned by unnumbered plumes."

Certainly not the least interesting of marine birds is the

Gull, belonging to a very numerous family, which includes also the squas, terns, petrels, shearwaters, albatrosses, noddies, skimmers, and others, all preying chiefly on fishes and mollusca, together with animal garbage of every kind. From the latter circumstance Buffon calls the gulls "the vultures of the ocean." Several of this family are the most oceanic of all birds, being seen hundreds of miles out at sea, apparently unwearied and restless. The gulls have very powerful wings, flying with ease against the roughest storms. In fine weather they fly high in the air, descending with great rapidity to seize the fishes on the surface of the water, or diving slightly for herrings and small fish within reach. Their plumage being close and thick, they are good swimmers. They have a close resemblance to the terns, or "sea-swallows," as they are sometimes called, but the bill is stronger, and the upper mandible much more curved towards the end. The symmetry and strength of gulls are remarkable, showing how Nature has adapted them in every particular for all the purposes of their predatory instincts.

"Let the reader," remarks Mr. Frank Buckland, "examine the pectoral or breast muscles of the next gull he kills: he will find them one solid mass of firm, hard muscle, admirably adapted to sustain and work the wings. What models of beauty and lightness are those wings! The bones are composed of the hardest possible kind of bone material, arranged in a tubular form, combining the greatest possible strength with the greatest possible lightness. If we make a section of the wing-bone of a gull, or, better still, of that of an albatross, we shall find that it is a hollow cylinder, like a wheat-straw; but, in order to give it still further strength, we see many little pillars of bone about the thickness of a needle extending across from side to side; these buttress-like pillars are in themselves very strong, and do not break easily under the finger. Again, at the top of the bone we find two or three holes, which communicate with the interior; through these, when the bird is alive, pass tubes, which are connected with the lungs; so that, when the bird starts for flight, he fills his wings and other bones with air, causing them to act something like a balloon on each side of him. This explains one of the chief reasons why man will never be able to fly: his arm-bones are filled with marrow, which he cannot by any means get rid of, should he be ever so anxious to fly like a bird."

Some of the larger gulls are very expert in breaking the shells of the mullusks on which they feed, by taking them up to a sufficient height in the air, and dropping them on a rock. Audubon, our famous naturalist, mentions an instance in which the gull, finding the shell not broken by the fall, carried it up a second and a third time, and dropped it from a loftier height, by which its purpose was effected. Gulls are able to endure hunger for a long time. An instance is related of one being kept without food for nine days, and yet retained a considerable degree of strength. When their prey is before them, they dart at it with such violence that they will swallow both bait and hook, and split themselves on the point placed by the fisherman under the fish which he presents to them.

The selfishness and rapacity exhibited by some larger members of the gull family has often been observed; the *Glaucous* is a notable instance, and is called by the Dutch sailors the "Burgomaster," from the tyranny which in virtue of its size and strength it exerts over most of the smaller birds of the Northern seas, compelling them to relinquish the fish they have taken; bad qualities, shared in a like degree by the *Parasiticus* Gull. Mr. Lamont describes these marine bashaws very amusingly:

"None of these birds ever seemed to take the trouble of picking up anything for themselves, but as soon as they observe any other gull in possession of a morsel which he is not able to swallow outright, they dash at him and hunt him through the air until the victim is obliged to drop whatever he has secured, and the ravenous burgomaster appropriates and swallows it himself. I have watched many of these nefarious transactions, and the result is always the same: the small gull turns, and twists, and doubles, and dodges, screaming all the time so pitifully that one would think he expected to lose his life instead of his dinner, but at last he is compelled to give up possession, and the burgomaster then ceases to molest him."

Sailors are very fond of playing off a joke upon the gulls, which are always hovering about ships. They take three or four pieces of sail-twine about six feet in length; these are tied together in the middle, and to the end of each a small piece of blubber or fat is attached tightly, and then thrown into the sea. A gull comes and swallows one piece, another then sees there is plenty to spare, and swallows the next; perhaps a third gull takes possession of another; but as they are all attached by the sail-yarns, whenever they try to fly away one or the other is compelled to disgorge his share; and this is continued, to the tantalizing suspense of the poor gulls, and the great fun of the sailors. This may be a confirmation of the old popular term applied to persons easily duped, but in most cases the gull shows great wariness and cleverness, especially in escaping from its insatiable enemy the heron.

The glaucous gull is an occasional visitor to English shores from its habitat in Northern Europe. One was shot at Galway during the "famine" year in Ireland, 1846. A soup kitchen had been established within some distance of the coast, and each day the stately-looking fellow left its maritime domain, and attracted by the smell, sailed about the vicinity of the soup. Many of the poor famished peasants regarded it with an unfavorable eye, not being accustomed to observe a white bird of such dimensions floating in the air, and uttering its hoarse cries overhead, as if laughing at their misery.

Another inhabitant of the cold regions is the Iceland Gull, smaller in size, and elegant in shape. Some species of this family are remakably beautiful: one of the smallest, the "Little Gull," from the Arctic shores, has a lovely roseate tint overspreading the white under-plumage. The Black-headed Gull abounds on English shores during autumn and winter, and is a fine bird, familiar and unsuspicious in his habits, and additionally interesting from the circumstance that this species was protected by the Druids, and was figuratively adopted as an emblem connected with the Deluge, and formed an important feature in their ceremonies.

The Great Black-backed Gull, distinguished also by the appellations of the Goose Gull, Gray Gull, and Parson Gull, the latter name arising from the contrast between the black back with the snow-white of the under-plumage, is a large and handsome bird. To every frequenter of the coast the stately and graceful form of this bird is well known, and whether observed in summer, when quietly sunning itself on the strand, or in winter amidst the conflicting war of elements steadying itself in the eddying blast, it cannot fail to excite admiration. At no time more attractive than when observed during hazy, foggy weather, a black-backed gull, looming through a cloud, with its immense sweep of wing (often exceeding five feet), increased by the state of the atmosphere to a giant size, almost reminds us of the albatross.

The Herring or Silvery Gull, is distinguished by the spotless purity of its plumage, and ranks among the most beautiful of the gulls that frequent our shores, and has been called the feathered dervish of the air from its rapid and gyratory mode of flying.

The Kittiwake is, with the exception of the "black-headed," the smallest of our common gulls, and during the summer the most frequent visitor on our coasts. Almost exclusively maritime in its habits, it never ventures inland like the other species, but contents itself with the food that it obtains on the sea.

The Skuas are ranked by naturalists in successive order after the gulls, who find in them determined antagonists. Armed with a powerful bill, the skua is capable of doing much mischief. It is related that one of these birds, which had received a slight injury in the wing-joint, was taken, and sent by the captain of a vessel on shore, in charge of a sailor, with instructions that the bird should be killed and stuffed. The sailor opening the basket in which it was confined rather hastily, the skua dashed ferociously at him, striking with its bill and buffeting with its wings, drawing blood with every successive stroke it made, until at last the sailor drew out his clasp-knife in self-defence, but so determined was the bird, that had not a table-cloth been thrown over it, the contest would have been of long duration.

The pugnacity of the skua is remarkable. No sooner does a skua observe an eagle within its domains than it makes a violent attack upon him. Mr. Drosier relates a very interesting anecdote on this subject. He was standing at the foot of the loftiest hill in Foula, Shetland: "an eagle was returing to his eyrie, situated on the face of the western crags, in appearance perfectly unconscious of approaching so near to his inveterate foe, as, in general, the eagle returns to the rocks from the sea without even crossing the smallest portion of the island. As I was intently observing the majestic flight of the bird, on a sudden he altered his direction and descended hurriedly, as if in the act of pouncing. In a moment five or six skuas passed over my head with astonishing rapidity, their wings partly closed and perfectly steady, without the slighest waver or irregularity. The gulls soon came up with the eagle, as their descent was very rapid, and a desperate engagement ensued. The short bark of the eagle was clearly discernable above the scarcely distinguishable cry of the skuas, who never ventured to attack their enemy in front, but taking a short circle around him, one made a desperate sweep or stoop, and striking the eagle on the back, darted up again almost perpendicularly. This cowardly attack was imitated by each of the other gulls, and continued some time, the eagle wheeling and turning as well as his ponderous wings would allow, and evidently harassed unmercifully, until I lost sight of the combatants among the rocks."

The Petrels are among the most interesting of marine birds. The name is said to be derived from the circumstance that besides the faculty of swimming, they possess that of supporting themselves on the water by striking very rapidly with their feet, which has caused them to be compared to St. Peter walking upon the water. These birds are to be seen in all seas of the globe from one pole to the other, and are the inseparable companions of mariners during their long navigations, following the vessels in great flocks to pick up any garbage thrown into the water. Their flight is almost always performed by hovering, and without presenting apparent vibrations. They drop promptly on their prey, which seems to consist chiefly of the blubber or fat of whales, mollusca, marine worms, and the spawn of fish. Neither the habits of the petrels, nor the structure of the bill adapt them for fishing. They have the faculty of spouting oil, as a means of defense, in the face of any one who may attempt to take them. Persons not aware of this fact have lost their lives by falling into the sea or down precipices.

The Storm-Petrel, the bird of ill omen among mariners, as has been already mentioned in another chapter, is about the size of a house-swallow, in length six inches, and the extent of the wings thirteen inches. The whole body is black except near the tail, some feathers of which are white. The ancients believed that the petrel hatched its eggs beneath its wing, as at all seasons and in every sea they had been remarked flying, while there appearance on land was never noticed:

> "The bird of Thrace,
> Whose pinion knows no resting-place."

It is true that the petrels do not quit the sea except at the time of laying, and for the purpose of making their nests upon very precipitous rocks, where they feed their young on half-digested animals. They retire there during the night, and utter a most disagreeable cry, resembling the croaking of a reptile.

The Terns or "Sea-swallows" have remarkably long wings and slender bills; the tail is forked, and the plumage generally is of a delicate pearl-white, with more or less black upon the head. The terns are continually on the wing, and although web-footed, are not seen to swim; they rest but seldom, and only on the land, feeding for the most part on small fish and mollusca, which they seize upon the surface of the water, but they also catch aërial insects. In flying they send forth sharp and piercing cries. The most elegantly formed of the terns is that called the "Roseate," the mantle of which is a pale tint, the under-parts of a rosy hue. Mr. Selby tells us that on the Farne Islands it breeds abundantly. "When intruding on the nest, the bird showed great anxiety, approaching so near that we knocked one or two down with a fishing-rod used by the keeper of the lighthouse for fishing from the rocks. All the terns are very light, the body being comparatively small, and the expanse of wings and tail so buoys them up that when shot in the air they are sustained, their wings fold above them, and they whirl gently down like a shuttlecock." The species are numerous and occur in both hemispheres.

The Skimmers, although possessing much of the general habits of the terns, are distinguished by the singular form of the bill, the upper mandible of which is considerably shorter than the other. They skim over the surface of the ocean with great swiftness, and scoop up small marine insects.

The Albatross, whose habitual dwelling is the Austral Ocean, from the Cape of Good Hope as far as New Holland, belongs to the genus *Diomedia*, and is the most powerful and bulky of the whole family. The extent of their outspread wings is enormous, yet their flight, except in stormy weather, is by no means lofty: like all the rapacious birds of the ocean, they are most voracious. They devour fish with so much gluttony that often one-half of the body remains outside of the bill until the part which is swallowed, being dissolved by digestion, leaves a passage for the rest. They are often gorged to such a degree as to be unable to fly, or escape the boats whlch pursue them. Although the flesh of the Albatross is hard and rank, yet sailors contrive to render it eatable, when they are in want of fresh provisions, by taking off the skin, and soaking the body in salt for twenty-four hours, then boiling it, and eating it with some strong sauce.

In spite of the strength and powerful bill of the albatross, it is by no means warlike, and will remain on the defensive against some of the gull tribe which harass them, and to escape such attacks they plunge their body into the water. They experience some difficulty in rising to their flight, and then strike the water rapidly with their feet and clap with their wings; but after this impulsion the wings remain developed, and they do nothing but balance themselves alternately from right to left, shaving the surface of the water with rapidity, and plunging in their heads now and then in search of food to a certain depth.

The divers are great destroyers of fish, and expert in their method of getting supplies, as their name would suggest. Indeed, they are said to dive with such celerity that they often evade a shot directed against them, sinking at the very moment the flash appears. These birds cannot support themselves on land except in a position nearly vertical, and by the assistance of their wings, which thus act as oars. Sometimes they fall with their stomach flat on the ground, and have some difficulty in raising themselves up. They are seen in our climates only when the rivers and ponds of cold countries are frozen, and they return to their homes in the north after the thaw. They undergo a periodical change of plumage in one form or another. The Red-throated Diver is tolerably common around the coasts, entering the mouths of rivers after shoals of sprats, etc. The Great Northern Diver, a remarkably handsome bird, occurs on our shores during winter, frequenting the vicinity of the oyster-scalps, and is there well known to the fisherman from its loud and monotonous call. Leemius remarks of the Laplanders, that if a person hears the cry of any of the divers in spring, and while fasting, the milk from his flocks will not curdle for the whole year. Vigilant and shy, if pursued, it exerts its admirable locomotive powers, and advances with immense speed. Nature has provided means of escape and safety to the divers in the flattened form of the body and the wonderful mechanism of the foot, the membrane of which can be closed preparatory to each stroke.

From the divers we are easily led to the family of the Auks, by means of the Guillemots, ocean birds to which the attribute of stupidity has been applied, but probably without sufficient reflection on their peculiar conformation, the wings being short and narrow so that the bird can scarcely flutter; the legs also from their position are quite unfit for the purpose of walking; and the natural element of the bird is only on the bosom of the sea, where it swims with the greatest swiftness, and even dives below the ice.

The common is the only one of the British guillemots that can be called abundant, the others being comparatively rare, and some only straggling visitants. It is found around English coasts, to the Shetland and Orkney islands, and also around the shores of temperate Europe. When near their breeding-places at the proper season, they assemble in thousands, at times blackening the sea.

Sitting closely along a ledge of rock, no matter how elevated above the sea, they impart all the appearance of being ranged in file, or, as they have been compared by the Manxmen, resembling an apothecary's shop—the even ledges of the rock, the shelves, and the birds the pots; while on the least alarm the entire range of the birds sweep downward in a line to the sea. Such successful divers are they, and rapacious feeders, that twenty-five herring fry have been counted in the stomach of a single bird. Congregated in parties of from eight to thirty, they evince the umost amiability towards each other, fishing and winging their way in small flocks to and from their breeding haunts.

The Great Auk is an inhabitant of Northern Europe, and has been rarely captured on our coasts. Of considerable size, its power of progression is limited only to the water, the shortness of its wings rendering it incapable of flight, and from the backward position of its legs, it stands erect and stately. Breeding in remote northern latitudes, the eggs are obtained with great difficulty. The length of the bird is said to be from thirty inches to three feet; the bill four inches long, is black with transverse furrows, the grooves white. In the dress of winter the chin, throat, and sides of the neck are white. The Razor-bill Auk is nearly equally abundant with the guillemot on all our coasts, breeding in the same manner together on rocks, and appearing off our shores during the winter in small parties.

The Puffin, or "Sea-Parrot," so named from the bill, which, in comparison with the size of the bird, is strongly developed, is a summer visitant to English shores, repairing to them for the purpose of lucubation. It sometimes breeds in fissures of the rocks; but its most general resort is in holes and burrows, either formed by itself or supplied by rabbits, if they happen to be inhabitants of the same locality. On the Bass Rock, the holes in the ruins of the old fortifications afford a retreat. The Puffin is used as an article of food by various islands and northern tribes in whose vicinity they breed. They are caught by stretching a piece of cord along the stony places where they chiefly assemble, to which nooses are attached.

The Penguins occupy habitually the most northern points and islands of Europe, of Asia, and of America; but they cannot remain at sea, except in calm weather. When the tempest surprises them far from shore, great numbers of them perish. Though they usually only shave the surface of the water in flying, they can elevate themselves to a certain height. By night they retire into the clefts of rocks and caverns. In their tottering walk they seem to rock from one side to the other. Their food consists in crustaceous animals, and they also live on shell mollusca and small fish, which they take in diving. They make their nests in holes on the sea coasts, which they enlarge with their bills and feet. These birds are singular in their habits. Darwin relates:

"One day, having placed myself between a penguin and the water, I was much amused by watching its habits. It was a brave bird, and until reaching the sea it regularly fought and drove me backwards. Nothing less than heavy blows would have stopped him: every inch gained he firmly kept, standing close before me erect and determined. When thus opposed he continually rolled his head from side to side in a very odd manner, as if the power of vision lay only in the anterior and bassel part of each eye. This bird is commonly called the 'jackass penguin,' from its habit while on shore, of throwing its head backwards, and making a loud, strange noise, very much like the braying of that animal; but while at sea and undisturbed, its note is very deep and solemn, and is often heard in the night time. In

BOOBIE.

CORMORANT.

ALBATROSS.

GREAT AUK.

diving, its little plumeless wings are used as fins, but on land as front legs. When crawling (it may be said on four legs) through the tassocks, or on the side of a grassy cliff, it moved so very quickly that it might readily have been mistaken for a quadruped. When at sea and fishing, it comes to the surface for the purpose of breathing with such a spring, and dives again so instantaneously, that I defy anyone at first sight to be sure that it is not a fish leaping for sport."

One of the greatest distroyers of fish is the Cormorant, belonging to the family of Pelicans, and the common species of which is widely distributed, extending around the whole coasts of our mainlands and islands, constructing their nests, on the summits of rocks most generally, of sea-weeds or materials collected on the waters. The bird is not easily approached at sea, but gets out of harm's way by flight, not by having recourse to diving, like so many of the true aquatic tribes: the flight, powerful and overland, is performed at a great height. When swimming it is easily distinguished by its long upright neck. So keen in fishing is the cormorant that advantage has been taken of the circumstance to train it for that purpose in the manner hawks are trained for fowling, a tight collar being put around the throat to prevent the swallowing of the prey. A bird of this species kept by a Colonel Montague was extremely docile, of a grateful disposition, and by no means vindictive. He received it by coach after it had been twenty-four hours on the road; yet, though it must have been hungry, it rejected every sort of food he could offer to it, even raw flesh; but as he could not procure fish at the time, he was compelled to cram it with meat, which it swallowed with evident reluctance, though it did not attempt to strike him with its formidable beak. After seeing it fed he withdrew to the library, but was surprised in a few minutes to see the stranger walk boldly into the room, and join him at the fireside with the greatest familiarity, where it continued, dressing its feathers, until it was removed to the aquatic menagerie. It became restless at the sight of water, and when set at liberty, plunged and dived without intermission for a considerable time, not capturing, or even discovering, a single fish; and, apparently convinced there were none to be found, it made no further attempt for three days.

The dexterity with which the cormorant seizes his prey is incredible. Knowing its own powers, if a fish is thrown into the water at a distance, it will dive immediately, pursuing its course under water in a direct line toward the spot, never failing to take the fish, and that frequently before it falls to the bottom. The quantity it will swallow at a meal is astonishing: three or four pounds twice a day are readily devoured, the digestion being excessively rapid. If, by accident, a large fish sticks in the gullet, it has the power of inflating that part to the utmost, and while in that state the head and neck are violently shaken, in order to promote its passage. In the act of fishing it always carries its head under water, in order that it may discover its prey at a greater distance and with more certainty than could be effected by keeping its eyes above the surface, which is agitated by the air, and rendered unfit for visional purposes. If the fish is of the flat kind, it will turn it in the bill, so as to reverse its natural position, and by this means only could such be got within the bill. If it succeeds in capturing an eel—which is its favorite food—in an unfavorable position for gorging, it will throw the fish up some height, dexterously catching it in a more favorable position as it descends. The cormorant lives in perfect harmony with the wild swan, goose, various sorts of duck, and other birds; but to a gull with a piece of fish it will instantly give chase.

A writer relates: "Several years ago I took a pair of these birds from a nest among the rocks of Howth (Ireland), and kept them for nearly two years, by which time they had attained their full growth. They were pleasant pets enough, unless when pressed by hunger, when they became outrageous and screamed most violently; when satisfied with food, they slept, roosting on a large trough placed for holding water. But woe to the man or beast attempting to approach them when hungry. It happened once that a gentleman's servant went to look at them while in this state: he wore a pair of red plush breeches that immediately attracted the attention of the birds, which I had been in the habit of feeding with livers and lights; the consequence was they made such a furious attack that I had to run to his assistance with a stick, and could not beat them off without the greatest difficulty. Their attack on cats, dogs, and poultry, if unprotected, was always fatal. They fought at once with their bills, wings, and claws, screaming frightfully all the time. In fact, the cause of my parting with them was their having destroyed a fine Spanish pointer: he had incautiously strayed into the place where I kept them, and they immediately flew at and attacked him in front and rear. His loud howling brought me to his aid, when I was astonished to find they had got him down, and before I could rescue him from their fury, they had greatly injured him in one of his shoulders, so much so that he afterwards died of the wound."

The Druids believed the appearance of a cormorant during the celebration of their mysteries was an evil omen. Milton describes the arch-fiend, who—

> "On the Tree of Life—
> The middle tree, the highest there that grew—
> Sat like a cormorant.

The Pelican, being furnished with a peculiar organ for storing up its prey, would seem to be still better adapted than the cormorant for being trained to fish. Labat mentions that the Indians adopt this practice, and dispatch a pelican in the morning, after having stained it red, and that it returned in the evening with its bag full of fish, which it was made to disgorge.

The sac or bag of the pelican is an elastic flesh-colored membrane, which hangs from the lower edges of the under mandible, reaching the whole length of the bill to the neck, said to be capacious enough to hold about four gallons of water. The bird has the power of contracting the bag by wrinkling it up under the mandible, so that it is scarcely visible; but after a successful fishing, it is incredible to what extent it is frequently distended. It preys chiefly on the larger fish, with which it fills its capacious pouch in order to digest them at leisure.

The great stretch of wing in the pelican, extending to eleven or twelve feet, and consequently double that of the swan or the eagle, enables it to support itself a long time in the air, where it balances itself with great steadiness, and only changes its place to dart directly downwards on its prey, which rarely escapes; for the violence of the dash, and its wide-spread wings, by striking and covering the surface of the water, make it boil and whirl, and at the same time stun the fish, and deprive it of the power of escape. When the pelicans are in flocks they act in concert, and, forming a great circle which they diminish by degrees, they thus enclose the fish, and all, at a certain signal, strike the water at the same moment, and amidst the disorder thus occasioned they plump in and seize their prey. These birds spend in fishing the hours of the morning and evening, when the finny tribe are most in motion, and they choose the places where they are most plentiful.

The pelican belongs more to warm than cold climates. It is very common in Africa and in some parts of Asia; it is met with also in this country and in the southern parts of Australia. It perches on trees, but does not nestle there, constructing on the ground a nest a foot and a half in diameter, furnished with soft sea-plants.

The flesh of the pelican was forbidden to the Jews as unclean. It has an ill taste, and in our country is used for its oil. The pouches of these birds have also been used to hold tobacco, and this skin, when dressed, is very soft.

To the pelican tribe also belongs the Gannet, Solan Goose, much larger than the gulls, from which they may be distinguished at a distance by a greater length of neck, the intense whiteness of the plumage, and the black tip of their wide-spread wings. The mode in which the Gannet fishes is peculiar. "In flight," remarks the Rev. C. A. Johns, "it circles round and round, and describes again and again the figure of eight, at a varying elevation above the water, in quest of herrings, pilchards, and other fishes, whose habit it is to swim near the surface. When it has discovered a prey, it suddenly arrests its flight, probably closes its wings, and descends with a force sufficient to make a jet of water visible two or three miles off, and carry it many feet downwards. When successful it brings its prize to the surface, and devours it without troubling itself about mastication. If unsuccessful, it rises immediately and resumes its hunting. It is sometimes seen swimming, perhaps to rest itself, for I did not observe that it ever dived on these occasions. My companion told me that the fishermen on the coasts of Ireland say that if this bird be chased by a boat when seen swimming, it becomes so terrified as to be unable to rise. The real reason may be that it is gorged with food. He was once in a boat on the Lough, when a gannet being seen a long way off, it was determined to give chase, and ascertain whether the statement was true. As the boat drew near, the gannet endeavored to escape by swimming, but made no attempt to use its wings. After a pretty long chase the boatmen secured it, in spite of a very severe bite which it inflicted on his hand. It did not appear to have received any injury, and when released on the evening of the same day, swam out to sea with great composure. A fisherman at Islay told me that in some parts of Scotland a singular method of catching these birds is adopted. A herring is fastened to a board, and sunk a few feet deep in the sea. The sharp eye of the gannet detects its prey, and the bird, first raising itself to an elevation sufficient to carry it down to the requisite depth, pounces on the fish, and in the effort penetrates the board to which it is attached. Being thus held fast by the beak, it is unable to extricate itself. Frequently also gannets are caught in the herring-nets at various depths below the surface. Diving after the fish, they become entangled in the nets, and are thus captured in a trap not intended for them. They perform good service to fishermen by indicating at a great distance the exact position of the shoals of fish."

Some idea may be formed of the fishing exploits of the gannet from what Buchanan states, that one hundred and five millions of herrings are destroyed annually by these birds at St. Kilda. They are summer visitants to the English coasts, and although from their power of flight they seem to be widely scattered, yet their real stations or breeding-places are few and local. The Bass Rock, St. Kilda, and Ailsa Craig have long existed as Scotch localities; while Lundy Island on the coast of Devon, and the Skelig Isles in Ireland, are less-known English and Irish stations.

It is on the Bass Rock, in the Firth of Forth, that they assemble in countless multitudes, and present an extraordinary sight to the beholder, nestling upon their eggs, greeting their mates on their arrival from the sea, or quarrelling if one happens to intrude a little to near another. Troops of birds in adult, changing, and first year's plumage, pass and repass, sailing in a smooth, noisless flight. The great proportion build on the ledges of the precipitous face of the rock, but a considerable number also place their nests—generally made carelessly of a few dried stalks of seaweed, rudely put together—on the summit near the edge, where they can be walked among; there the birds are very tame, allowing a person to approach them, but when a foot is held out aggressively they will bite at it.

Most, if not all, of these breeding stations are rented from the proprietors, the rent being paid chiefly by the feathers. The young geese are killed and cured. The inhabitants of St. Kilda, the most western of the Hebrides, are said to consume twenty-two thousand of the young birds every year, besides eggs. The gannet is easily kept in confinement, though the required supply of fish renders its keeping expensive. It is indifferent alike to cold or stormy weather; the air-cells which give lightness to the body are developed in an extraordinary degree. Montague remarks "the gannet is capable of containing about three full inspirations of my lungs, divided into nearly three equal portions, the cellular parts under the skin on each side holding nearly as much as the cavity of the body. In the act of respiration there appears to be always some air propelled between the skin and the body, as a visible expansion and contraction is observed about the breast, and this singular conformation makes the bird so buoyant that it floats high on the water, and does not sink beneath the surface, as observed in the cormorant and shag."

The Hooper or Wild Swan is the most common of its species in England and America, being a general winter visitant. The length to the end of the toes is five feet; to that of the tail, four feet ten inches; extent of wings, seven feet three inches; and weight from thirteen to sixteen pounds. The lower part of the bill is black; the base of it, and the space between that and the eyes, is covered with a naked yellow skin; the whole plumage in the old birds is of a pure white, the down being very short and thick. The cry of the wild swan is very loud, and may be heard at a great distance, from which the name of "Hooper" is derived. When they fly high, and numbers of different ages and sexes are mingled together, their notes are far from disagreeble.

Belonging to the family of the *Fulconidæ* are birds of the eagle kind, which fish on their own account, robbing others of their prey when they can, and pursuing nearly the same method of dashing from a height upon the fish in the water. The Great Sea-Eagle is a distinguished member of this predatory family, measuring in length three feet, and in extent of wings six feet six inches. This bird often presents a fine feature in the wild and desolate landscape. Its most favorite haunts in Britain are the northern coasts of Scotland, where the headlands reach a stupendous height, are perpendicular on the face, and where the shelves and ledges selected for breeding or roosting-places are secure from aggression either from above or beneath. Here the sea-eagle resides constantly at one season, or he finds a safe shelter during the night, after his more extended hunting excursions. Here he is monarch of all he surveys; amidst the numerous sea-fowl, his companions, his pale gray-tinted plumage and outspread tail being conspicuous when opposed to the dark green sea or the deep and rich shades of many of these splendid precipices. Although of great size and imposing aspect, it is less elegant than the golden eagle, and inferior in courage and activity to many of the smaller species of the tribe. When standing, its postures are by no means graceful, but the keenness of its bright and fierce eye enlivens its appearance, and under excitement it throws itself into beautiful and picturesque attitudes, drawing back its head, and erecting the narrow and pointed feathers of the neck.

Besides a fondness for fish—in capturing which, however, the sea-eagle is not half so dexterous as the osprey—the bird is such a predaceous intruder on the farm-yard, that in the Hebrides a fierce war is waged against him.

The farmers of the isles of St. Kilda proceed to their extermination, some carrying coils of rope, others bundles of dry heath and burning peat, and ascend to the brow of the mountains, where the fissured and shelved precipice hangs over the foamy margin of the Atlantic. Strings of gannets, cormorants, and guillemots are seen winding round the promontories, while here and there over the curling waves is seen hovering a solitary gull. They have reached the brink of the cliffs, over which the more timid scarce dare venture to cast a glance, for almost directly under their feet is the unfathomed sea, heaving its heavy billows some hundred feet below the place to which they cling. The eagles are abroad, sailing at a cautious distance in circles, uttering wild and harsh screams, and as they sweep past displaying their powerful talons. One of them fastens the rope to his body, passing it under his arms, and securing it under upon his breast by a firm knot. The rest dig holes with their heels in the turf, and sitting down in a row, take firm hold of the cord. The adventurer looks over the edge of the cliff, marks the projecting shelf which overhangs the eagle's nest, and is gradually lowered towards it, bearing in one hand the bundle of heath with the cord attached to it, and the peat burning in the middle, and with the other pushing himself from the angular projection of the rock. At length he arrives on the shelf, and calls to those above to slacken the rope, but keep fast hold of it. Then creeping forwards and clinging to unstable tufts of vegetation, on the sides of the rock, he looks downwards and ascertains the precise position of the nest, in which are two eagles covered with down, skeletons of fishes, birds, and lambs heaped around them. At sight of the human face—which to their imagination is anything but divine—the young eagles shrink back in terror, cowering beneath the projecting angle that partly roofs the nest. Their enemy now retreats, disposes the bundle of heath in a loose manner, blows the peat into a flame, and partially encloses it. Once more he approaches the brink, casting an anxious eye towards the old eagles which are wheeling in short circles and uttering confused and piercing cries; then blowing the flame, kindles the bundle of combustibles, and rapidly lowers it right into the nest. The young birds scream and hiss, throwing themselves into attitudes of defence. The heath smokes and crackles, and at length blazes into full flame; then the sticks, sea-weeds, wool, and feathers of the nest catch fire, and the ascending column of smoke indicates to the ropemen above that the deed is doing. Flames and smoke conceal the young birds from the avenger's gaze, but he stirs not until they have abated, and he sees the huge eyrie and its contents reduced to ashes. He then calls to his friends, who tighten the rope, and preparing himself for the ascent, is hauled up, encountering no small danger from the fragments which are loosened from the rock, and the difficulty of keeping his face and breast from the ragged points which project from the cliff. Birds have feelings as well as men, and those of the eagle are doubtless acute, for the old birds wheel and scream along the face of the rock for many days in succession, and as by this time the summer is far advanced, they form no new nest.

But the king of winged fishers is the famous Osprey, the Fishing Eagle *par excellence*, or Fishing Hawk, as it has been variously named, a bird remarkable among the rapacious kind for the peculiar adaption it enjoys for fishing. The wings of the male osprey are sixty inches in length, the body being twenty-three; the female, however, is larger, but does not differ much in color, which is generally in the upper parts a deep brown, beautifully glossed with light purple, the margins and tips of the feathers being pale brown or brownish-white. The osprey finds a worthy antagonist in the white-headed eagle.

Elevated on the high dead limb of some gigantic tree that commands a wide view of the neighboring shore and ocean, the white-headed eagle seems calmly to contemplate the motions of the various feathered tribes that pursue their busy avocations below: the snow-white gulls slowly winnowing the air; the busy sand-pipers coursing along the sands; trains of ducks streaming over the surface; silent and watchful cranes, intent and wading; clamorous crows; and all the winged multitude that subsist by the bounty of this vast liquid magazine of Nature. High over all these hovers one whose actions instantly arrest all the attention of the observer. By his wide curvature of wing and sudden suspension in air he knows him to be the osprey, the "fish-hawk," settling over some devoted victim of the ocean. His eye kindles at the sight, and balancing himself with half-opened wings on the branch, he watches the result. Down—rapid as an arrow from heaven—descends the distant object of his attention, the roar of its wings reaching the ear as it disappears in the deep, making the surges foam around. At this moment the eager looks of the eagle are all ardor; and levelling his neck for flight, he sees the osprey once more emerge struggling with his prey, and mounting in the air with screams of exultation. These are the signals for our hero, who, launching into the air, instantly gives chase, soon gains on the fish-hawk; each exerts its utmost to mount above the other, displaying in the struggle the most elegant and sublime aërial evolutions. The unencumbered sea-eagle rapidly advances, and is just on the point of reaching his opponent, when, with a sudden scream, probably of despair and honest execration, the osprey drops his fish; the eagle, poising himself for a moment, as if to take a more certain aim, descends like a whirlwind, snatches it in his grasp before it reaches the water, and bears his ill-gotten booty to the woods.

The osprey on leaving its nest, usually flies direct until it reaches the sea, then sails round in easy curving lines, turning sometimes in the air as on a pivot, apparently without the least exertion, rarely moving its wings. Suddenly it checks its course as if struck by a particular object, which it seems to survey for a few moments with such steadiness that it appears fixed in the air, flapping its wings. This object, however, it abandons, and is again seen sailing round as before. Now its attention is again arrested, and it descends with great rapidity, but before it reaches the surface shoots off on another course, as if ashamed that a second victim had escaped. It now sails at a short distance above the surface, and by a zig-zag descent, and without seeming to dip its feet in the water, seizes a fish, which, after carrying a short distance, it drops and probably yields up to the bald eagle, and again ascends by easy spiral circles to the higher regions, where it glides about with all the ease and majesty of its species. From hence it descends like a perpendicular torrent, plunging into the sea with a low rushing sound, and with the certainty of a rifle. In a few moments it emerges, bearing in its claws the struggling prey, which is always carried head-foremost, and having risen a few feet above the surface, shakes itself as a water-spaniel would do, and then seeks land. If the wind blows hard, and its nest be in a quarter from which it comes, it is amusing to see with what judgment the osprey beats up to windward; not in a direct line, but like an experienced navigator, making several successive tacks to accomplish its purpose.

The ospreys watch and pursue fish with as much avidity as the true eagles hunt their game on land; and Nature, as we have remarked, has provided them with the means for so doing. Fish are slippery, and therefore its claws are long and much curved, its toes nearly of equal length, and capable of being applied in the most effectual manner, in pairs, two and two opposite each other. It must also possess considerable power, and therefore its legs are strong and muscular, and to prevent its being inextricably entangled the claws are smooth and rounded, so that they can, if necessary, be readily withdrawn. The animals on which it feeds live in the water, ordinarily beyond its reach, coming occasionally to the surface; the bird, therefore, has a comparatively slender form, with very long wings, so as to enable it to remain without fatigue sailing or hovering over the water until an opportunity of pouncing occurs. To prevent its plumage from being injured by its sudden immersion into the water, the feathers of the lower surface are rather more compact and considerably shorter than in eagles and most other birds of the family, and those of the leg are short all round, while most other species have a large tuft of short feathers. The structure of the wings is also curious: in the osprey they are very long, yet length is not of itself an indication of great speed so much as the power of easy suspension in the air and of continued flight. The osprey requires to hover long over the waters, often over the open sea at some distance from land, sometimes for hours together before an opportunity for pouncing on its prey occurs. Its form, therefore, is as light as is compatible with strength.

> "True to the season, o'er our sea-beat shore,
> The sailing osprey high is seen to soar,
> With broad unmoving wing, and circling slow,
> Marks each loose straggler in the deep below—
> Sweeps down like lightning, plunges with a roar,
> And bears his struggling victim to the shore."

We have now to notice another family, the Phaeton or Tropic Birds, so named because, from their habitual residence under the burning zone, bounded by the tropics, they seem attached to the chariot of the sun, to use a classical metaphor. From this climate they remove but little, and their appearance indicates to seamen their approaching passage under this zone, from whatever side they may arrive. Still, they advance seaward many hundreds of miles.

The Frigate-Bird is the representative of this species, the swiftest ranger of the ocean, whose extended wings measure a width of seven feet. How this bird treats the unfortunate "booby" (also a fish-hunter) is described by a writer, who says:

"Every one who has read the romantic naratives of the old voyagers is familiar with the name of the booby, so termed by seamen from its apparent stupidity and familiarity, suffering itself to be knocked down by a stick, or taken by the hand when it alights, as it often does, on the spars or shrouds of a vessel. This habit seems quite unaccountable. Many birds have manifested a similar fearlessness of man when first discovered, but have soon learned the necessity of precaution; but the booby will manifest the same unnatural tameness after being long accustomed to the cruelty of man. It does not arise from helplessness, as it is a bird of powerful wing, like its relative the common gannet; neither is it a sufficient explanation to affirm, as is sometimes done, that it arises from a peculiar difficulty in rising to flight after alighting, because it is not unfrequently caught in the air by the hand, so incautiously does it approach man. Notwithstanding this apparent stupidity, the booby is a dexterous fisher. Hovering over a shoal of fishes, he eagerly watches their motions, turning his head from side to side in a very ludricous manner. He presently sees one of the unwary group approach the surface: down he pounces like a stone, plunging into the waves, which boil into foam with the shock. Nor fails he to seize the scaly victim, with which he emerges into the air, and soon it is lodged whole in his capacious stomach. But the frigate-bird has watched the proceeding, and instantly betakes himself to the pursuit. Sweeping down upon the unfortunate booby, he compels him to disgorge the fish which he has just swallowed, and which, long before it can reach the water, is seized and again devoured by the oppressor.

"The frigate-bird neither swims nor dives; the seamen even believe that it sleeps on the wing: whether this be so or not, there is good evidence that the same individuals will remain in the air for several successive days; they are never known to alight on a vessel. Though the chase of the booby is so usual as to be considered one of its constant means of dependence, yet it also fishes for itself; precluded, however, from plunging into the sea, it can only take such as, like the flying-fish, leap into another element. With such success, however, does it attack these, that it has been seen to snap up three in succession in the course of a few minutes."

The frigates fly with great rapidity, and brave the tempest by shooting above their region and remaining balanced in the air until they can alight upon some rock to rest, for the length of their wings would prevent them from rising either from the waves or the ground. Their sight must also be remarkably piercing to enable them to discover, at such distances as quite escape our vision, the places where pass the flying-fishes, their chief relish. Instead of precipitating their head foremost, like birds which have the faculty of diving, the frigate holds its head and feet in a horizontal direction, striking the upper column of air with its wings, then, raising and fixing them one against the other above its back, it darts on its prey with such address and velocity that it rarely escapes. The tropic birds, like the cormorant, perch on the highest trees, and make their nests in the holes of precipitous rocks or in the hollows of trees. The young, while in the nest gathered up in a ball and covered with a down of the most brilliant white, have a resemblance to powder-puffs. Of the long tail-feathers—sometimes twenty-four inches—the Otaheitans make plumes for their warriors.

The Boobies have been met with in every sea and in every quarter of the globe. They fly with the neck extended, the tail spread out, and the wings almost motionless. Their cry participates of those of the goose and raven. They remove much less from land than the frigate-birds.

CHAPTER XIX.

SUPERSTITIONS CONNECTED WITH THE OCEAN.

"I saw the new moon late yestreen
With the old moon in her arm;
And if we go to sea master,
I fear we'll come to harm."—*Old Ballad.*

IT is not surprising that men accustomed to the monotony of a seafaring life, remote from the educational influences afforded to those on land, with the many wonders of the vast ocean around them, full of strange mystery, which science only can partially unveil; with minds thus generally untutored, and consequently more susceptible to superstitious fancies, it is not astonishing that such persons should be among the most credulous of mankind. It is true that the spread of knowledge in modern times has removed many of the absurd notions peculiar to seamen; but, as a class, they may still be considered among the foremost believers in the supernatural.

From the earliest times the sea has been regarded as the region of fabulous marvels. The ancient mariners performed their voyages in a vague mist of capricious doubts and fancies, omens and prognostics, which excited terror or inspired confidence. Every object that met their gaze was endowed by them with some miraculous agency for good or for evil. Their course over unknown waters, peopled by their mythology with imaginary creatures, would naturally create awe and suspicion.

Horace, lamenting at Virgil's departure for Athens, rebukes the impiety of the first mariner, who ventured, in the audacity of his heart, to go afloat, and cross the briny barrier between nations. He esteems a merchant favored specially by the gods should he twice or thrice return in safety from a distant cruise. He tells us he himself had known the terrors of the dark gulf of the Adriatic and had exoerienced the treachery of the western gale.

Ancient writers are diffuse in the description of prodigies witnessed by mariners at sea, many of which, doubtless originating from simple causes, received the addition of a divine interposition. The sudden breaking up of a dense fog, and the sun shining in undimed splendor, was attributed to the appearance of Apollo himself, as the saints in later ages were supposed to miraculously intervene for the protection of seamen. Apollonius of Rhodes, the Greek poet, describes the Argonauts (Greek heroes who, under the command of Jason, went in search of the Golden Fleece) as suddenly benighted at sea in broad daylight by a dense black fog. They pray to Apollo, and he descends from heaven, and alighting on a rock, holds up his illustrious bow, which shoots a guiding light farther to an island. The delusions of these pagan times continued through succeeding ages, modified only by the change of religion and a better knowledge of navigation. These notions, under various forms, still prevail in some foreign countries, where the divine light of evangelical truth has not pierced, while other phases of superstition still linger among our own sailors as regards omens, gook luck, and a number of other senseless notions.

Legends of a ridiculous character abound in most all of the old writings, but we will now pass on to later superstitions. You have no doubt heard of the "Phantom Ship," which was supposed, when seen by sailors—or rather present in their imaginations only—to foretell disaster. This story originated with the Dutch, and found believers among seamen of all countries. Sir Walter Scott alludes to this spectral illusion as a harbinger of woe:

"The phantom ship whose form
Shoots like a meteor through the storm,
When the dark scud comes driving hard,
And lower'd is every topsail-yard,
And canvas wove in earthly looms
No more the brave the storm presumes
Then 'mid the roar of sea and sky,
Top and top-gallant hoisted high,
Full spread and crowded every sail,
The demon frigate braves the gale,
And well the doom'd spectators know
The harbinger of wreck and woe."

Water-spouts at sea were regarded in olden times with great terror. Sailors were accustomed to discharge artillery at these moving columns to accelerate their fall, from a fear lest the vessel should be sunk by them. The principal danger, however, arises from the wind blowing in sudden gusts in the vicinity of the spout from all points of the compass, sufficient to capsize small vessels carrying much sail. Another practice was to cut the air with a knife, while reciting some prayers, by which simple enchantment it was supposed the water-spouts would be reduced to submission. If it happened, however, to be in an obstinate mood, two sailors would draw their swords, and strike at each other, in true gladiator style, taking care between each blow to make the sign of the cross.

It is a cheering instance of human progress that, by the introduction of lightning-conductors into ships, the fearful electric currents which destroyed many noble vessels is now placed under control and rendered powerless.

Among the ancients it was believed that certain persons had the power of raising tempests at sea. In the "Odyssey," Æolus is described as possessing these attributes, and

Calypso, in the same work, is said to have been able to control the winds.

The belief in human agency to influence the ocean was prevalent in the fifteenth century. A curious confession was made in Scotland about the year 1469, by one Agnes Sampson, a reputed sorceress, who avowed that "at the time His Majesty (James VI.) was in Denmark, she took a cat and christened it, and afterwards bound to each part of that cat the chiefest parts of a dead man, and several joints of his body; and that in the night following, the said cat was conveyed into the midst of the sea, by herself and other witches, sailing in their baskets, and so left the said cat right before the town of Leith in Scotland. This done, there arose such a tempest in the sea as a greater hath not been seen, which tempest was the cause of the perishing of a boat or vessel wherein were sundry jewels and rich gifts, which should have been presented to the new Queen of Scotland at Her Majesty's coming to Leith."

Such was the language of a silly old woman, probably extorted by torture from a weak imagination.

King James, in his "Demonology," states "that witches can raise stormes and tempestes in the aire, either on sea or land," which was in answer to Reginald Scot, who in his "Discoverie of Witchcraft" ridiculed the "black art" severely, and he had the advantage of his royal master, the "British Solomon," as he had been equivocally termed, in this and many other statements.

The Evil One was supposed to have a direct influence on the winds and waves.

Some sailors have a strange opinion of satanic power and agency in stirring up winds, and that is the reason they so seldom whistle on shipboard, esteeming it to be a mockery, and consequently an enraging of the devil.

We should scarcely expect that the mere turning of a stone was supposed to have had an effect in procuring favorable breezes, yet we read that the inhabitants of some parts of the Western Islands had implicit faith in this charm. In the chapel of Fladda Chuan there was a blue stone fixed in the altar, of a round form, which was always moist. It was the custom of any fishermen who were detained on the island by contrary winds to wash this blue stone with water, expecting by this to obtain a favorable wind. So great was the regard paid to this stone, that any oath sworn before it could never be broken. Another mode of these primitive islanders to secure auspicious winds was of a bucolic character, and consisted in hanging a he-goat to the mast-head.

A similar feeling with regard to the efficacy of stones, though for another object, existed among the fishermen of Iona This took the shape of a pillar, and the sailor who stretched his arm along it three times in the name of the Trinity could never err in steering the helm of a vessel. The Finlanders are said to have used a cord, tied with three knots, for raising the wind: when the first was loosed, they could expect a good wind; if the second, a stronger; and if the third, such a storm would arise that the sailors would not be able to direct the ship, or avoid rocks, or stand upon the decks. The French seaman in former days had a comical notion that the spirit of the storm was propitiated by flogging unfortunate midshipmen at the mainmast.

Particular seasons of the year and saints' days were held in superstitious regard among mariners, and peculiar customs were attached to them. The old practice of setting the nets at Christmas Eve was general among Swedish fishermen. The sailors at Folkestone, in Kent, chose eight of the largest and best whitings out of every boat, when they came home from the fishery. Out of the profit arising from these they made a feast every Christmas Eve. On Allhallow's Even, or the vigil of All Saints' Day, the fishermen of Orkney sprinkled what was called *fore-spoken* water over their boats when they had not been successful. They also made the sign of the cross on their boats with tar. The sailors in the Island of St. Lewis had an ancient custom of sacrificing to a sea-god called *Shony*, at Hallow-tide. They came to the church of St. Malvay, each seaman having his provisions with him. Every family furnished a peck of malt, and this was brewed into ale. A fisherman was selected to wade into the sea, carrying a cup of ale in his hand and crying, "Shony, I give you this, hoping you will send us plenty through the year."

The fishermen of Finland believed that any among them who created a disturbance on St. George's Day would provoke storms and tempests. At Dieppe, in Normandy, even to a late period, All Saints' Day was religiously observed by the sailors of that port. Those who ventured out to sea on that anniversary were supposed to have the "double sight;" that is, each one beheld a living likeness of himself seated in close contact, or when engaged in any work, doing the same. If the nets were cast out, they were found, on drawing them in, to contain nothing but bones. On the same day toward midnight, a funeral car was heard driven slowly by a team of eight white horses, preceded by dogs of the same color. Those who listened might hear the voices of those sailors who had died in the course of the year, Those persons who dared to look at this fearful scene were doomed to die shortly afterwards; so, as the hour approached, every house was barred and windows closed.

The Russian Twelfth day (18th of January) is devoted to the singular custom of blessing the waters of the Neva, there being no parallel ceremony in any other couutry, except the practice once observed at Venice, of the Doge espousing the sea. On the same day at Constantinople, the Greek Patriarch performs a similar custom by throwing a cross into the sea, and it is said that skillful divers generally succeed in obtaining it before reaching the bottom.

The fishermen who dwell on the coasts of the Baltic never used their nets between All Staints' Day and St. Martin's Day, believing that any infraction of this rule would prevent them from getting fish through the whole year. A similar observance, for the same reason, was held on St. Blaise's Day. They also considered sneezing on Christmas Day a favorable omen for the ensuing year.

The fishermen of Hartlepool preserve many old customs, such as Carling and Palm Sundays, and Easter Day. At Christmas the children sing carols, and sword-dancers go about the streets; and on the first Monday after the Epiphany, the stot or fool-plough (a small anchor) is dragged through the town, and donations requested.

Sailors have always had their prejudices with regard to certain days of the week. That ominous day, Friday, so dark-lined to so many weak-headed individuals—not only at sea but on shore—was and is still considered by many mariners a blank day for sailing. A Cornish saying places Candlemas Day as ill omened for sailing. Bishop Hall, speaking of a superstitious man, observes, "he will never set to sea except on Sunday." At Preston-Pans, it seems, that holy day was usually selected for sailing to the fishing grounds: a clergyman of the town preached against this Sabbath-breaking, and the sailors, to prevent any ill befalling them in consequence, made a small image of rags, and burnt it on the top of their chimneys.

Apparitions have always been a fruitful source of terror to seamen. A few years ago half a dozen sailors on board a man-of-war took it into their heads that there was a ghost in the ship, and declared *they smelt him.* The captain laughed at them, and called them a parcel of lubbers. A few nights afterwards they were in great terror, saying the ghost was behind the beer-barrels. The captain, annoyed at their folly, ordered a dozen lashes to each of them, which effectually stopped all talk about the spirit. When the barrels were removed some time afterwards, a dead rat was found, which had given rise to the story. Brand mentions that the cook of a vessel belonging to Newcastle died on a homeward passage. One of his legs was shorter than the other, which had given him an odd appearance when he walked. A few nights after the body had been committed to the deep, the captain was alarmed by his mate assuring him that the man was walking on the sea before the ship. The captain certainly saw something that seemed to move as the cook had walked, and ordered the ship to be moved towards the object. The seamen were greatly terrified; but it was soon found that the cause of all the commotion was part of a main-top, the remains of some wreck, floating before them. In the campaigns of the French fleet at Mitylene, the crew of a brigantine are said to have seen the figure of a montrous and hideous seaman descend in the waters at Zante, with one of the crew who had defied the Virgin while playing at dice on board.

A belief is still widely entertained in the virtues of a child's caul as a preservative against drowning and shipwrecks. In a Plymouth (England) paper, the *Western Morning News*, of the 9th of February, 1867, there was a notice to mariners: "For safeguard at sea: a child's caul for sale, price five guineas." The pretended virtues of this object reminds us of what is told of Augustus Cæsar, that he carried a seal's skin about him as a preservative against lightning.

Rats leaving a ship are considered indications of misfortune, probably from the same idea that crows will not build their nests upon trees that are likely to fall. A droll story is told of a cunning Welsh captain, whose ship was infested by rats, which he was anxious to get rid of. The vessel was lying in the Mersey at Liverpool, and hearing that there was another, laden with cheese, in the basin, he got alongside of her about dusk, and soon saw all the rats attracted by the rich smell of the cheese into his neighbor's ship, when he quietly had his own removed to a safe distance.

Omens for good or evil were derived from birds and marine animals. Shakespere alludes to the halcyon when he says:

> "Disown, affirm, and turn their halcyon beaks
> With every gale and vary of their masters."

The osprey is abundant during the summer along our coasts, and its presence is hailed by the fisherman as the harbinger of summer, with the same feelings of satisfaction as the appearance of the gannet on the English shores.

> "The osprey sails above the sound,
> The geese are gone, the gulls are flying,
> The herring-shoals swarm thick around,
> The nets are launched, the boats are plying;
>
> "'Yo, yo, my hearts! let's seek the deep,
> Raise high the song, and cheerly wish her,
> Still as the bending net we sweep,
> God bless the fish-hawk and the fisher!'"

The tern is considered in the same favorable light as the osprey and the kingfisher; but the stormy petrels, the "Mother Carey's chickens" of early times, bring apprehensions of fearful dangers to the seamen, owing probably to the appearance of the birds when several hundred miles from land, apparently untired, and seldom seen resting or eating, together with its ominous color. The petrel does actually caution mariners of an impending tempest by collecting under the stern of the ship.

> "Thus doth the prophet of good or ill
> Meet hate from the creatures he serveth still;
> Yet he ne'er falters; so, petrel, spring
> Once more on the waves with thy stormy wing."

It is curious to find crows employed in the early ages as guides to mariners. We are told that when Flok, a famous

Norwegian navigator, was going to start from Shetland to Iceland, he took on board some crows, because the mariner's compass was not then in use. When he thought he had made a considerable part of his way, he threw up one of his crows, who, seeing land astern, flew to it, thus indicating the route. Such was the simple mode of keeping a reckoning and steering their course pursued by the bold navigators of the stormy Northern Ocean.

It is still believed that sea-gulls retiring to land foretell a storm; but the migration of sea-birds generally arises from their security in finding food, such as earth-worms and larvæ, driven out of the ground by severe floods. The fish on which they prey in fine weather in the sea leave the surface and go deeper.

Bourne says that, " seeing *three* magpies augurs a successful voyage;" but this will scarcely hold good with the superstitions respecting the same bird formerly held by seamen. Sir Walter Scott relates that a friend on a journey to London found himself in company with a seafaring man of middle age, in the same mail coach, who announced himself as master of a vessel in the Baltic trade. In the course of conversation the seaman observed, "I wish we may have good luck on our journey; but there is a magpie!" "And why should that be unlucky?" said my friend. "I cannot tell you," replied the sailor; "but all the world agrees that one magpie bodes ill luck; two are not so bad; but three are the Evil One himself. I never saw three magpies but twice, and once I nearly lost my vessel, and afterwards, when I was on land, I fell from my horse and was much injured."

The swan was an omen of fair weather to mariners. Coleridge has immortalized the albatross, as the harbinger of good fortune, in the "Ancient Mariner:"

> " At length did cross an albatross,
> Through the fog it came;

As if it had been a Christian soul,
 We hail'd it in God's name.

"It ate the food it ne'er did eat,
 And round and round it flew;
The ice did split with a thunder-fit,
 The helmsman steer'd us through.

"And a good south wind sprang up behind;
 The albatross did follow;
And every day, for food or play,
 Came to the mariners' halloa!

"In mist or cloud, or mast or shroud,
 It perch'd for vespers nine,
While all the night, through fog-smoke white,
 Glimmer'd the pale moonshine.

"'God save thee, Ancient Mariner,
 From the fiends that plague thee thus;
Why look'st thou so?' 'With my crossbow
 I shot the albatross!'

* * * * *

"And all averr'd I had kill'd the bird
 That made the breeze to blow:
'Ah, wretch!' said they, 'the bird to slay,
 That made the breeze to blow?'"

The albatross is remarkable for the extent of its wanderings; indeed, it may almost be said to pass from pole to pole, and is seen at a greater distance from land than any other bird. Hence sailors regarded this companion of their voyages with superstitious fondness.

Dolphins, as well as porpoises, when they play about a ship, are supposed to foretell storms. The ancient navigators, however, regarded them in a different light, and believed that they conveyed shipwrecked seamen to shore in times of peril. The story of Arion is well known; and Spenser, in his "Marriage of the Thames and Medway," alludes to this romantic fiction, at the sight of which

"All the raging seas for joy forgot to roar."

Like many other old pagan fictions, this story was invested by the earlier Christian converts with a deeper, holier meaning; and the dolphin, so constantly recognized in sculptures and frescoes, points, not to the deliverer of Arion, but to Him who, through the waters of baptism, opens to mankind the path of deliverance.

We need scarcely be surprised at the superstitions of seamen in former days, when instances of such gross ignorance and credulity are found among the writers of those times. A belief long prevailed that the barnacle, a well-known kind of shell-fish, found adhering to the bottom of ships, would, when broken off, become a species of goose. Several old writers assert this, and more than one from personal observation. The numerous tentacles or arms of the animal inhabiting the barnacle-shell, which are disposed in a semi-circular form and have a feathery appearance, seem to have been all that could reasonably be alleged in favor of this strange supposition.

Carrying dead bodies in ships has always been a sore point with sailors, and the sight even of an empty coffin works upon their prejudices. Such Nelson found was the case, when one was sent to him by a brother officer made from the mainmast of the French ship "L'Orient," to remind the illustrious hero that amidst all the glory that surrounded him, he was but mortal. Nelson received the present in the proper spirit, and had the coffin placed in his own cabin in the "Vanguard," but the crew could not bear to have the obnoxious memorial in sight, and it was accordingly ordered to be sent below.

In the Orkneys, mariners, on going to sea, would consider themselves in the greatest danger if by accident they turned their boat in opposition to the sun's course. In Sweden it is considered a bad omen to turn the prow of a vessel toward the shore, and for anyone to say, "Good luck" to the fishermen when starting; also, that pins found in a church, and made into hooks, get the best fish. Tackle, they affirm, stolen from a friend or a neighbor, secures better luck than when purchased for money—a species of larceny more profitable to the fisherman than comfortable to his friends.

Sneezing—a potent omen in ancient times—had its portent for good or evil among seamen in former days: a sneeze on the left side, at the moment of embarking, foreboded evil, while a fortunate sneeze on the right side betokened a favorable voyage.

"Good luck" is as much the creed of the fisherman as it is of many superstitious persons on land. Only a few years ago, it was related that the herring fishery being very backward, some of the fishermen of Buckie dressed a cooper in a flannel shirt with burrs stuck all over it, and in this position he was carried in procession through the town in a handbarrow. This ridiculous ceremony was done to procure "better luck." It happened, also, in a district where there were several churches, chapels, and schools. The fishermen of the Firth of Forth believe that if they chance to meet a woman bare-footed who has broad feet, when they are going to their vessels, they will have "bad luck," and the same fatality attends the sale of fish for the first time in the day to a person having broad thumbs. It is considered "unlucky" to lose a water-bucket or mop at sea. Children on board are regarded with favor by seaman as likely to bring good luck; not so a cat, which is sure to turn the scale of chance in the wrong direction. Whittington, however, the renowned "thrice Lord Mayor of London," could not have shared in this superstition, if some old stories are true. To play at cards on board is considered unlucky; at some places boats crew's are changed from time to time for the same reason.

Very curious and quaint are the reflections and expressions of those who go down to the sea in ships and spend their lives therein. They see so much that is sublime and wonderful, in fact, that it is not extraordinary that their imagination and memories should be a storehouse of the fabulous which is scarcely more surprising than what they actually see and feel often during their monotonous and yet vivid experiences. The water is their world, and the ship that rides it is not only a castle to them, but a thing endowed with life. Personification is not restricted to their floating homes, but everything breathes, lives, moves; and the great deep and every fleeting shadow seems peopled.

CHAPTER XX.

MARINE PRODIGIES.

"God quickened in the seas and in the rivers
So many fishes of so many features,
That in the waters we may all see creatures,
Even all that on the earth are to be found,
As if the world were in deep waters drown'd.
For seas, as well as skies, have sun, moon, and stars
As well as air, swallows, rooks, and stares,
As well as earth, vines, roses, nettles, melons,
Mushrooms, pinks, gillyflowers, and many millions
Of other plants more rare, more strange than these,
As very fishes living in the seas:
As also rams, calves, horses, hares, and hogs,
Wolves, urchins, lions, elephants, and dogs;
Yea, *men and maids*, and which I most admire,
The mitred bishop and the cowlèd friar;
Of which examples, but a few years since,
Were shown the Norway and Polanian prince."—DU BARTAS.

UNBOUNDED are the regions of fable; and probably no department of Nature is so prolific in supplying food for the wildest fancies of the imagination as the great unfathomed ocean depths, which conceal so many mysteries.

The ancients had their sea-divinities and monsters in profusion. It is true that the powerful mind of Aristotle, the great father of early philosophy, rejected with disdain the credulous tales and fabulous stories of his age in regard to natural history; but the writings of Pliny, the natural historian, abound in prodigies and absurdities, as also those of Ælian, and other ancient authors. For many centuries a mist of doubt, error, and fanciful credulity prevailed with regard to the inhabitants of the ocean. Even in 1554 a work on fishes by Rondelet, a physician at Montpellier, although written with tolerable exactitude in some particulars, concludes with a chapter illustrated by grotesque figures of certain marine monsters; among others, a fish dressed as a monk, and a "bishop-fish" in full pontificals. Where the extraordinary originals, from which these cuts were taken, came from, is not known; but they were probably fabricated in the true Barnum style from the skins of some large species of sharks or rays, by the priests of that period, to excite the superstitious veneration of the people, and pursuade them, as Du Bartas, in the quotation at the head of this chapter, wishes us to believe, that even the sea contains bishops and monks.

Until the commencement of the seventeenth century, nothing like a dawn of true light with regard to natural history seemed to strike upon the popular mind. A laborious but very credulous professor at Bologna, Aldrovandus, wrote no less than fourteen folio volumes on the subject, published in 1640; but the true and the false, fable and nonsense, are strangly intermixed. Some of the engravings in these books are very curious, and give an idea of the exaggerative style of their contents. These we have, magnified in every extent, by succeeding writers.

The Kraken, described by Pontoppidan in his "Natural History of Norway," is one of the most extraordinary of these wonderful sea-monsters, and claims the peculiar privilege of the wide domain of the Norwegian waters.

"Our fishermen usually affirm," says this writer, "that, when they row out several miles to sea, particularly in hot summer days, they are informed by various circumstances that the kraken is at the bottom of the sea. Sometimes twenty boats get together over him, and when, from well-known indications, they perceive it is rising, they get away as fast as they can. When they find themselves out of danger, they lie upon their oars, and in a few minutes they see the monster come to the surface. He there shows himself sufficiently, though only a small part of his body appears. Its back, which appears to be a mile and a half in circumference, looks at first like a number of small islands, surrounded with something which floats like sea-weeds; here and there a large rising is observed like sand-banks; at last, several bright points or horns appear, which grow thicker the higher they rise, and sometimes they stand up as high and as large as the masts of middle-sized vessels. It seems these are the creature's arms, and it is said, if they were to lay hold of the largest man-of-war, they would pull it down to the bottom. After the monster has been a short time on the water, he begins slowly to sink again; and then the danger is as great as before, because the motion of the sinking causes such a swell and such an eddy or whirlpool that he carries everything before it."

Such is the description of the fabulous kraken. Divested of its supernatural powers and dimensions, there may be some foundations for these exaggerations in the occasional appearances of huge cephalopods (molluscous animals having the head covered with tentacula or feelers, serving as feet), to the general characters of which the description given of its form and monstrous arms sufficiently agrees. Many such animals are known to exist in some seas, and there are reasons for believing that much larger creatures of the same species exist.

It is a favorite notion of Pontoppidan that from the appearance of the kraken originate those dim traditions of floating islands being observed in the North seas. It has been sought to identify the cuttle-fish of enormous size with the kraken, and stories are told of men having been drawn over the sides of vessels by their enormous arms. In 1834, Captain Neill, of the ship *Robertson*, of Greenock, saw the head and snout of a great sea-monster, of which a sketch was taken at the time. It appeared like a vessel lying on her beam-ends. The *Robertson* was hauled up so as to near it, and it was discovered to be the head and snout of a great fish swimming to windward: immediately above the water, its eye was seen like a large deep hole. The part of the head which was above the water measured about twelve feet, and its width twenty-five feet. The snout was about fifty feet long, and the sea occasionally rippled over one part, leaving other parts dry and uncovered. Several records exist in Scotland of the appearance of similar animals to that which have been noticed, but the result appears to infer the existence of some enormous cuttle-fish, possessed of characters which distinguish it essentially from every other creature with which we are familiar. Pliny's vast animal, with prodigious arms, which impeded the navigation of the Straits of Gibraltar, would seem to have had a family likeness to the kraken.

The great Sea-Serpent appears to have some analogy to the same monster as we have described, and here again the sapient Pontoppidian raises our eyes in astonishment at his description of this marine prodigy, which he describes as six hundred feet in length, lying in the water in many folds, and appearing like many hogsheads floating in a line at a considerable distance from each other. Such a creature is said to have been seen on the coast of Norway, in 1819, for a whole month, seeming to doze in the sunbeams. In 1822 and 1837, it is said to have reappeared in the same waters. Our sailors, not to be outdone by the Norwegians, relate several cases in which prodigious sea-serpents have been seen in the Atlantic, opinions varying as to the length of the animal, averaging from eighty feet to two hundred and fifty yards, making curves "perpendicular to the water, and with eyes brilliant and glistening."

British sailors have also their account in the prodigy witnessed in 1848, in the South Atlantic Ocean, not far from the coast of Africa, by the officers and crew of H.M.S. "Dædalus."

According to the account forwarded to the Admiralty, the animal was seen not in bright and fine weather, but with a murky atmosphere, and with a long ocean swell. It was swimming rapidly, and with its head and neck above water, and appeared an enormous serpent, with head and shoulders kept about four feet constantly above the surface of the sea. It passed so close under the lee-quarter of the vessel, that its features were easily recognized. The diametre of the serpent was about fifteen or sixteen inches behind the head, and the animal was never, during the twenty minutes that it continued in sight, once below the surface of the water. The color was dark brown, with yellowish-white about the throat. It had no fins but something like the mane of a horse, or rather a bunch of sea-weed, washed about its neck.

England's great anatomist, Dr. Owen, has expressed much doubt as to the existence of a great sea-serpent, on the ground that no bones or other remains of such recent animal have been found. Notwithstanding this high authority, it may be, however, that many animals in the ocean depths, of this character, are still unknown to us.

It has been supposed that large fishes of the Ribbon family may have given rise to some of these stories about the great sea-serpent. One was lately captured at the Bermudas, apparently an immature fish, but more than sixteen feet in length, and with a long row of flexible filaments, or slender threads on the back of the head and anterior parts of the back, which might well represent the mane alluded to as an appendage of the serpent prodigy. The fishes of this kind are inhabitants of great depths in the ocean, and this may account for the rarity of their appearance. One species belongs to the Northern Seas, where the appearance of the sea-serpent has been particularly recorded; others belong to the warmer regions. It may be that these fishes attain to a length that would corroborate the assertions of those who have seen the sea-serpent, making a due allowance for exaggeration under such novel circumstances.

Mr. Adams, naturalist of the English ship "Samarang," writing of Sooloo and the molucca Archipelagoes, remarks: "I have often witnessed the phenomenon which first gave origin to the marvelous stories of the great sea-serpent, namely, lines of rolling porpoises, resembling a long string of buoys often extending a hundred yards. These account for the so-named protuberances of the serpent's back. They keep in close single file, progressing rapidly along the surface of the water by a succession of leaps, part only of their uncouth forms appearing to the eye."

The belief in Mermaids and Mermen, prevalent through the remotest ages, was also especially strong in the Scandinavian countries, and some traces of the delusions still linger on some of the out-of-way coasts of the Northern seas. A very high antiquity is claimed for these mythic creatures. Ancient history abounds with notices of them. One was called by the Babalonians *Odakon*, and is regarded as identical with Dagon, the national god of the Philistines, so frequently mentioned in the Scriptures. It is always represented on medals as half fish and half woman, but the Hebrew writers speak of it as a masculine being. In the excavations of Khorsabad, M. Botta found a figure of Odunes, a creature half man and half fish. At the excavations at Nimroud, Mr. Layard discovered a gigantic figure with a fish's head as a cap, and the body of the fish depending over the shoulders. On the coins of Ascalon is figured a goddess, above whose head is a half-moon, and at her feet a woman with her lower extremities like a fish. It is singular how the prevalence of the tales of mermaids exists among Celtic nations, indicating these water-nymphs as having been originally deities of the people. The Peruvians had also their semi-fish gods. These form the types of those imaginary creatures, the subjects of ancient poetry, the Tritons, who were represented as half men and half fish, having power to calm tempests; and probably, too, of the Syrens, whose songs were said to lure the unhappy seamen to destruction.

Innumerable are the stories that are told of mermaids and mermen: they have been made the subject of numberless songs by ancient and modern bards. Shakespere alludes to the vocal powers of these mythic creatures:

> "I heard a mermaid on a dolphin's back
> Uttering such dulcet and harmonious breath
> That the rude sea grew civil at her song."

Laureate inquires:

> "Who would be
> A mermaid fair?
> Singing alone,
> Combing her hair
> Under the sea,
> In a golden curl,
> With a comb of pearl,
> On a throne?"

In the "Speculum Regale," an Icelandic work of the twelfth century, is the following description of a mermaid: "A monster is seen also near Greenland, like a woman as far down as the waist; long hands and soft hair, the neck and head in all respects like that of a human being. The hands seem to people to be long, and the fingers not to be parted, but united by a web like that on the feet of water-birds. From the waist downwards this monster resembles a fish, with scales, tail, and fin. This prodigy is supposed to show itself more especially before heavy storms. The habit of this creature is to dive frequently, and rise again to the surface with fishes in its hands. When sailors see it playing with the fish, or throwing them toward the ship, they fear they are doomed to lose several of the crew; but when it casts the fish, or, turning from the vessel, flings them away from her, the sailors take it as a good omen that they will not suffer loss in an impending storm, The monster has a very horrible face, with broad brow and piercing eyes, a wide mouth, and double chin."

Many of the so-called mermaids exhibited in a stuffed condition from time to time have proved sometimes clever, but more frequently bungling "shams." Among the latter may be classed the exhibition of the famous P. T. Barnum, a few years since, which proved to be the combination of the head of a monkey with the tail of a fish! The probability is that all the stories about these prodigies have originated in the appearance of seals, walruses, to which we have already alluded, and to what are called the herbivorous cetacea, from their living on sea-plants, and which consists among others of the manatee of the West Indies, the dugong of the Eastern seas, and the stellerus, an inhabitant of the Polar regions.

The best-known species of the Manatee, or Sea-Cow, is found in the West Indies and on the western coasts of tropical America. These sometimes attain a length of twenty feet, and a weight of three or four tons, and they live chiefly in shallow bays and creeks, and in the estuaries of rivers. The skin is very thick and strong, and is almost destitute of hair. The fingers can be readily felt in the swimming paws, and, connected together as they are, possess considerable power of motion, whence the name manatee (from the Latin *manus*, "a hand"). This animal is usually found in herds, which combine for mutual protection when attacked, placing the young in the centre. When one is struck with a harpoon, the others try to tear it out. The females show great affection for their young.

The Dugong—numbers of which frequent the coasts of

Ceylon, allured by the still waters and the abundance of sea-weeds—is, perhaps, one of the most likely representatives of what is considered a "mermaid" that could be found. There is a rude approach to the human outline in the shape and attitude of the mother dugong while suckling her young, holding it to her breast by one flipper while swimming with the other, the heads of both being above water; and when suddenly disturbed, diving and displaying her fish-like tail. These, together with her habitual demonstrations of strong natural affection, might readily give rise to the fable of the mermaid.

Megasthenes records the existence of a creature in the ocean near Taprobane (Ceylon), with the aspect of a woman; and Ælian, adopting and enlarging on his information, peoples the seas of Ceylon with fishes having the heads of lions, panthers, and rams; and stranger still, in the form of satyrs! Statements such as these must have had their origin in the hairs which are set round the mouth of the dugong, somewhat resembling a beard. The Portuguese cherished for a long time their belief in the mermaid; and the historian of the proceedings of the Jesuits in India gravely records that seven of these monsters, male and female, were captured at Ceylon and carried to Goa, where they were dissected by the physician to the Viceroy, and "their internal structure found to be in all respects similar to the human!" A dugong, killed at Ceylon in 1847, measured upwards of seven feet in length, but specimens considerably larger have been taken.

The female dugong, or sea-cow of Sumatra, will follow her young to the death, and is usually taken with them. The sea-calves have a short, sharp, pitiable cry, which they frequently repeat, and, like the stricken deer, are also said to shed tears, which, according to Sir Stamford Raffles, were carefully bottled by the common people, and preserved as charms to secure affection.

Only one species of the Stellerus—of the same genus as the two mentioned—has been known, about twenty-five feet in length, a native of the polar seas, and never observed since the middle of last century, so that it is supposed to be extinct. The characteristic features of this animal would lead one to suppose, also, that it may have contributed to the misconceptions about the mermaid.

An eminent writer remarks that the exhibition of strange fishes appears to have been at its height in the reign of Elizabeth. Shakespere twice alludes to it: once in the "Winter's Tale" (Act IV., Scene 3), where Autolycus says: "Here's another ballad of a fish that appeared upon the coast on Wednesday, the fourscore of April, forty thousand fathoms above water, and sung this ballad against the hard hearts of maids. It was thought she was a woman, and was turned into a cold fish, for she would not exchange flesh with one that loved her. The ballad is very pitiful, and as true;" and again, in the "Tempest" (Act II., Scene 2). A printed notice, dated 1566, has for its title "The Description of a Rare or rather Most Monstrous Fishe, taken on the East Coast of Holland, the 17th November, Anno 1566," with a woodcut of the fish, and underneath the following lines:

> "The workes of God, how great and strange they be!
> A picture plaine, behold, heare you may see."

Two years later there is another printed notice of a "moste true and marvellous strange wonder, the lyke hathe seldom been seene, of xvii monstrous fishes, taken in suffulke, at Downam Brydge, within a myle of Ipswiche, the xi daye of October, in the yeare of our Lorde God, 1568." Stow, in his "Annales," gives a particular description of this "wondrous draught of fishes," some of them being "eight and twentie foote in length at least."

CHAPTER XXI.

MONSTERS OF THE DEEP.—SEA DRAGONS.

"A world of wonders, where creation seems
No more the works of Nature, but her dreams."

MONTGOMERY.

WE have chosen for the present chapter a subject of the deepest interest, for it carries our thoughts to ages beyond the human mind to conceive, when the ocean, covering an immense expanse of our globe, swarmed with gigantic reptiles in the highest state of development, living in the open sea, and seeking the shore occasionally, crawling along the beach in search of prey. These reptiles were inhabitants of the ocean thousands of ages past, and from their strength and voracity must have been fearful scourges in that element.

Lyme Regis, a seaport in Dorsetshire, became remarkable at the commencement of the present century for the number of fossils embedded in the limestone cliffs, and a native of the place, Mary Anning, who had been engaged from her childhood in searching for what were then popularly called "curiosities," saw in 1811 an immense bone projecting from the ledges of the rocks, and having traced the remains of what she considered an enormous crocodile, she employed some laborers to dig out the blocks of stone in which they were petrified. These remains, placed together, proved to be the skeleton of a marine monster about thirty feet in length, with jaws nearly six feet feet long. This was a "fish-lizard," of which seven or eight different species may be seen in the British Museum.

Many have, no doubt, often been amused and perhaps frightened in their childhood, by the descriptions of dragons and other fearful monsters, which found their way, we are sorry to say, much too frequently in story-books; but one could not imagine a more dreadful-looking creature than this huge sea-lizard, with a head like a crocodile, and the jaws provided with a great number of immense teeth (in some cases one hundred and eighty); the eyes in volume exceeding the size of a human head, and so constructed as to afford a wonderful magnifying power in tracing their prey through the darkness and depth of the ocean.

The body was like that of a fish, with a broad and long tail, and four paddles (instead of feet, like those of the lizard or crocodile), similar to those of the whale tribe, enabling the animal to move, as *they* do, with rapidity through the water; and with such a construction of the breast-arch as enabled it to descend to the bottom of the sea in search of food, which consisted of fishes and reptiles, the remains of which have been discovered with the bones of the animal, thus showing upon what it subsisted. When we discover in the body of a fish-lizard the food which it has engulfed an instant before its death—when the intervals between its sides present themselves still filled with the remains of fishes which it had swallowed some ten thousand years ago, or at a time even twice as great—all these immense intervals vanish, and we find ourselves, so to speak, thrown into immediate contact with events which took place in epochs immeasurably distant, as if we occupied ourselves with the affairs of the previous day.

Another huge fossil marine animal, the *Plesiosaurus*, a species of the fish-lizard, somewhat allied in its structure to the animal just mentioned, was first discovered at Lyme Regis, about the year 1823. This was a most extraordinary

MONSTERS OF THE OCEAN BEFORE THE FLOOD.

creature, with the head of a lizard, the teeth of a crocodile, a neck of enormous length, resembling, on a very enlarged scale, that of a swan, the ribs of a chameleon, a body rounded like that of a great marine turtle, a tail shorter, in comparison with the length of the body, than the fish-lizard, acting the part of a rudder in directing the course of the animal through the water.

That the habits of this huge reptile were aquatic, is evident from the form of its paddles; that it was marine is equally so from the remains with which it is universally associated; that it may have occasionally visited the shore, the resemblance of its extremities to those of the turtle may lead us to conjecture: its motion, however, must have been awkward on land, and its long neck may have impeded its progress through the water. May it not, therefore, be concluded that it swam upon or near the surface, arching back its long neck like the swan, and occasionally darting it down on the fish which happened to float within its reach? It may perhaps have lurked in shoal water along the coast, concealed among the sea-weed, and raising its nostrils to a level with the surface from a considerable depth, may have found a secure retreat from the assaults of dangerous enemies.

The Plesiosaurus was scarcely as large as the "sea-lizard;" some of the species, however, measure from eighteen to twenty feet.

The two animals just described do not seem to have had, as far as can be judged, a scaly covering; but another monstrous reptile of the primitive seas, the *Teleosaurus*—a kind of fossil crocodile inhabiting the seas and rivers of the Old World, the great pirate of the ocean—was armed to the teeth, and clothed with an impenetrable coat of mail both on the back and stomach. This fearful animai was thirty feet in length, the head measuring from three to four feet, with enormous jaws, well defended beyond the ears, sometimes with an opening of six feet, through which they could swallow animals the size of an ox.

These animals were, from their enormous size and voracity, the terror of the primitive seas. After them we have the *Mœsasaurus*, a creature whose remains were first discovered at Maestricht, on the banks of the Meuse, in 1780. This occurred when the knowledge of these ancient prodigies was still in its infancy. One saw in it the head of a crocodile, another that of a whale. Among those interested in the discovery of these ancient vestiges of creation was an officer of the garrison at Maestricht, named Drouin. He purchased the bones of the animal as the workmen disengaged them from the rock, and formed a collection of fossil rarities at Maestricht, which excited great curiosity. The head, which exceeded six feet in length, was sent to France and placed in the Museum of Natural History at Paris.

To show the wonderful anatomical knowledge of scientific men in modern times, we may mention that Cuvier was able to ascertain the character of the entire skeleton of this huge animal from the examination of the jaws and teeth alone, and even from a single tooth! This told a history in itself: being without a root, not hollow, as in the crocodile, but solid throughout, and joined to the sockets by a broad bony basis, it became an instrument of enormous strength, and proved how formidable the animal must have been. It had sufficient velocity to overtake and capture fishes of immense size, with which the ancient seas abounded. In length it was about twenty-five feet; the tail was flattened on each side, but high and deep, forming a straight oar of great strength to propel the body.

CHAPTER XXII.

SUBMARINE SCENERY.

" The water is calm and still below,
 For the winds and waves are absent there,
And the sands are bright as the stars that glow
 In the motionless fields of upper air.
There, with its waving blades of green,
 The sea-flag streams through the silent water,
And the crimson leaf of the dulse is seen
 To blush like a banner bath'd in slaughter !"

PERCIVAL.

AS will be seen in the following pages, the earth has its counterpart in the ocean, and this not only as respects the physical conformation or the portions of water having the same form as parts of the land, but what is more signficant, the sea is full of breathing life, beautiful flowers and magnificent forests. In the tropical regions on the earth we have most luxuriant vegetation, while the polar regions are nearly barren. The order seems reversed in the ocean; for in the warmer waters of the Indian ocean and China seas, there is no vegetation, but unnumbered varieties of beautiful fishes and aquatic animals. In the polar waters we have countless varieties of beautiful flora, growing and flourishing in luxuriant wantonness.

We do not purpose to try to account for this apparently strange order of arrangement, but shall merely endeavor to describe it as it is, beginning with the submarine animal life of the tropical waters.

It is in the warm sea regions that the glory of submarine scenery is developed, the great transparency of the water in various places affording an ample view of the magnificent objects which gem the ocean's depths. The poet Moore, writing of the Bahamas (the earliest discovery of Columbus), a chain of islands in the Atlantic, remarks on the singular clearness of the water, so that the rocks are seen to a very great depth. "As we entered the harbor," he observes, "they appear so near to us that it seemed impossible to avoid striking on them." Addressing the Marchioness of Donegal, he says:

> "Believe me, lady, when the Zephyrs bland
> Floated our bark to this enchanted land—
> These leafy isles, upon the ocean thrown
> *Like studs of emerald o'er a silver zone*—
> Never did weary bark more sweetly glide,
> Or rest its anchor in a lovelier tide."

Dr. Collingwood describes a scene of marvelous submarine beauty in the China seas. He speaks of Fiery Cross Reef on a day when the sea was so calm that the ship's anchor could be distinctly seen sixty or seventy feet from the surface. Rowing over a two-fathom patch, he allowed the boat to drift slowly, and gazed on the sea treasures beneath him.

"Glorious masses of living coral strewed the bottom; immense globular madrepores (zoophytes); vast overhanging mushroom-shaped expansions; complicated ramifications of interweaving branches, mingled with smaller and more delicate species, round, finger-shaped, horn-like, and umbrella forms, lay in a wondrous confusion, and these were painted in every shade of delicate and brilliant coloring—grass-green, deep blue, bright yellow, pure white, rich buff, and more sober brown; altogether forming a kaleidioscope effect of form and color unequaled by anything I ever beheld. Here and there was a large clam-shell, wedged in between masses of coral, the gaping zigzag mouth covered with a projecting mantle of the deepest Prussian blue; beds of dark purpled, long spined sea-urchins, and the thick black bodies of sea-cucumbers varied the aspect of the sea bottom. In and out of these coral groves, like gorgeous birds in forest trees, swam the most beautifully colored and grotesque fishes—some of an intense blue, others bright red, yellow, black, salmon colored, and every hue of the rainbow, curiously barred, and bound, and bearded."

All the deepest colors we are acquainted with are those of hot climates, and all the lightest those of cold ones. The brilliant color of fishes, shells, and sea-weeds of the tropics, and especially of the Indian and Caribbean Seas, are spoken of with admiration by every navigator.

Wondrously beautiful is the fish, the imperial *Chœtodon*, spiny-finned fishes inhabiting the southern seas of China. The singular splendor of this animal will give an idea of the the generic name of a family of marvels that exist in the bosom of the deep. Its body is deep blue, marked all over by about thirty-two narrow bands of orange-yellow. The pectoral fins are black, and the entire tail a bright yellow. It is rather a large fish of its kind, sometimes attaining the length of fifteen inches. The tribe to which this fish belongs seems to have been particularly favored, for Nature seems to have bestowed her brighest ornaments on them with a most lavish hand.

We may remark here, though—however, diverging somewhat from our present subject—that the glorious beauty of these and other inhabitants of the warm seas, while it pleases the eye and excites admiration, has one drawback, and a very important one too. For the nourishment of man they are not to be compared to the far less showy but more wholesome fishes of the colder waters, which produce the species best suited for food, and very far superior in flavor. Maury states from his own knowledge, that seamen, even after long voyages, prefer their salt beef and pork to a mess of fish, resplendant with all the hues of the rainbow, caught in the warm seas; reminding us of what the poet so aptly says:

> "It is the flavor forms the test of merit,
> Which, when with wholesome qualities combined
> Forms the intrinsic value of all food.
> If mere exterior is to claim the palm,
> Then must the woodcock to the parrot yield,
> The spotted leopard supersede the deer,
> And dories to the blue-striped wrasse give place."

To the eye of the experienced naturalist, how many, varied, and beautiful are the forms which meet his gaze in the transparent depths of the ocean! Dr. Collingwood describes a magnificent spectacle which he witnessed and declares to be truly a wonder of the deep. This consisted of five or six large gelatinous bodies, forming an oblique line, each one of a bright and delicate green color, and with a large rich ruby spot, which shone in the water like carbuncles. Another consisted of a long and delicate chain, which might be compared to a necklace of diamonds set with brilliant rubies, the whole waving gracefully in the currents of the water.

Among these marine gems of the "purest water" which add such splendor to the submarine scenery of the tropics, we may mention, also, the ruby-colored *Etelis*, a fish allied to the perch tribe—though differing, from possessing strong and long teeth—so named from its color, which Cuvier compared to the tints of the ruby. The eye of this splendid fish is a conspicuous object, and of a golden orange. The color of the etelis is bright ruby-red, relieved by stripes of bright golden yellow, which run along the ridges of the scales. But there are numbers of such glorious fishes—shoals

> "Of fish that with their fins and shining scales

SUBMARINE SCENERY IN INDIAN OCEAN.

Glide under the green wave, in sculls that oft
Bank the mid-sea;
Or, sporting with quick glance,
Show to the sun their waved coats dropp'd with gold,
Or in the pearly shells at ease attend
Moist nutriment, or under rocks their food
In painted armor watch."

The Indian Ocean, one of the five grand divisions of the universal ocean, is especially rich in its submarine scenery.

We dive into the liquid crystal of its waters, and it opens to us the most wondrous enchantments of the fairy tales of our childhood's dreams. The strangely branching thickets bear living flowers. Dense masses of *Meandrinus* (a genus of polyps), and *Astreas* ("a star;" animalculæ which form coral), contrast with the leafy cup-shaped expansions of the *Explanarius*, the variously ramified *Madrepores*, which are now spread out like fingers, now rise in trunk-like branches, and now display the most elegant array of interlacing branches. The coloring surpasses everything: vivid green alternates with brown or yellow; rich tints of purple, from pale red-brown to the deepest blue. Brilliant rosy, yellow, or peach-colored *Nullipores* overgrow the decaying masses, and are themselves interwoven with the pearl-colored plates of the *Retipores*, resembling the most delicate ivory carvings. Close by wave the yellow and lilac fans, perforated like trellis-work, of the *Gorgonius*. The clear sand of the bottom is covered with the thousand strange forms and tints of the sea-urchins and star-fishes. The leaf-like *Flustras* and *Escharas* adhere like mosses and lichens to the branches of the corals; the yellow, green, and purple-striped limpets cling like monstrous cochineal insects upon their trunks. Like gigantic cactus-blossoms, sparkling in the most ardent colors, the *Sea-Anemones* expand their crowns of tentacles upon the broken rocks, or more modestly embelish the flat bottom, looking like beds of variegated ranunculuses. Around the blossoms of the coral shrubs play the humming-birds of the ocean—little fish sparkling with red or blue metallic lustre, or gleaming in golden green, or in the brightest silvery tints.

Softly, like spirits of the deep, the delicate milk-white or bluish bells of the jelly-fishes float through this charmed world. Here the gleaming violet and gold-green Isabelle, and the flaming yellow, black, and vermilion-striped coquette chase their prey; there the band-fish shoots snake-like through the thicket, like a long silver ribbon, glittering with rosy and azure hues. Then comes the fabulous cuttle-fish, decked in all colors of the rainbow, but marked by no definite outline; appearing and disappearing, inter-crossing, joining company and parting again, in most fantastic ways; and all this in the most rapid change, and amidst the most wonderful play of light and shade, altered by every breath of wind and every slight curling of the surface of the ocean. When day declines, and the shades of night lay hold upon the deep, the fantastic garden is lighted up with new splendor. Millions of glowing sparks, little microscopic medusas and crustaceans, dance like glowworms through the gloom. The sea-feather, which by daylight is vermilion-colored, waves in a greenish phosphorescent light. Every corner of it is lustrous. Parts which by day were dull and brown, and retreated from the sight amidst the universal brilliancy of color, are now radiant in the most wonderful play of green, yellow, and red light; and to complete the wonders of the enchanted night, the silver disc, six feet across, of the moon-fish, moves, slightly luminous, among the crowd of little sparkling stars.

How like a dream of romance and fairy beauty is this vivid description of submarine scenery in the tropics! What exquisite loveliness exists in those still, transparent waters! far exceeding in richness and coloring the most attractive objects that meet the eye on land. And while only a very small portion of these ocean wonders are unfolded to human gaze, what vast and countless glories are hidden in the great ocean depths to all save Him.

The most luxuriant vegetation of a tropical landscape cannot unfold as great wealth of form, while in the variety and splendor of color it would stand far behind this garden landscape, which is strangely composed exclusively of animals, and not of plants; for, characteristic as the luxuriant development of vegetation of the temperate zones is of the sea bottom, the fullness and multiplicity of the marine Fauna is just as prominent in the regions of the tropics. Whatever is beautiful, wondrous, or uncommon in the great classes of fish and *Echinoderus* (animals which include the sea-urchin and star-fish), jelly-fishes and polyps, and the mollusks of all kinds, is crowded into the warm and crystal waters of the tropical ocean—rests in the white sands, clothes the rough cliffs, clings, where the room is already occupied, like a parasite, upon the first comers, or swims through the shallows and depths of the elements; while the mass of the vegetation is of a far inferior magnitude.

The Asterias, or star-fishes, so frequently alluded to in the description of submarine scenery by naturalists, belong to a genus of molluscous worms, and some species that have been often observed on the sea-shore. The most curious of the sea-stars, perhaps, is that called Caput Medusæ, or basket-fish, which inhabits most seas, and consists of five central rays, each of which divides into two smaller ones, and these are again divided into two others; the same kind of division and subdivision being continued to a vast extent, and every ray regularly decreasing in size, until at length the ramifications amount to many thousands, forming a beautiful network spread over the water. The color of the worm varies, being sometimes pale, sometimes reddish, white and brown. The arms of the star-fishes are furnished on their lower surfaces with suckers, which enable them to crawl along the smoothest rocks.

The madrepores, millipores, and nullipores are polypi, classed in the third family of the Coralliferi, including all the numerous species which were for a long time regarded as marine plants, and in which numerous individuals are so united as to form compound animals, for the most part fixed, like plants, by a branched stem, or by simple expansions of a solid substance at the base or in the middle of the group.

No more lovely ornaments of submarine gardens could be imagined than the Anemones, a name thus applied about a century ago by the indefatigable naturalist, Ellis, who made them the subject of some remarkable investigations, and who remarks that "their tentacles being disposed of in regular circles, and tinged with a variety of bright lively colors, very nearly represent the beautiful petals of some of our most elegantly fringed and radiated flowers, such as the carnation, marigold, and anemone," reminding us of what Du Bartas says, in his quaint poem on the birth of the world, that seas have

> "Pinks, gilliflowers, mushrooms, and many millions
> Of other plants."

The reader may have seen some of the smaller species of anemones on the rocks of our sea-coast and in aquariums, but to observe these animals in their full bloom of loveliness, we must gaze into the transparent waters of the tropical seas, where they attain their greatest size and beauty, spreading out their delicate tentacles or "feelers," and displaying all the vivid colors which render them so remarkable. The similarity of some of these animal "flowers" to the flora of the earth is very singular. Hughes describes some of them as found in a submarine rock-basin:

"In the middle of it there is a fixed stone or rock which is always under water. Round its sides, at different depths, seldom exceeding eighteen inches, are seen at all times of the year, issuing out of little holes, certain substances that have the appearance of fine radiated flowers, of a pale yellow or a bright straw-color, slightly tinged with green, having a circular border of thick-set petals, about the size of, and much resembling, those of a single garden marigold, except that the whole of this seeming flower is narrower at the setting on of the leaves than any flower of that kind."

> "Each following billow lifted the last foam
> That trembled on the sand with rainbow hues;
> The living flower that, rooted to the rock,
> Late from the thinner element,
> Shrank down within its purple stem to sleep,
> Now feels the water, and again
> Awakening, blossoms out
> All its green anther necks."

This reads like a gardener's description of some new and rare plants.

But the elegance and beauty of the anemones belong only to their native element; when left dry by the receding tide, they contract into a jelly-like mass, and the glorious hues that shone through the clear waters of the ocean fade away.

"I once cut off," adds Mr. Hughes, "with a knife which I held for a long time out of sight near the mouth of a hole out of which one of these animals appeared, two of the seeming leaves. These, when out of the water, retained their shape and color; but being composed of a membrane-like substance, surprisingly thin, it soon shrivelled up and decayed."

Each species generally selects a peculiar haunt, but they are found in every sea. Some appear suspended from the vaults of submarine reefs; others cover the more exposed sides of rocks with a sort of flower-like tapestry.

One species commonly found on the English coasts, and a gem of the aquarium, is named *Mesembryanthemum*, after the fig-marigold, an annual of English flower gardens. If you look attentively at one of these animals, you will be struck with its remarkable beauty. Around the margin of the mouth there is a circle of little azure knobs or knots, like turquoise beads. Another British anemone, called the *Crassicornis*, exhibits the most attractive colors—red, varied with white, orange, green and yellow.

The term applied by naturalists to these very interesting "animal flowers" is *Actinia* ("a ray"), subdivided into a number of genera, and is now the type of a family called *Actiniadæ*.

The sea-anemones are a hungry class, preying especially on small crabs, which they clasp in a fond embrace, and eventually devour. Another peculiarity in these strange and beautiful marine animals is their power of reproducing organs of their own bodies that may have been broken off. Mr. Bennett relates:

"I had once brought to me a specimen of the crassicornis, that might have been originally two inches in diameter, and that had, somehow, contrived to swallow a valve of *Pecten Maximus* (a genus of two-shelled mollusks) of the size of an ordinary saucer. The shell, fixed within the stomach, was so placed as to divide it completely into two halves, so that the body, stretched tightly over, had become thin and flattened like a pancake. All communication between the interior portion of the stomach and the mouth was of course prevented; yet, instead of emaciating and dying of an atrophy, the animal had availed itself of what had undoubtedly been a very untoward accident, to increase its enjoyments and its chances of double fare. A new mouth furnished with two rows of numerous tentacula, was opened up at what had been the base, and led to the under-stomach: the individual had become a sort of Siamese twin, but with greater intimacy and extent in its unions."

The anemones I need scarcely further describe to you, except as soft fleshy bodies, with mouths surrounded by several rows of tentacles or feelers, which they expand, contract, and move at will with wonderful ease, shrinking when touched into a solid round mass with a slippery surface, which renders it difficult to remove them without injury.

The singular clearness of the waters of the Red Sea has often been noticed by travelers, as presenting views of submarine scenery of the greatest beauty.

It has been said that, when leaning over the edge of a boat on the smooth surface of the sea, the pebbles and the pure white sand can be seen very distinctly at the depth of even one hundred and eighty feet. Through the body of the water can be discerned the minutest objects at an immense depth. The secrets of the deep thus laid open afford the most magnificent spectacle that can be conceived. In one part the whole forests of pale pink and red coral spread forth their luxuriant branches, and impart a blush to the element in which they grow. How varied, how beautiful is their coloring! a brilliant red or blue, or gorgeous with orange or the deepest black. In one spot they are of a dead white or living purple, in another a bright yellow or crimson, and everywhere fancifully diversified.

It is in the Red Sea also that the strange family to which the Sea-Slug and the Sea-Cucumber belong are found in great abundance—many of the species exhibiting splendid colors, and making the bottom of the sea, particularly among coral formations, gay and lovely as a garden. Here also are seen the zoophyte *Gorgonias*, the stem of which is usually brown or black, while the fleshy parts often exhibit colors of great brilliancy. One species, the Sea-Fan, is brought to us as a curiosity. Here are also the *Serpula* a species of ringed animals, like worms, inhabiting a limy tube like that of mollusks, which they attach to rocks and shells in the sea. The wide end of the tube is open, and from this the animal protrudes its head and gills, which expand as beautiful fan-like tufts. They are generally most splendidly colored. And here also are the *Sertularia*, a genus of zoophytes attached to stones, shells, and sea-weeds, and very beautiful.

DRAG NET.

The waters of the North Sea, along the west coast of the Scandinavian peninsula, have been remarked by all observers for being of an extraordinary transparency. "As we passed" (says a writer) "slowly over the surface, the bottom, which here was in general a white sand, was clearly visible with its minutest objects, where the depth was from one hundred and twenty to one hundred and fifty feet. During the whole course of the tour I made, nothing appeared to me so extraordinary as the inmost recesses of the deep thus unveiled to the eye. The surface of the ocean was unruffled by the slightest breeze, and the gentle splashing of the oars scarcely disturbed it. Hanging over the gunwale of the boat, with wonder and delight I gazed on the slowly moving scene below. Where the bottom was sandy, the different kinds of asteriæ, echini, and even the smallest shells, appeared at that great depth conspicuous to the eye, and the water seemed in some measure to have the effect of a magnifier, by enlarging the objects like a telescope, and bringing them seemingly nearer. Now, creeping along, we saw, far beneath, the rugged sides of a mountain, rising toward our boat, the base of which, perhaps, was hidden some miles in the great deep below. Though moving on a level surface, it seemed almost as if we were ascending the height under us; and when we passed over its summit, which rose in appearance to within a few feet of our boat, and came again to the descent, which on this side was suddenly perpendicular, and overlooking a watery gulf, as we pushed gently over the last point of it, it seemed almost as if we had thrown ourselves down this precipice, the illusion, from the crystal clearness of the deep, actually producing a sudden start. Now we came again to a plain, and passed slowly over the submarine forests and meadows which appeared in the expanse below; inhabited, doubtless, by thousands of animals, to which they afford both food and shelter—animals unknown to man; and I could sometimes observe large fishes of singular shape gliding softly through the watery thickets, unconscious of what was moving above them. As we proceeded the bottom became no longer visible; its fairy scenes gradually faded to the view, and were lost in the dark green depths of the ocean."

Having briefly described some of the beautiful fishes and animals of the transparent tropic waters, let us consider the vegetation and submarine forests of the arctic waters.

Columbus, in his search for a new world, encountered one of the greatest marvels of ocean vegetation—a garden of enormous extent in the waste of waters, which perplexed and terrified his timid seamen. "When," as Robertson relates, "about four hundred leagues to the west of the Canaries, he found the sea so covered with weeds, that it resembled a meadow of vast extent, and in some places they were so thick as to retard the motions of the vessel. This strange appearance occasioned new alarm and disquiet to the sailors." They imagined that they had now arrived at the utmost boundary of the navigable ocean; that these floating weeds would obstruct their further progress, and concealed dangerous rocks, or some large tract of land that had sunk, they knew not how, in that place. Columbus endeavored to persuade them that what alarmed ought rather to encourage them, as it was a sign of their approaching land. At the same time a brisk gale arose, and carried them forward; several land-birds were seen hovering about the ship, and directed their flight towards the west; a whale, also, was seen heaving up his huge form in the distance, which Columbus affirmed was a favorable indication of the neighborhood of land. The desponding crew resumed some degree of spirit, and began to entertain fresh hopes.

The marine vegetation that threatened to impede the course of the adventurous Columbus was the Gulf-weed, so termed from its great abundance in the Gulf of Mexico. The Portuguese called the waters thus covered the Grassy Sea, for the surface, during several days' sailing, is literally carpeted with the weed. Here the beautiful fishes of the warmer latitudes,

"with fry innumerable, swarm,"

and find a refuge from their relentless pursuers in the ocean; and the whole mass, extending many miles in space, affords food and shelter to an infinity of small marine animals.

In the Atlantic Ocean these sea-weeds cover an expanse of two hundred and sixty thousand square miles, a vast mass of vegetable matter that no other similarly furnished tract of open water is known to produce.

These sea-weeds are occasionally thrown up by currents on our own shores, and may be known by the cluster of air-vessels that the Gulf-weed bears, and which, from their appearance, have given them the name of tropical grapes.

How marvelous is this vast provision of Nature in the ocean depths for the wants and nourishment of animal life, all created for wise purposes by the Great Being

> "Who sleeps not—is not weary; in whose designs
> No flaw deforms, no difficulty thwarts,
> And whose beneficence no change exhausts!"

Innumerable animalculæ (visible or invisible to the naked eye), the chief food of the whale and also of many species of fish eaten by man, derive their sustenance from sea-weeds. Myriads upon myriads of eggs of fishes find security in this tangled mass of sea-plants, and the young fish are sheltered there until they acquire strength to commit themselves to the water. It has been remarked that "the vegetable kingdom in the sea is no barren spot in the garden of Nature, but in usefulness and abundance it is not inferior to the most favored spots on land." But the character of sea-weeds is very different to land-plants: the former, supported in a liquid of greater specific gravity than themselves, do not require the woody fibres which are necessary for land-plants, except such as support themselves by climbing; and, as they derive their nourishment from the water which covers them, they do not need the continuous vessels which are so necessary to land-plants for their growth and life. This is explained by the simple experiment of placing one portion of sea-weed in water, and exposing the other part to the air, when the latter will speedily dry and wither, while the former retains its freshness.

Again, the trees, and flowers, and shrubs which adorn

our gardens require the bright beams of the sun to warm them into life and beauty; but the plants that thrive in the depths of the ocean are not dependent for their existence on light, for only a feeble ray can reach many of them in their rocky homes far beneath the surface.

Humboldt mentions the fact of a sea-weed of a fine grass-green color being brought up from a depth of one hundred and ninety-two feet, where it had vegetated, though the light that had reached it could not have been more than that afforded by half the light of an ordinary candle.

Who can conceive the mighty operations of the All-Powerful Creator in the depths of the ocean? What transcendent wonders lie hidden in the waste waters! Let us imagine to ourselves vast submarine forests, which we know to exist —an almost boundless extent of vegetation, which lives, thrives, and decays, unseen by mortal eyes—how insignificant is human comprehension!

Viewing these tribes of sea-weed in the most careless way, as a system of subaqueous vegetation, we see the depths of ocean shadowed with submarine groves, often of vast extent, intermixed with meadows, as it were, of the most lively hues, while the trunks of the larger species, like the great trees of the tropics, are loaded with innumerable minute kinds, as fine as silk, or transparent as a membrane.

How singular the contrast, also, between the gigantic weeds that line the ocean depths, and spread forth their knotted shoots upon the surface of the water, and the small, beautifully colored, delicate plants that cling to the rocks, and rival in lovliness the choicest flowers of our gardens!

On the shores of the North Pacific may be seen the *Nereocystus*, with a slender stem, upwards of three hundred feet long, bearing at its extremity a large air-vessel six or seven feet in length, shaped as a barrel, and crowned with a tuft of upwards of fifty forked. leaves, each thirty to forty feet long, forming the fishing-grounds of the sea-otter, who can seek his prey with greater certainty amidst the shade of the enormous leaves.

In the Antartic regions the growth of sea-weeds is remarkable. The Tree sea-weed rises from the ocean with a huge stem or trunk eight or ten feet in height, and the thickness of a human thigh. The ends of the branches give out leaves two or three inches broad, which, when in the water, hang down like the boughs of a willow. Thousands of these aquatic trees, uprooted by the currents, are often mistaken for driftwood, and are collected for fuel. Darwin mentions some sea-weeds that grow on the rocks in the Arctic seas, which, though of prodigious length, instead of being spread along the bottom of the ocean, are in part floated on the surface by means of the numerous air-vessels they contain. These gigantic sea-plants are sometimes fifteen hundred feet in length. So full of air-vessels are they, that they look like a honeycomb.

One species of sea-weed in the Antarctic regions, in its horizontal growth at the surface of the ocean, ranges between two hundred and seven hundred feet in length; and that at the Falkland Islands the beach is lined for miles with entangled cables of this weed, much thicker than the human body.

Opposed to these gigantic marine plants we have multitudes of smaller growth, combining the most delicate, beautiful, and curious characteristics of form and color, among which we may mention the Water Flannel, which waves backwards and forwards like the pendulum of a clock; and some of us no doubt have often, when at the sea-side, enjoyed the fun of cracking the air-vessels of the Bladder-weed, and pulled to pieces the thready weeds that children call sea-silk. The sea-coasts present an exhaustless variety of pleasure derived from sea-weeds. There one may find the Whip-lash, which grows from thirty to forty feet in length, and is used for fish-lines in Scotland; the Net-weed, which spreads its delicate interlaced threads like a web in the water; the feathery *Callithamnion*, one of the most lovely of sea-weeds, of a bright, fine, rosy-red color; the branches divided like the teeth of a comb. Then there is the Fern-leaf sea-weed, another attractive plant, and the splendid Fan-weed, representing a collection of hundreds of beautiful little fans, every one of which, if minutely examined, is of exquisite workmanship. On the southern coast of England is found a common shore-plant of the tropical seas, the Peacock's Tail, another lovely sea-weed. When growing, the fronds are rolled up into cups, while the delicate fibres with which they are bordered reflect the most glorious tints. Then there are the curious Sea-thongs or Girdles, which are often seen on the coast, and which, when taken out of the water and held by the stem, resemble a flag-staff and streamers.

The varieties of form and substance in sea-plants are also highly interesting subjects for comtemplation: some are like masses of jelly, others are elastic like India-rubber; many are tough as leather, others firm as wood; some have delicate transparent leaves, others have thick, finely-veined, or nerveless leaves.

The plants of the ocean gardens can vie also in glowing tints with many of our most attractive land-flowers.

The natural colors of many sea-plants are exceedingly beautiful when viewed in their native element; but exposure to the sun and air—unless they are preserved with the greatest care and delicacy—causes them to fade. Those of the red species, which abound chiefly in the temperate zone, acquire their richest tints in the deepest water. The plants of an olive color are mostly found in the neighborhood of the tropics, while the green species principally inhabit the polar seas. But, besides the colors mentioned, there are countless of other shades.

Having alluded to the beauty and richness of ocean vegetation, we will now mention its usefulness, in addition to the shelter and nourishment it affords to the inhabitants of the deep—

> "Invisible,
> Amid the floating verdure millions stray."

They soften the currents of rolling waters, and lessen the violence with which the waves would, otherwise, break upon the shores of the land.

Darwin alludes to the value of sea-weeds to those who traverse the ocean: "I believe, during the voyage of the *Beagle* and *Adventure* not one rock near the surface was discovered which was not buoyed by this floating weed. The good services it thus affoads to vessels navigating near this stormy land (Terre del Fuego) is evident, and has certainly saved many from being wrecked."

And now let us consider the use of sea-weeds for food. The value of these in many parts of the world is very great; the Chinese, especially are the largest consumers of any nation, and have various ingenious methods of preparing them for the table. Ceylon Moss, formerly much esteemed, is the product of an esculent sea-weed gathered on the western coast of Ceylon, and possesed many nutritious qualities. Carrageen Moss is a sea-weed much used for food in Ireland; it is also frequently employed instead of isinglass for making soups and jellies. In Bavaria it serves for clarifying beer. The young stalks of the Tangle-weeds, when well-boiled and served up with pepper and vinegar, are very wholesome. One species grows to the length of twenty feet.

There may be found on the sea-shore a pretty weed resembling in shape the palm of a hand, with leaves, like fingers growing around it. This is properly called Dulse, and is eaten both raw and roasted, the taste resembling that of cooked oysters. This is also a favorite food of lobsters, crabs, and other shell fish. The Icelanders have a particular relish for this sea-weed, and prepare it by drying, when it gives out a white powdery substance, which is sweet and palatable. Cattle are also very fond of dulse, especially sheep, for which reason it is often called sheep's dulse. These animals seek it eagerly on the sea-shore, and are sometimes carried away by the tide in their eagerness to obtain it. In Kamtchatka it is used for making a fermented beverage.

The marine vegetable called Laver, so much esteemed in various parts of England as a relish for the table, is a species of sea-weed, stewed and served as a sauce.

The uses to which sea-weeds are applied are, indeed, numerous and important. We will merely mention a few. The ashes of marine plants afford a large quantity of soda salts, and especially the carbonate, such as Kelp which is prepared by merely burning certain species of weeds suitable for the purpose, and this was formerly in great demand for the manufacture of glass, but now there is a better and cheaper means of getting soda from salt.

According to Pliny, the value of soda in making glass was discovered by a mere accident. A vessel loaded with soda was once driven ashore on the coast of Palestine. The crew landed, and made a fire upon the sands to boil their kettle. They took some lumps of the soda for the kettle to rest upon, without the least idea of what would result. The soda was melted, and, uniting with the sand, formed a rough kind of glass.

But kelp, although superseded in this respect, is valuable from the circumstance that iodine (discovered in 1811, by Courtois, in the waste liquors produced in the manufacture of carbonate of soda from the ashes of sea-weeds), which is so necessary in medicine, in photography, and various processes connected with the arts, is chiefly derived from it. Iodine exists in the waters of the ocean and mineral springs, marine shelly animals, and sea-vegetation generally, but not to the same extent as in kelp. When heated, iodine rises in a vapor of a violet color and this is condensed and solidified by a chemical process.

Iodine is found in large quantities in the sea-weeds which cover the rocks for miles round the west coast of Ireland. The average yield of British kelp is said to be ten thousand tons yearly, of the value of forty thousand pounds.

In some of the countries bordering on the Baltic, sea-weeds are used for packing materials and for stuffing articles. The Ulva Marina is extensively employed in our own country for the latter purpose. Attempts have been made to manufacture paper from sea-weeds; marine sugar is obtained from several species. The Chinese derive from them a gum for making their lanterns and transparencies, also a varnish, and a size for the manufacture of silk and paper.

CHAPTER XXIII.

THE BED OF THE OCEAN.—DEEP SEA SOUNDINGS.

"WE dive," says Schleiden, "into the liquid crystal of the Indian Ocean, and it opens to us the most wondrous enchantments, reminding us of fairy tales in childhood's dreams. The strangely branching thickets bear living flowers, the coloring, beauty, and variety of which surpasses everything. The clear sand of the bottom is covered with the thousand strange forms and tints of sea-urchins and star-fishes. The leaf-like flustras and escharus adhere like mosses and lichens to the branches of the corals; the yellow, green, and purple-striped limpets cling like monstrous cochineal insects upon their trunks. Like gigantic cactus-blossoms, sparkling in the most ardent colors, the sea-animones expand their crowns of tentacles upon the broken rocks, or more modestly embellish the flat bottom. Around the blossoms of coral shrubs play the humming-birds of the ocean, little fish sparkling with red or blue metallic glitter, or gleaming in in golden green, or in the brightest silvery lustre. Softly, like spirits of the deep, the delicate milk-white or bluish bells of the jelly-fishes float through this charming world. Here the gleaming violets and gold-green Isabelle, and the flaming yellow, black, and vermillion-striped coquette chose their prey; there the band-fish shoots, snake-like, through the thicket, like a long silver ribbon, glittering with rosy and azure hues. Then comes the fabulous cuttle-fish, decked in all colors of the rainbow, but marked by definite outlines, appearing and disappearing, intercrossing, joining company and parting again in most fantastic ways; and all this in the most rapid change, and amid the most wonderful plays of light and shade, altered by every breath of wind, and every slight curling surface of the ocean. When day declines, and the shades of night lay hold upon the deep, this fantastic garden is lighted up with a new splendor. Millions of glowing sparks, little microscopic medusas and crustaceans dance like glowworms through the gloom. The sea-feather, which by daylight is vermillion-colored, waves in a green phosphorescent light. Every corner of it is lustrous.

"Ports which by day were perhaps dull and brown, and retreated from the sight amid the universal brilliancy of color, are now radiant in the most wonderful play of light; and, to complete the wonders of the enchanted night, the silver disc, six feet across, of the moon-fish, moves, slightly luminous, among the cloud of little sparkling stars. The most luxuriant vegetation of a tropical landscape cannot unfold as great wealth of form, while in the variety and splendor of color it would stand far behind this garden landscape, which is strangely composed exclusively of animals, and not of plants; for, characteristic as the luxuriant development of vegetation of the temperate zones is of the sea bottom, the fullness and multiplicity of the marine Fauna is just as prominent in the regions of the tropics. Whatever is beautiful, wondrous, or uncommon in the great classes of fish and Echinoderms, jelly-fishes and polyps, and the mollusks of all kinds, is crowded into the warm and crystal waters of the tropical ocean—rests in the white sands, clothes the rough clifs, clings, when the room is already occupied, like a parasite, upon the first comers, or swims through the shallows and depths of the elements, while the mass of the vegetation is of a far inferior magnitude."

What is the average depths of the sea? The time has been when it was difficult to answer this question with any degree of certainty, because of the difficulty met with in taking soundings, caused chiefly by the deviations of submarine currents. When the "deep soundings" were first undertaken such a thing as the submarine current, or a system of oceanic circulation was unknown by navigators. As a consequence of this ignorance many marvelous soundings were reported; indeed it was supposed by some that the sea in many places was without bottom, as, apparently, the longest line could not reach it. But when it was definitely determined that there were many and deflecting under-currents running through the ocean, it was seen why the line would keep paying out indefinitely long after the weight attached might have reached the bottom. The question now arose as to how the difficulty was to be surmounted. Many and ingenious plans were tried; for it was judged, and rightly so, too, as the sequel proved, that it was well worth the time, means, and patience employed in arriving at the result. Fortunately we were living in an era of the world's history when the dungeon and stake were not the reward prescribed to discoverers and benefactors. Naval officers in the service of their respective governments were striving in honest emulation, as to whom should be the first to win the coveted prize. It remained for the United States Navy to solve the problem, and that, too, by a very simple contrivance, which will be best described by noting its operation. We will suppose a ship in the mid Atlantic. A small boat is put off from the vessel, containing the necessary crew and implements with which to take the soundings. As the little boat cannot be readily anchored, it is kept stationary by the use of the oars. In a convenient place in the boat lies coiled perhaps three miles of good stout rope. Attached to this rope is a sinker weighing sixty-four pounds. All is made ready, and it is finally cast overboard. It is carefully watched. By the shock felt it is decided that the weight has just reached the bottom. At this moment the line is cut. Now, how much of this line remains on the reel? One half a mile, we will say, by way of illustration. Since there were three miles of rope on the reel at first, and there remains one-half of a mile, the remainder, two and one-half miles, is in the water—the sea is two and one-half miles deep here.

By this simple method the depths of the sea can be determined. The plan may seem an easy and novel one, but it has been far-reaching in its results. It opened the way for a series of successful experiments that have been of great practical benefit, and the end is not yet. It will be our purpose to describe as fully as the limited space will allow, the various steps of discovery, dependent upon this first step, and the controversies to which they gave rise.

"If the waters of the Atlantic were drawn off, what a sight would be exposed to our view. The very ribs of the earth, with the foundations of the sea, would be brought to light, and we should have presented to us in one view, in the empty cradle of the ocean, a thousand fearful wrecks with that array of dead men's skulls, great anchors, heaps of pearls, and inestimable stones, which, in the poet's eye, lie scattered in the bottom of the sea, making it hideous with the sight of ugly death."

To lay down on our maps the mountains of the earth is regarded as an important thing. May it not be equally of as much importance to bring regularly within the domain of science the physical geography of the sea? By the aid of the deep sea soundings previously described, the greater part of the bed of the Atlantic has been carefully mapped out by the late Capt. Maury; it is to his persevering and well directed labors in this new field of discovery that we owe some of the conclusions recorded in this chapter.

We had now arrived at a period of anxiety and expectation, especially on the part of mariners and scientists. Simply to know that the sea was just so deep in this or that place was of but little importance standing alone. Exploration was investing the subject with an ever increasing interest. Better directed attempts began to be put forth. The question agitating thoughtful minds was, "of what is the bottom of the ocean composed?" And again, "Can we determine this, and if we do, of what importance will it be to the world?" The outcome of this agitation was the sounding apparatus invented by midshipman J. M. Brooks, U. S. N. With the aid of this instrument, specimens of the bottom of the ocean were brought up from a depth of four miles. One of the questions was now answered, and now succeeded another period of anxious expectation. The thoughtless might have asked: "Of what use is all this trouble just to procure a handful or two of mud, clay, sand, or of something else from the bottom of the sea?"

The law of gravitation has existed since time began. Was the discovery of it of any importance? What important impulses of activity and utility came into existence with this discovery? Every natural law discovered, at once becomes the basis of great reorganization and new invention. The philosophers asked: "Where should this discovery be ranked," or in other words, "what was to be the use of these deep-sea soundings?" These questions were difficult of answer categorically, but every physical fact, every expression of nature, the work of any and all of those agents which make the face of the world what it is, and as we see it, is interesting and instructive. Until a group of these facts are grasped, it cannot be determined what practical bearings they may have, though the thoughtful know that they contain many precious jewels which the experts of philosophy will not fail to bring out, polished and bright, and beautifully adapted to man's purposes.

The first practical answer to the question concerning the use of deep sea soundings was in an attempt to utilize the knowledge in the laying of the submarine cable across the Atlantic.

There is, at the bottom of this sea, between Cape Race, in Newfoundland, and Cape Clear, in Ireland, a remarkable steppe, now known as the *telegraph plateau.* In August, 1858, a cable was laid upon it from Valencia in Ireland, to Trinity Bay in Newfoundland, and a few messages were passed through it when it ceased to work. This cable has since become a perfect success, and many others have been constructed, as we shall have occasion to notice in another chapter. Probably the first, or 1858 cable would have been an immediate success, had its projectors better availed themselves of the light which the deep sea soundings cast upon the bed of the ocean. This will be better understood as we get further along in this subject.

It was upon this very plateau that Brooks' sounding apparatus brought up its first trophies from the bottom of the sea. These specimens were judged to be clay. They were carefully preserved and labeled and eventually sent to the proper bureau for microscopic examination. It was ascertained that all these specimens contained microscopic shells, calcareous and silicious, and that, too, unmixed with sand or gravel. Of all the specimens brought up, all were animal, and not one of the mineral or vegetable kingdom. The ocean teems with life we know, and of the four elements, fire, earth, air, or water, the sea most of all abounds with living creatures. These little mites of shells had not a particle of gravel or sand among them, and it may be inferred from this fact that the sea there at least is at *rest.* There was not motion enough to abrade these very delicate organisms, nor currents enough to sweep them about and mix up with them a grain of the finest sand, nor the smallest particle of gravel torn from the loose beds of debris that here and there strew the bottom of the sea.

This plateau is not too deep for the wire to sink down

and rest upon, yet is not so shallow that currents, or icebergs, or any abrading force can derange the wire after it is once lodged in its bed.

The analysis of the treasures recovered from the rich bottom of the sea show us that the roaring waves and the mightiest billows repose upon cushions of still waters; that in the "blue water" the skeleton of the earth is protected, as with a soft garment, from the abrading action of its currents, and that the lower stratum of its waters is so nearly at rest that it can neither disturb nor move the lightest bit of drift once lodged there.

"The tooth of running water is very sharp. See how the Hudson has eaten through the Highlands, and the Niagara cut its way through layer after layer of solid rock. But what are the Hudson and Niagara, with all the fresh water courses of the world, by the side of the Gulf Stream, and other great rivers in the ocean? And what is the pressure of fresh water upon river-beds in comparison with the pressure of ocean water upon the bed of the deep sea? It is not so great in contrast as the gutters in the street are to the cataracts. Then why have not the currents of the sea worn its bottom away? Simply because they are not permitted to get down to it."

A very simple calculation would show us the enormous weight of a simple foot square of water from the surface to the bottom. In round numbers such a quantity of water would weigh over one million pounds. Now let us multiply the number of solid feet indefinitely, and then take into consideration the fact that some of the ocean currents run at the rate of four miles an hour. Of what would the bottom of the sea have to be composed to stand such abrasion? The hardest adamant could stand no such friction. Why then has not the bottom of the sea been worn away? Why have not the currents cut through the solid crust in which its bottom is rocked, and ripped out from the bowels of the earth the masses of incandescent, molten matter which geologists tell us lie pent up and boiling there? If the currents of the sea, with their four-mile velocity at the surface, and this immense pressure in its depths, were permitted to chafe against its bed, the Atlantic, instead of being two miles deep and three thousand miles broad, would, we may imagine, have been long ago cut down into a narrow channel that might have been as the same ocean turned up on edge, and measuring two miles broad, and three thousand miles deep; but had it been so cut, the proportion of land water-surface would have been destroyed, and the winds, for lack of area to play upon, could not have sucked up from the sea vapors for the rains, and the face of the earth would have become as a desert without water.

Now there is a reason why such change should not take place, why the currents should not uproot nor scour the deep bed of the ocean, why they should not throw out of adjustment any physical arrangement whatever in the ocean; it is because that in the presence of everlasting wisdom a compass was set upon the face of the deep; because the waters were measured in the hollow of the Almighty hand; because bars and doors were set to stay the proud waves; and because, when He gave to the sea His decree that its waters should not pass, He laid the foundations of the world so fast that they should not be removed forever."

No indications of running water have been found beyond the depths of about three thousand feet. The various cable companies, in determining their routes, select such as will enable them to lay their cables below the lower of the dual currents, where they will be as free from disturbance as though they were resting upon dry land. Another interesting fact developed that may be noticed here, is that beyond a certain depth neither animal decomposition nor vegetable decay takes place. The most frail and delicate organisms of the sea can remain in its depths for an indefinite length of time without showing a single trace of decay. The entertainment of this fact suggests many beautiful fancies, some touching thoughts, and a few useful ideas; and among these last are found reasons for the conjecture that the gutta-percha or other insulating material in which the cables are encased, become, when lodged beyond a certain depth, impervious to the power of decay; that with the weight of the sea upon them, the destructive agents which are so busy in the air cannot here find room for play. Very curious it is that destruction and decay should be imprisoned and rendered inoperative at the bosom of the great deep!

How inadequately can we, with the limited space necessarily allotted to this chapter give with that detail which might be expected to interest the general reader, the many facts and fancies that ought to be recorded in dealing with so important a subject.

Thus far in our treatment of the subject we have described with some minuteness the successful determinations of the actual depths of the sea with the mapping of its bottom; the fact that below certain depths the sea was at *rest,* and the other important fact, that in the deepest ocean decay does not take place. It has also been pointed out that so soon as these discoveries were made known they were at once seized and practically utilized in the laying of submarine cables for telegraphic communication to all important parts of the hospitable globe accessible by water communication. Deep sea soundings are still being prosecuted in various parts of the world of waters, and of the laying of cables there seems no end. But, notwithstanding the prompt utilization of the discoveries already unfolded, there remained another important consideration in the successive order of the new determinations and that was whether or not there was *life in the depths of the sea.*

Astronomers had measured the volumes and weighed the masses of the most distant planets. The " ooze " and bottom of the ocean had remained a sealed volume, rich with ancient and eloquent legends, and suggestions of many an instructive lesson that might, when rightly read, be useful and profitable to mankind.

The sea, with its myths, has suggested attractive themes to all people and all ages. The human mind longed to know more of its wonders and understand its mysteries. There were, in its bosom, untold wonders; therefore the contemplative mariner, as in mid ocean he gazed upon its mirrored surface, continued to experience sentiments akin to those that fill the mind of the devout astronomer, when, in the stilly night, he looks at the stars, and wonders.

The specimens from the bottom of the sea brought up by Brooks' sounding apparatus, were carefully labeled, and a part was sent to Professor Ehrenberg, of Berlin, and the remainder to the late Professor Baily, of West Point, for microscopic examination. After such examination the latter decided that the animalculæ, whose remains Brooks had brought up from the bottom, probably did not live or die there. "They would have had no light there, and, had they lived there, their frail little texture would have been subjected, in its growth, to the pressure of a column of water twelve thousand feet high. They probably lived near the surface, where they could feel the genial influences of both light and heat, and were buried in the lichen coves below after death." Maury says, in this same connection: "Brooks' lead and the microscope, therefore, it would seem, are about to teach us to regard the ocean in a new light. Its bosom, which so teems with animal life, its face, upon which time writes no wrinkles, makes no impression, are, it would now seem, as obedient to the great law of change as is any department whatever, either of the animal or vegetable kingdom. It is now suggested that henceforward we should view the surface of the sea as a nursery teeming with nascent organisms, its depths as the cemetery for families of living creatures that outnumber the sands on the sea shore for multitude. Where there is a nursery, hard by there will also be found a grave-yard—such is the condition of the animal world. But it has never occurred to us before to consider the surface of the sea as one wide nursery, its every ripple a cradle, and its bosom one vast burial place."

Several years ago the French Academy sent out bottles, and caused specimens of air from various parts of the world to be collected and brought home to be analyzed. The nicest tests which the most skillful chemists could apply were incapable of detecting any, the slightest, difference as to ingredients in the specimens from either side of the equator, so thorough in the performance of their office are these agents. Nevertheless, there are a great many more demands on the atmosphere by the organic world in one hemisphere than in the other, and consequently a great many more inequalities for these agents to restore in one than in the other. Of the two, the land of our abiding place most abounds with life, and consequently here there is a heavier burden on the air. Here the hearthstone of the *human* family has been fixed. Here, with our fires in winter and our crops in summer, with our work-shops, steam engines, and fiery furnaces going night and day, with the limitless demands which the animal and vegetable kingdoms are making upon the air above us, we cannot detect any difference in the make-up of the atmosphere in different parts of the world; and yet, notwithstanding the perfect balances between the kingdoms of the organic world, there are causes at work in every hamlet which would produce a difference of adjustment were it not for these rising and falling columns of air; for the obedient wind; this ground system of circulation, these little cogs and ratchets which have been provided for its perfect working.

"On those parts of the solid portions of the earth's crust which are at the bottom of the atmosphere, various agents

DIVER AT WORK.

are at work, leveling both upward and downward. Heat and cold, rain and sunshine, the wind and the streams, all, assisted by the forces of gravitation, are unceasingly wasting away the high places on the land, and as perpetually filling up the low. But it would seem that these same leveling agents are powerless in the deep sea. There are no abrading processes at work there; neither frost nor rains are felt, and the force of gravitation there is so paralyzed that it cannot use half its power, as on dry land, in tearing the overhanging rock from the precipice and casting it down into the valley below. Though the same agencies are not at work in leveling off the floor of the sea, as are changing the face and form of the earth, let us not forget the myriads of anamalculæ that make the surface of the sea sparkle and glow with life: they are secreting from its surface solid matter for the seeming purpose of filling up these cavities below. These little marine insects build their habitations at the surface, and when they die, their remains, in vast multitudes, sink down and settle upon the bottom. They are the atoms of which the mountains are formed—plains spread out. Our marl-beds, the clay in our river bottoms, large portions of many of the great basins of the earth, are composed of the remains of just such little creatures as these, which the ingenuity of Brooks has enabled us to fish up from the depth of nearly four miles below the sea-level. These *Foraminifera*, therefore, when living, may have been preparing the ingredients for the fruitful soil of a land that some earthquake or upheaval, in ages far away in the future, may be sent to cast up from the bottom of the sea for man's use."

By bringing up specimens from the depths of the ocean, and studying them through the microscope, it has been ascertained that the bed of the ocean is lined with the remains of its own dead: with marine feculences which lie on the bottom as lightly as rests the gossamer in a calm at the bottom of the atmospheric ocean.

The streams of running water in the oceanic machinery play rather about its surface than in its depths; the causes that produce currents reside at and near the surface. These causes are changing heat and alternating cold with their powers of contraction and expansion—wind and sea-shells with evaporation and precipitation; and it would appear that none of these agents seem capable of reaching very far down into the depths of the great sea with their influences—apparently not much farther down than the light can reach.

"Does," says Maury, "any portion of the shells which Brooks' sounding-rod brings up from the bottom of the ocean live there, or are they all the remains of those that lived near the surface in the light and heat of the sun, and were buried in the bottom of the sea after death? Philosophers were divided in opinion upon this subject. The facts, as far as they went, seemed at first to favor the one conjecture nearly as well as the other. Under these circumstances, I inclined to the anti-biotic hypothesis, and chiefly because it would seem to conform better with the Mosaic account of creation. The sun and moon were set in the firmament before the waters were commanded to bring forth the living creature; and hence we infer that light and heat are necessary to the creation and preservation of marine life; and since the light and heat of the sun cannot reach to the bottom of the deep sea, my own conclusion, in the absence of positive evidence upon the subject, has been that the habitat of these mites of things hauled up from the bottom of the great deep is at and near the surface. On the contrary, others maintained, and perhaps with equal reason, the biotic side of the question. Professor Ehrenburg, of Berlin, is of this latter class." We refer to and quote thus extensively from Maury, as at the time this interesting and important discussion began, he was the great authority on all matters connected with the physical geography of the sea, having done as much, and perhaps more, than any other one person, in clearing up or elucidating the mysteries of the sea. Many of his conclusions, as novel and strange as they may have seemed, were at once accepted as indisputable facts, and passing time has proved their correctness. Additional research and investigation has shed new light upon some subjects that at that time (1858) may have been supposed to be settled, and this subject of life in the deep sea may be one of those questions. We shall present briefly both sides of the question, with the evidence adduced, and leave the reader to form his own conclusion.

As soon as the deep-sea specimens were mounted on the slides of the microscope, the two great masters of that instrument—Baily, of West Point, and Ehrenburg, of Berlin, discovered the greater part of the small calcareous *carapaces* to be filled with a soft pulp, which both admitted to be fleshy matter. From this fact the German argued that there is life at the bottom of the deep sea. Ehrenburg further contended that, owing to the great quantity of peculiar forms and of soft bodies existing in the innumerable carapaces, and also the large number of unknowns, *increasing with the depth,* he was firmly of the opinion that stationary life existed in the deep sea.

The anti-biotics, considering the number of unknowns increasing with the depths, contended that the tides, the currents, and the agitation of the waves all reach to the bottom in shallow water; that they sweep and scour from it the insects, and bear them off into the deep water; that after reaching a certain depth, this sediment passes into the stratum of quiet waters that underlie the roaring waves and tossing currents of the surface, and through this layer of quiet waters these organic remains slowly find their way to the final place of repose as ooze at the bottom of the deep sea; that, through such agencies, this same deposit ought to be richer than that of the shallow water with infusorial remains.

Concerning the soft, pulp or fleshy matter contained in these calcareous remains, the anti-biotics held that the little creatures were preserved for a while after death, and until they reached a certain depth, by salt and afterward by pressure. We have already referred to the fact established that, beyond a certain depth, animal decomposition or vegetable decay does not take place. This fact was brought forward as evidence. The point claimed for it was that the little animals need not have *lived* in the depths of the sea, in order that their remains might be brought up from the bottom, still containing the fleshy matter, but that owing to the antiseptic properties of sea water, this same fleshy matter might be preserved for a long time after death, almost in a perfect state of preservation. Another argument, as regards the necessity of light and heat in the preservation of animal life, we referred to in the beginning. Finally, many of the forms brought up from the deep sea, were recognized as coming from fresh water sources, forced down by the rivers into the sea where they were carried along by the ocean rivers and finally deposited into the sea.

The unabraded appearance of these shells, and the almost total absence among them of any detrius from the sea or foreign matter, suggests most forcibly the idea of perfect repose at the bottom of the deep sea. Some of the specimens are as pure and as free from the sand of the sea, as the snow flake that falls, when it is calm, upon the lea, is from the dust of the earth. Indeed, these soundings suggest the idea that the sea, like the snow-cloud with its flakes in a calm, is always letting fall upon its bed showers of these microscopic shells; and we may readily imagine that the "sunless wrecks," which strew its bottom, are, in the process of ages, hid under this fleecy covering, presenting the rounded appearance which is seen over the body of the traveler who has perished in the snow storm. The ocean, especially within and near the tropics, swarms with life. The remains of its myriads of moving things are conveyed by currents, and scattered and lodged in the course of time all over its bottom. This process, continued for ages, has covered the depths of the ocean as with a mantle, consisting of organisms as delicate as the macled frost, and as light in the water as is down in the air.

But whence came the little silicious and calcareous shells which Brooks' lead has brought up, in proof of its sounding, from the depth of over two miles?" Did they live in the surface water immediately above? or is there a *habitat* in some remote part of the sea, whence, at their death, the currents were sent forth as pall-bearers, with the command to deposit the dead corpses where the plummet found them.

It will be remembered that these discoveries and ascertainments were made in the period from 1850 to 1860. Much was crowded in this brief span of time. Philosophers and scientists were startled and invigorated; all were in a state of earnest looking-forward-to and "What next?" and "What next?" was continually being asked by the thinkers. The period immediately succeeding this and continuing to the present time, has been one of remarkable activity in this field of research. Brooks' sounding rod and Maury's discoveries only opened the way. Later mechanical appliances have made that way much easier, and the results ascertained by the means of better conducted contrivances are more satisfactory. It was very confidently believed for a considerable length of time by most of those who thought at all on the subject, that Baily and the anti-biotics were right indisputably in their opinion that there was no life in the depths of the ocean. This conclusion is now seriously questioned, and perhaps with reason.

Following Maury's researches under the auspices of the American navy, were several "Deep Sea Dredging Expeditions," sent out by various governments. While a detailed account of all these expeditions would be instructive and interesting, we shall have to content ourselves with but a brief glance at one or two only, and those sent out by the British government. We may refer especially to the cruises of the *Lightning* and *Porcupine*, and the "Voyage of the *Challenger*." For an account of these cruises and their important determinations we are particularly indebted to Dr. C. Wyville Thomson, Director of the Civilian Scientific Staff of the *Challenger* Exploring Expedition.

We shall only be able at this time to refer to so much of the work of these expeditious as treats of "Life in the Depths of the Sea." The dredging operations of the *Lightning* and *Porcupine* were done during the years 1868, 1869, and 1870. Preceding these years this question of "life in the depths" had remained an open one, the majority of the naturalists and scientists probably inclining toward the antibiotic theory. We have given at some length the opposite opinions of Ehrenburg and Baily. The next high authority who expressed an opinion was Professor Huxley, and he was very guarded. He says, in his report to the British admiralty concerning an examination he had made of samples procured by Captain Dayman in the *Cyclops* in 1857: "How can animal life be conceived to exist under such conditions of light, temperature, pressure, and aëration as must obtain at these vast depths? To this one can only reply that we know for a certainty that even very highly organized animals do contrive to live at a depth of three hundred or four hundred fathoms, inasmuch as they have been brought up thence, and that the difference in light and heat at four hundred and at two thousand fathoms is probably, so to speak, very far less than the difference in complexity of organization between these animals and the humble protozoa and Protophyta of the deep-sea soundings. I confess, though, as yet, far from regarding it proved that the *Globigerinæ* live at these depths, the balance of probability seems to me to incline in that direction."

In 1860, the celebrated naturalist, Dr. Wallich, took part in a British sounding expedition to Iceland, Greenland and Newfoundland. During the cruise soundings were taken, and specimens of the bottom were brought up from depths from six hundred to two thousand fathoms; many of these were the now well-known gray "Globigerina ooze." On the return voyage, about midway between Cape Farewell and Rockall, thirteen star-fishes came up from a sounding of one thousand two hundred and sixty fathoms, "convulsively embracing a portion of the sounding line which had been payed out in excess of the already ascertained depth, and rested for a sufficient period at the bottom to permit of their attaching themselves to it." Dr. Wallich warmly advocated the view that the conditions of the bottom of sea were not such as to preclude the possibility of the existence of even the higher forms of animal life, and discussed fully the arguments advanced by the other side. However, his was the expression only of an individual opinion, and it was thought that no new fact had been elicited. Star-fishes had come up on several previous occasions adhering to sounding lines, but the certain proof was still wanting that they had lived upon the ground at the depth of the sounding.

In 1857, a cable was laid between Sardinia and Bona. In 1858, it became necessary to repair it, and a length of about thirty miles was picked up and successfully replaced. In 1860, it completely failed. In picking up a portion of this cable from a depth of about one thousand two hundred fathoms, a great many specimens of animal forms were found adhering to it. The condition of these forms was such as to indicate pretty clearly that they had become attached to the cable while it was at rest at the bottom. "Up to this time," says Dr. Thomson, "all observations with reference to the existence of living animals in extreme depths had been liable to error, or at all events to doubt, from two sources. The appliance and methods of deep sea soundings were imperfect, and there was always a possibility, from the action of deep currents upon the sounding-line, or from other causes, of a greater depth being indicated than really existed; and again, although there was a strong probability, there was no absolute certainty that the animals adhering to the line or entangled on the sounding instrument had actually come up from the bottom. They might have been caught on the way." The animal forms upon which these conclusions were based were not sticking loosely to the cable, under circumstances which might be accounted for by their having been entangled upon it during its passage through the water, but they were moulded upon its outer surface or cemented to it by calcareous or horny excretions, and some of them from what is known of their history and life, must have become attached to it as minute germs, and have grown to maturity in the position in which they were found. This may be regarded as the best proof yet furnished of the existence of highly organized beings living at a depth of more than one thousand fathoms.

During the cruises of the *Lightning* and *Porcupine*, already referred to, fifty-seven hauls of the dredge were taken in the Atlantic at depths beyond five hundred fathoms, and sixteen at depths beyond one thousand fathoms; and in all cases life was found in abundance. Two casts were taken in depths greater than two thousand fathoms. In both of these cases life was abundant; and with the deepest cast, two thousand four hundred and thirty-five fathoms, off the mouth of the Bay of Biscay, living, well-marked, and characteristic examples of all the five invertebrate sub-kingdoms, were taken. And thus the question of the existence of abundant animal life at the bottom of the sea may be considered finally settled for all depths, for there is no reason to suppose that the depths anywhere exceeds four thousand fathoms; and if there be nothing in the condition of a depth of two thousand five hundred fathoms to prevent the full development of a varied fauna, it is reasonable to suppose that even an additional thousand fathoms would not make any great difference.

With Brooks' sounding-lead only about a handful of the bottom could be brought up at a time. With the dredging appliances used in later expeditions as much as two hundred pounds would sometimes be brought up from the depths at a single haul. The operation of casting the dredge, and bringing it up again with its specimens from the bottom, is a delicate and protracted one: it sometimes requires from eight to ten hours to make a single haul.

The conditions which might be expected to effect animal life at great depths of the sea are pressure, temperature, and the absence of light which apparently involves the absence of vegetable food. The conditions of pressure are certainly very extraordinary. At two thousand fathoms a man would bear upon his body a weight equal to twenty locomotive engines, each with a good long train loaded with pig iron. We sometimes find when we get up in the morning, by a rise of an inch in the barometer, that nearly half a ton has been quietly piled on us during the night; but we experience no inconvenience, rather a feeling of exhilaration and buoyancy, since it requires a little less exertion to move our bodies in the denser medium. Water, we know, is very incompressible, and, together with this fact, we may conclude that these animals in the depths are so constituted that they can bear whatever pressure there is, though it is by no means certain that these same highly organized beings could survive the change of condition involved in the sudden removal of this pressure.

As yet we have very little exact knowledge as to the distance to which the sun's light penetrates into the waters of the sea. It is certain that beyond the first fifty fathoms plants are barely represented, and after two hundred fathoms they are entirely absent. The question of the mode of nutrition of animals at great depths becomes, therefore, a very singular one. The practical distinction between plants and animals is, that plants prepare the food of animals by decomposing certain inorganic substances which animals cannot use as food, and recombining their elements into organic compounds, upon which animals can feed. This process is, however, so far as we are at present aware, constantly affected under the influence of light. There seems to be no light at the bottom of the sea, and there are certainly no plants there except such as may sink from the surface, and yet the bottom of the sea is a mass of animal life.

The only practical explanation relative to the maintenance of this .life is, that all sea-water contains a certain quantity of organic matter, in solution and suspension. Its sources are obvious. The rivers bring in a considerable quantity. Every shore is surrounded by a fringe which averages nearly a mile in width, of olive and red sea-weed. Then there is the "Sargasso-sea" or marine prairie, in the middle of the Atlantic, covering over three million square miles. The sea is full of animals which are constantly dying and decaying. Nearly all the animals at extreme depths belong to one sub-kingdom, the Protozoa; whose distinctive character is that they have no special organs of nutrition, but absorb nourishment through the whole of their jelly-like bodies. Most of these animals secrete exquisitely formed skeletons, some of silicia, some of carbonate of lime. There is no doubt that they extract both of these substances from the sea-water; and it seems more than probable that the organic matter which forms their soft parts is derived from the same source. It is thus quite intelligible that a world of animals may live in these dark abysses, but it is a necessary condition that they must chiefly belong to a class capable of being supported by absorption through the surface of their bodies of matter in solution, developing but little heat, and incurring a very small amount of waste by any manifestation of vital activity.

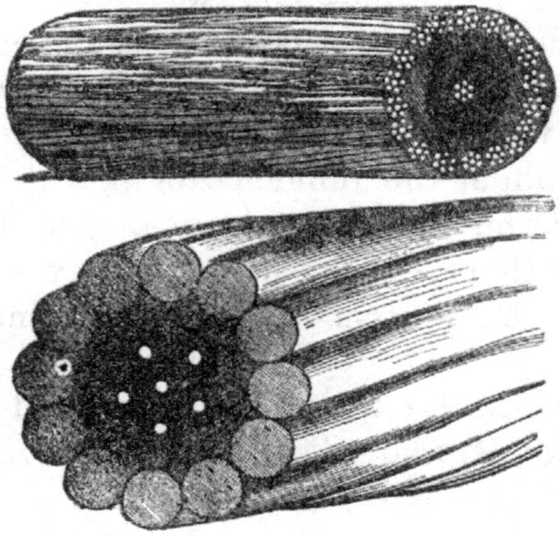

SECTIONS OF OCEAN CABLE.

According to this view, it seems probable that in all periods of the earth's history some form of the Protozoa-rhizopods, sponges, or both, predominated greatly over all other forms of animal life in the depths of the warmer regions of the sea. The rhizopods, like the corals of a shallower zone, form huge accumulations of carbonate of lime, and it is probably to their agency that we must refer those great bands of limestone which have resisted time and change, and come in here and there with their rich imbedded lettering to mark like milestones the progress of passing ages.

CHAPTER XXIV.

PHENOMENA OF THE OCEAN.

NAVIGATORS in the Northern seas have the opportunity of witnessing to perfection some curious phenomena, among which may be mentioned the Mirage, a name given by the French to an optical deception in the atmosphere by which a ship appears as if transferred to the sky. These appearances were regarded by the credulous, in former times, as supernatural; but they are referred to the refractive and reflective properties of the atmosphere. Not only is there an increase in the vertical dimensions of the objects affected, so that low coasts frequently assume a bold and precipitous outline, but objects sunk below the horizon are brought into view with their natural position changed and distorted.

Dr. Hayes gives the following vivid description of the optical delusion:

"These Arctic skies do sometimes play fantastic tricks, and on no occasion have I witnessed the exhibition to such perfection. The atmosphere had a rare softness, and throughout almost the whole day there was visible a most remarkable mirage, or refraction, an event of very frequent occurrence during the calm days of the Arctic summer. The entire horizon was lifting and doubling itself continually, and objects at a great distance beyond it rose, as if by strange enchantment, and stood suspended in the air, changing shape with each changing moment. Distant icebergs

WATER-SPOUT.

MIRAGE.

and floating ice-fields, and coast-lines and mountains, were thus brought into view—sometimes preserving for a moment their natural shapes, then widening and lengthening, rising and falling, as the wind fluttered or fell calm over the sea. The changes were as various as the dissolving images of a kaleidoscope, and every form the imagination could conceive stood out against the sky. At one moment a sharp spire, the prolonged image of a distant mountain-peak, would shoot up, and this would fashion itself into a cross, or a spear, or a human form, and would then die away, to be replaced by an iceberg, which appeared as a castle standing upon the summit of a hill, and the ice-fields coming up with it, flanked it on either side, seeming at one moment like a plain, dotted with trees and animals; again, as rugged mountains, and then breaking up after awhile, disclosed a long line of bears, and dogs, and birds, men dancing in the air and skipping from the sea to the sky. There was no end to the forms which appeared every instant, melting into other shapes as suddenly. For hours we watched the 'unsubstantial pageant,' until a wind from the north ruffled the sea, when, with its first breath, the whole scene melted away as quickly as the 'baseless fabric' of Prospero's vision."

Another writer, during a voyage to the eastern coast of Greenland, was amused by the singular refractive power of the Polar atmosphere. The rugged surface of the coast assumed the form of castles, obelisks, and spires, which here and their were linked together, so as to present the appearance of an extensive city. At other times it resembled a forest of naked trees, and it was easy to conceive colossal statues, porticoes of rich and regular architecture, shapes of lions, bears, horses, etc. Ships were seen inverted, and suspended high in the air, and their hulls often so magnified as to resemble huge edifices. Objects really beneath the horizon were raised into view in a most extraordinary manner. It seems positively ascertained, that points on the Greenland shore, not above three or four thousand feet high, were seen at the distance of one hundred and sixty miles. The extensive evaporation of the melting ice, with the unequal condensation produced by streams of cold air, are considered as the chief sources of this extraordinary refraction.

The same navigator relates that when in the Polar Sea, his ship had been separated for some time from that of his father, which he had been looking out for with great anxiety. At length, one evening, to his astonishment, he beheld the vessel suspended in the air in an inverted position, with the most distinct and perfect representation. Sailing in the direction of this visionary appearance, he met with the real ship by this indication. It was found that the vessel had been thirty miles distant, and seventeen beyond the horizon, where her appearance was thus elevated into the air by this extraordinary refraction.

Sometimes two images of a vessel are seen, the one erect and the other inverted, with their topmasts and their hulls meeting, according as the inverted image is above or below the other.

The most remarkable instance of mirage that has been seen was that in which a vessel, with all sails set, at one moment looked like an immense black chest, no sails or masts being visible. On observing her for a time, the black body seemed to separate horizontally into two parts, and two sets of mingled sails occupied the intervening spaces, with one set of very small sails above. The figures afterwards became more distinct, and three images were clearly discerned. Another vessel changed, also, from the form of a great square flat-topped chest, to five distinct images, the upper with the sails erect, and the two lower double images with their sails rather confusedly intermingled.

Another phenomenon which is seen in its highest perfection in the Polar seas is the *Aurora Borealis*, or the "Northern Daybreak," so named from its appearance in that part of the heavens, and its close resemblance to the aspect of the sky before sunrise.

AURORA BOREALIS IN ARCTIC REGIONS.

During the winter of the Northern Hemisphere, the inhabitants of the Arctic zone are without the light of the sun for months together, and their long dreary night is relieved by the light of this meteor, which occurs with great frequency in those regions, and the exceeding beauty of which those who have seen it only in our latitudes can hardly conceive. It is generally described as an immense curtain, waving its folds like the canopy of an ample tent agitated by the wind, and fringed with a border of light of the richest colors and most vivid brilliancy. It is sometimes seen for a few minutes only, or an hour, or through the whole night, and through several nights in succession.

A dingy aspect of the sky in the direction of the north is generally the precursor of the Aurora, and this gradually becomes darker in color, and assumes the form of a circular segment, surrounded by a luminous arch, and resting at each end on the horizon. Sometimes the blue sky is seen between the cloud and the horizon. After shooting a number of rays or streamers, the dark part of the cloud generally changes and becomes very luminous. The rays continue to be shot from the upper edge, sometimes at some distance, or very close to each other. Their light is very dazzling; bright columns slowly issue upwards from openings in the main cloud, becoming broader as they proceed. When the Aurora attains its full brightness and activity, rays are projected from every part of the arch, and if they do not rise too high, it presents the appearance of a comb furnished with teeth. When the rays are very bright, they assume a green, violet, purple or rose color, giving to the whole a variegated and brilliant effect. The height of the Aurora has been differently estimated, but it has been seldom found to exceed ninety miles; but its geographical extent is enormous. The origin of the phenomenon is yet unexplained, but it is generally supposed to be electrical. Franklin regarded it as the result of a slow and continual discharge of electric fluid from the atmosphere about the poles to the air above; and Sir Humphrey Davy and other electricians noticed the striking similarity between the Northern Lights and electricity discharged through rarefied air.

The Aurora has been observed in almost every part of the world. The ancients regarded its appearance with great terror, as the precursor of dire events; and there is no doubt that the fiery meteors, representing to their imaginations armies fighting in the heavens, and described by many writers as having preceded remarkable occurrences, must have been this phenomenon. The Indians also regarded these lights as the spirits of their fathers roaming through the land of souls. This idea may have originated from the long streaks of light which spread out with inconceivable swiftness, but always appearing to move to and from a fixed point, somewhat like a ribbon held in the hand and shaken.

Other luminous meteors are seen by the navigators of the Northern Ocean to perfection, arising, apparently, from the refraction caused by the minute and highly crystalized particles of ice floating in the atmosphere. The sun and moon are often surrounded by Halos, circles of vapor, tinted with the brightest hues of the rainbow. Arctic voyagers frequently mention the fall of icy particles during a clear sky and a bright sun, so small as to be scarcely visible to the naked eye, and detected by their melting on the skin; and others larger, presenting a remarkably interesting appearance. M'Clintock, in his "Voyage of the Fox," observes: "The snow crystals of last night are extremely beautiful; the largest kind is an inch in length, and its form exactly resembles the end of a pointed feather. Stellar crystals, two-tenths of an inch in diameter, have also fallen; these have six points, and are the most exquisite things when seen under a microscope. In the sun, or even the moonlight, all these crystals glisten most brilliantly, and as our masts and rigging are abundantly covered with them, the 'Fox' was never so gorgeously arrayed as she now appears."

Parhelia, or mock suns, in the vicinity of the real orb, shine at once in different quarters of the firmament. They are most brilliant at daybreak, diminish in lustre as the sun ascends, but again brighten at his setting. Edward Parry describes a parhelion of remarkably gorgeous appearance which he saw during a winter's sojourn at Melville Island. It continued from noon until six in the evening. It consisted of one complete halo, with segments of several others, displaying in parts the colors of the rainbow. Besides these, there was another perfect ring, of a pale white color, which went right around the sky parallel with the horizon, and at a distance from it equal to the sun's altitude, and a horizontal band of white light appeared passing through the sun. Where the band and the inner halo cut each other, there were two parhelia, and another close to the horizon, directly under the sun, which formed the most brilliant part of the spectacle, being exactly like the sun slightly obscured by a thin cloud at his rising or setting.

A singular phenomenon observed on the Arctic seas by Mr. O'Reilly is mentioned in his account of Greenland. The atmosphere had been obscured by a fog, and the sunlight, falling on the mist, formed an ellipsis, strongly illuminated, apparently rising from the surface of the ocean to the upper edge of the mist. The inner edge was pearly white, with the faintest tinge of blue; the middle yellow, deepening into brown and purple; the outer edge a blackish blue. In the centre of this oval, Mr. O'Reilly, who had ascended into the hurricane house, saw reflected his whole figure, of a colossal size, the head surrounded by a circle of the brightest rainbow colors.

The sun, for some time before it finally departs for the Arctic winter, and also after its reappearance in spring, tinges the sky with hues of matchless splendor, which far outvie even the glory of an Italian sky. The edges of the clouds, near the sun, often present a fiery or burnished appearance, whilst the opposite horizon glows with a deep purple, gradually softening into a delicate rose color of inconceivable beauty.

Another phenomenon which meets the eye of the Arctic navigator is the Ice-blink; a peculiar brightness in the atmosphere which is almost always perceptible on approaching ice. It is a stratum of clear whiteness, occasioned evidently by the glare of light reflected obliquely from the surface of the ice against the opposite atmosphere. This shining streak, which looks always brightest in clear weather, indicates to the experienced navigator, twenty or thirty miles beyond the limit of direct vision, not only to the extent and figure, but even the quality, of the ice.

ICE BLINK.

Another phenomenon remarked by voyagers is the ***Tide Rip and Sea Drift.*** The first may be described as a commotion in the waters not unlike that produced by a conflict of tides or other powerful currents. These sometimes move along with a roaring noise, and the inexperienced navigator expects to find his vessel drifted by them a long way out of his course. But he generally finds after the commotion has subsided that his ship's course has not been really affected. This may be accounted for on the supposition of an equilibrium of opposing forces. The *Drift* is certainly a movement of the waters, though not of sufficient dignity to be denominated a current. If a bottle or other object be cast into the ocean near the equator, it will eventually bring up at the icy regions about the poles, and in time find its way back again to the tropical waters from whence it started, unless caught up by some of the known currents. This may be regarded as another evidence of the complete circulation of the ocean as affected by heat and cold.

If a river of comparatively warm water be found in a cold latitude, it must have been heated in a tropical region. This phenomenon is a fact, and not only this, but the cold sea rivers may be found near the equator. The effect upon climate we know, and, in some sense, the cause. What a beautiful compensation is this regulator of marine climate! This Drift is irrespective and in addition to the known currents. Its rate is nearly four knots a day. Though having some influence upon navigation, its effect is mainly climatic.

As we have already said, these currents, drifts, or floes are well defined as to their extent, direction and effect; and and it will probably not be uninteresting to examine to some little extent the causes that bring about these phenomena, phenominal in themselves. In other chapters we have referred to the tiny polyps and other marine animals as agents in disturbing the equilibrium of the ocean—each one producing a ripple, the total of which amounts to a current.

Evaporation and precipitation are other agencies we will now discuss. The better to understand their operation, let us imagine a district in the tropical waters set apart. We will now suppose that we have a machine large and powerful enough to pump up the water one mile in depth in this district. It must not only force up this quantity of water but it must bear it away and discharge it elsewhere. We have limited this comparison to arrest the attention, for one hundred times and vastly more than this quantity is pumped up daily and discharged. It can readily be imagined that this cannot be done without greatly disturbing the equipoise of the waters; but it must also be remembered that every disturbance is compensated, and every force equally disturbed. Again having recourse to Maury, to whom we are indebted for many valuable hints relative to these matters, let us compute how much the fall of a single inch of rain over an extensive region in the sea, or how much the change even of two or three degrees of temperature over a few thousand square miles of its surface, tends to disturb its equilibrium, and, consequently, to cause an aqueous palpitation that is felt from the equator to the poles. As an example in illustration: the surface of the Atlantic Ocean covers an area of about twenty-five millions of square miles.

Taking one-fifth of this area, we will suppose a fall of rain one inch deep to take place over it. This rain would weigh three hundred and sixty thousand millions of tons; and the salt which, as water, is held in solution in the sea, and which, when that water was taken up as vapor, was left behind to disturb equilibrium, weighed sixteen millions more of tons, or nearly twice as much as all the ships in the world could carry at a cargo each. This rain might fall in an hour, or it might fall in a day; but, occupying what time it might in falling, this rain is calculated to exert so much force—which is inconceivably great—in disturbing the balance of the ocean. If all the water discharged by the Mississippi River during the year were taken up in one mighty measure, and cast into the ocean at one effort, it would not make a greater disturbance of the sea than would the supposed rain-fall. Now this is but for one-fifth of the Atlantic, and the area of the Atlantic is about one-fifth of the sea-area of the world; and the estimated fall of rain was but one inch, whereas the average for the year is sixty inches; but we will assume it for the sea to be no more than thirty inches. In the aggregate, and on an average, then, such a disturbance in the equilibrium of the whole ocean as is here supposed occurs seven hundred and fifty times a year, or at the rate of once in twelve hours. Moreover, when it is recollected that these rains take place now here, now there; that the vapor of which they were formed was taken up at still other places, we shall be enabled to appreciate the better the force and effect of these irregular movements in the sea.

FIRST STAGE OF WATER-SPOUT.

SECOND STAGE OF WATER-SPOUT.

The Greeks applied the term "*prester*" to the water-spout, which signifies a fiery fluid, from its appearance being generally accompanied with flashss of lightning and a sulphureous smell, showing the activity of the electrical principle in the air.

A few minutes suffices in general for the duration of this phenomenon, but several have been known to continue for nearly an hour. In the Mediterranean as many as sixteen water-spouts have been observed at the same time. The principal danger seems to be from the wind blowing in sudden gusts in their vicinity from all point of the compass, sufficient to overwhelm small vessels carrying much sail.

Mr. Ellis, in a cruise amongst the islands of the Pacific Ocean, had, on more than one occasion, a perilous escape from these phenomena. At one time the weather seemed clearing from a previous storm, when one of the native boatmen pointed to a large cylindrical water-spout, extending like a massive column from the ocean to the dark impending clouds:

"It was not very distant, and seemed moving toward our apparently devoted boat. The roughness of the sea forbade our attempting to hoist a sail in order to avoid it; and as we had no other means of safety at command, we endeavored to calmly await its approach. The natives abandoned themselves to despair, and either threw themselves along at the bottom of the boat, or sat crouching on the keel, with their faces downward and their eyes covered with their hands. The sailor kept at the helm. Mr. Barff sat on one side of the stern, and I on the other, watching the alarming object before us. While thus employed we saw two other water-spouts, and subsequently a third, if not more, so that we seemed almost surrounded with them. Some were well defined, extending in an unbroken line from the sea to the sky, like pillars resting on the ocean as their basis, and supporting the clouds; others assuming the shape of a funnel or inverted cone, attached to the clouds, and extending toward the waters beneath. From the distinctness with which we saw them, notwithstanding the density of the atmosphere, the farthest could not be many miles distant. In some we could imagine to have traced the spiral motion of the water as it was drawn into the clouds, which were every moment augumenting their portentous darkness. The sense, however, of personal danger and immediate destruction if brought within the vortex of their influence, restrained in a great degree all curious, and what, in other circumstances, would have been interesting observation on the wonderful phenomena around us, the mighty agitation of the elements, and the terrific sublimity of these wonders of the deep.

"The roaring of the tempest, and the hollow sounds that murmured on the ear as the heavy billows rolled in foam or broke in contact with opposing billows, seemed as if deep called unto deep, and the noise of the water-spouts might almost be heard, while we were momentarily expecting that the mighty waves would sweep over us. Our prayers were offered to Him who is a very present help in every time of danger, for ourselves and those who sailed with us; and under these or similar exercises several hours passed away. The storm continued during the day. At intervals we beheld, through the clouds and rain, one or other of the water-spouts, the whole of which appeared almost stationary, until at length we lost sight of them altogether, when the spirits of our native voyagers evidently revived."

The natives of the South Sea Islands, although scarcely alarmed at thunder and lightning are at sea greatly terrified by the appearance of water-spouts. They occur more frequently in the South than in the North Pacific, and although often seen among the Society Islands, are more rarely met with in the Sandwich group. But throughout the Pacific, water-spouts of varied form and size are among the most frequent of the splendid phenomena and mighty works of the Lord which those behold who go down to the sea in ships, and do business upon the great waters. They are sublime objects of unusual interest when viewed from the shore; but when beheld at sea, especially if near, and from a small and fragile bark, it is almost impossible so to divest the mind of a sense of personal danger as to contemplate with composure their stately movement, or the rapid internal circular eddy of the waters.

The Tornado—which, however is a general term employed to designate what is called a hurricane or whirlwind —is a sudden and violent storm of wind, accompanied by lightning and heavy torrents of rain, occurring frequently in the Indian Ocean, on the coasts of Africa, and other places in the tropics. While the tornado is passing over a ship, a loud creaking noise, occasioned by the electric fluid descending along the masts, is distinctly heard amongst the rigging. After the squall has passed beyond the ship, the lightning again appears to descend in sheets, as they did on its approach. Typhoons have their origin in the ocean to the east of China immediately about Formosa, Luzon, and the islands immediately to the south, and their course is generally along the coast of China. The body of the storm advances at the rate of twelve miles an hour. It is very probable that typhoons arise from opposing aërial currents, each highly charged with moisture which they have taken up from the oceans they have traversed; and their intensity is aggravated by the large quantity of heat disengaged in the condensation of the vapor of the atmosphere into the deluges of rain which fall during the storm.

The Trade-Winds, which are classed under the designation of "constant" winds, probably owe the origin of their name to the facilities afforded to trade and commerce by their constant prevalence and uniform course. They are perpetual in the torrid zone, blowing from the eastward with little variation. They were not known to the ancients, and seem to have been unknown even to modern seamen up to the time of Columbus, who had passed some time at the Canaries, to which the trade-winds extend in summer, and who seems to have conceived a just idea of their extent. On his first voyage, after leaving the Canaries, his crew were greatly alarmed at finding that the wind always blew from the north-east or east, and feared they would be prevented by it from returning to their native country. Columbus, however knew otherwise, and on his return from the newly discovered islands his tack was north of the trade-winds, in the region of the changeable winds. After the time of Columbus, European navigation extended rapidly in the Atlantic and Indian Oceans, and the trade-winds became gradually known.

Seamen dwell with delight in the region of the trade-winds, not only on account of the favoring gale, but its genial influence, the transparent atmosphere, the splendid sunsets, and the brilliancy of the unclouded heavens day and night. The origin of the trade-winds is ascribed to the rarefaction produced in our atmosphere by the apparent diurnal progress of the sun. It appears that the heat caused by the sun in the air is strong enough to produce this rarefaction to an extent of about sixty degrees of latitude, as the trade-winds, including what is termed the "region of calms," extend over such a portion of the globe. In this immense space the rarefied air is replaced by the colder and denser air which rests over the region contiguous to that of the trade-winds, and this transportation of air is the trade-wind.

It may not be entirely out of place here to say a few words respecting Atmospheric Currents, and to hazard a few conjectures respecting other causes than those mentioned that are among the agents producing the trade-winds and their "eastings."

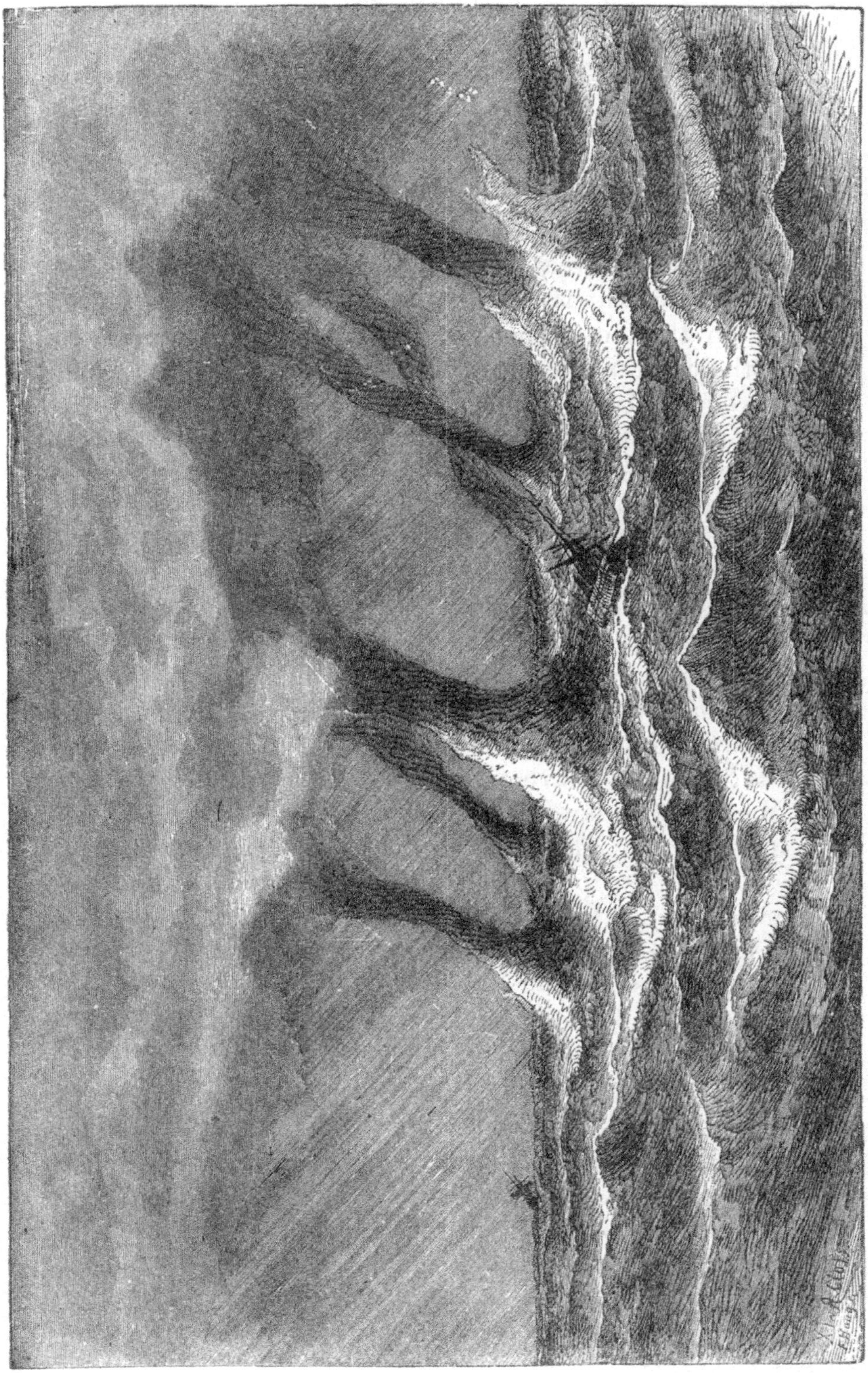

WATER-SPOUT.

The physical laws of the sea were created by God. To the devout even so simple a statement would be sufficient to awaken holy and reverential feelings; but the skeptic, in whom is no faith, wants to *see*. The more closely we examine the works of creation the more we must revere the Creator. It is impossible to escape the conviction that nothing in this admirable system is the result of accident, but on the contrary we are forced to admit that all is the result of wonderful and perfect design, and what is of still more consequence, this perfectly adjusted mechanism is arranged for the happiness and comfort of unthankful humanity.

The circulation of the currents of atmosphere is as well defined as are the currents in the seas. The wind strata constantly flow from the poles to the equator, and of course they must return there; therefore we may conclude that each current is dual. These direct and opposite currents are forced to move in a sort of a curve, turning *to* the west as they *leave* the poles for the equator, and in the opposite direction when they return—this turning being caused by the rotation of the earth on its axis. "Thus, in substance," says Maury, "the laws of motion, the force of gravity, and the shape of our planet all work together in compelling every wind either to force air up from the surface into the higher regions, or to draw it down to us from the azure or starry expanse above." By this everlasting agitation, or mixing up, the surface of the globe is kept in that condition so necessary and requisite to our well-being. Every exhaled breath, kindled blaze, and every atom of vegetation that grows and decays, adds something that is poisonous or takes away something that is healthful from the air around.

The agents of the air, appointed to maintain its chemical status, preserve its proportions, and adjust its ingredients to keep them in that state best calculated for the purpose intended, do not dally in the performance of their various offices.

If the air that is inhaled were not taken from us and renewed, warm-blooded life would cease to be; if the constituents of the air, carbon, oxygen, hydrogen, and water, were not in proper quantities given out by the ever-moving air to the flora of the world, all vegetation would perish. But we are saved from this calamity by this wind that brings down fresh supplies of life from above, and takes up from the earth that which has become stale. In the laboratory aloft the noxious gases are deflagrated among the clouds, purified and made over by ways known only to Him whom they obey. The cry going up from the crowded cities for fresh air from the mountains or the ocean, is a constant reminder of the life-giving virtues of circulation.

The study of its mechanism is good and wholesome in its influences, and the contemplation of it is well calculated. to excite in the bosom of right-minded philosophers the deepest and best of emotions.

"How minute, pervading and general, sure, and perfect must be that system of circulation which invests the atmosphere and makes the 'whole world kin.'" We can imagine the lither sky filled with crystal vessels continually ascending and descending between the bottom and the top of the atmospheric ocean full of life-giving air; these little buckets are let down by invisible hands from above, and as they are taken up again, they carry off from the surface, to be purified in the work-shop of the heavens, phials of mephitic vapors and noxious gases, with the dark and deadly air of marshes, ponds and rivers.

Among the many offices assigned by their Creator to the ocean and air currents, two may be particularly noticed in this connection: first, to act and react upon each other, and second, to distribute moisture over the surface of the earth, thus tempering the climate of different latitudes. When the northeast and southeast trades meet and produce the calms at the equator, the air, by the time it reaches this belt, is full of moisture, for in each part of the world it has traveled obliquely over a vast extent of ocean. In one sense these air currents may be said to terminate here—it is the end of their laterial direction. They have but one way to escape, and that is in an upward direction. As they turn about and go up, they expand and become cooler; a portion of the vapor that they have carried in their clasp is thus pressed together and comes down in the shape of rain. Therefore it is that it rains here almost continually. Old sailors tell us that the precipitation is so great and constant here, that they can frequently scoop up *fresh* water from the *sea* to drink. All the moisture that the air has taken up in its appointed journey across the waters, is not deposited in the calm belts: Let us enquire as to what becomes of the rest; for it must be understood that nature takes nothing from the earth that she does not give back again in some form or other. We see the great rivers constantly discharging great bodies of water into the ocean, and yet "the sea is not full." Whence do these waters come and whither go? That they come from their sources we know, but how are these beginnings supplied? The fountains must be renewed constantly or they would soon be exhausted. "The springs of these rivers are supplied from the rains of heaven, and these rains are form of vapors which are taken up from the sea, that "it be not full," and carried up to the mountains through the air. Note the place whence the rivers come, thither they return again. Picture now the mighty rivers of America, Europe and Asia, drawn up by the atmosphere, and rushing along in visible st¬eams back through the air to their beginnings in the mountains, and that, too, by the way of channels so regular, sure, and well-defined, that the amount thus carried year by year is nearly the same.

This apparently capricious atmosphere is a mighty machine. It performs its office with regularity and certainty, and is as obedient to law as is the steam engine to the will of its builder. "It, too, is an engine,"—the South seas being the boiler, and the northern hemisphere the condenser.

How admirable is the system of terrestrial adaptations! A little study of this system would show us why the proportions between the land and the water were arranged as we find them. With more water and less land we should have had more rain than we now have; with more land and less water we would have less rain; then, indeed, the climate would have been different over the whole face of the earth, and the inhabitants could not have existed as they are. The unthinking may marvel that God should number the very hairs of your head and note the sparrows fall; but what shall we say when we contemplate the earth as a whole, and from such points of view as we have presented? The sea, the air and the land, together with the mechanism of each, prove the most simple that each and all "have their origin in one "Omniscient idea."

The Monsoon is a term applied to periodical winds which prevail almost entirely in the northern part of the Indian Ocean. The force with which these winds blow is much greater than that of the trade-winds. It is frequently impossible to stem their violence in any way. Many vessels which have endeavored to force their way against them have been compelled to give in and to enter the nearest harbor. Other vessels are obliged to change their course, and to reach their destination by following a different track, wide of the straight route, and thus avoiding the monsoon. But although these winds, to vessels which miss the right season, render their voyages long and laborious, yet they greatly favor those of ships that arrive at the right period. It is chiefly by the assistance of the north-east and north-west monsoons that the voyages of merchant vessels bound from Canton to England are accomplished in short periods.

Monsoons, when compared with the trade-winds, exercise a most beneficial and important office in nature, especially in their relation to rain-fall, the fertility of the greater part of Southern Asia being entirely due to them. The shiftings of the monsoons is not all at once, and in some places the time of the changes is attended with calms, in others with variable winds; and particularly those of China, at ceasing to be westerly, are frequently very tempestuous, and such is their violence that they seem to be of the nature of the West India hurricanes.

The monsoon commences with great severity and presents an awful spectacle; the inclement weather continues, with more or less violence, from May to October. During that period the tempestuous ocean rolls from a black horizon, literally of "darkness visible," a series of floating mountains, heaving under heavy summits, until they approach the shore, when their stupendous accumulations flow in successive surges, and break upon the beach. Every ninth wave is observed to be generally more tremendous than the rest, and threatens to overwhelm the settlement. The noise of these billows equals that of the loudest cannon, and, with the thunder and lightning so frequent in the rainy season, is truly awful.

It is not easy to explain the origin of the monsoons: they appear to be only a modification of the trade-winds, produced by the peculiar form of the countries lying within and around the Indian Ocean.

The lightning attending these phenomena is described as fearfully vivid, realizing the description of a storm by Shakespeare:

"I have seen
The ambitious ocean swell, and rage, and foam.
To be exalted with the threat'ning clouds;
But never till to-night, never till now,
Did I go through a tempest dropping fire.
Either there is a civil strife in heaven,
Or else the world, too saucy with the gods,
Incenses them to send destruction.

Such sheets of fire, such bursts of horrid thunder,
Such groans of roaring wind and rain."

Other appalling phenomena that we may mention are Hurricanes or Cyclones; revolving winds, which appear to originate not far from the west coast of Africa, where they move in a more or less north-westerly course, until they reach our coast, where they begin to turn, and continue to the north-east, the vortex gradually widening, until they are lost in the ocean between Iceland and the British isles. These do not follow a straight line, but adopt a whirling course. The cyclones vary in breadth from fifty to five hundred miles, sometimes contracting, and in that case increasing fearfully in violence.

The direct causes of these terrible atmospheric disturbances are not precisely determined; they may arise from an electric condition of a portion of the atmosphere (indeed, atmospheric changes of every kind seem to be connected with electricity), or the sudden heating of the air over some insulated portion of the African mainland, although there is ground for believing that they are not occasioned by heat alone.

Hurricanes have been more terribly destructive than even earthquakes. The "Great" hurricane which commenced at Barbadoes in 1780, was so termed from its terrible results. The very bottom and depths of the sea seemed to be uprooted, and the waves rose to such a height that forts were washed away, and their guns carried about in the air like chaff; houses were raised and ships were wrecked, bodies of men and beasts were lifted up in the air and dashed to pieces. At the different islands not less than twenty thousand persons lost their lives on shore; while, farther to the north, the "Stirling Castle" and the "Dover Castle," men-of-war, went down at sea, and fifty vessels were driven on shore at the Bermudas, During a hurricane at

HURRICANE.

Guadaloupe, in January, 1825, a brig was whirled out of the water, and actually blown to pieces in the air. At St. Vincent's, in 1831, the water of the sea was raised to such a height as to flood the streets. The waves broke over cliffs seventy feet high. A very remarkable cyclone, or spiral hurricane, passed over a portion of England and the British and St. George's Channels during the autumn of 1859, destroying an enormous amount of shipping. The cyclone in Calcutta, in 1864, was one of the most awful events on record, of which there is no parallel except that in 1832; but even the latter was not so disastrous in its effects. Upwards of three hundred ships were wrecked or irreparably damaged, and the results on shore were appalling.

The noise of the wind in hurricanes is described by a seaman "as the most tremendous unearthly screech he had ever heard." The electric phonomena that sometimes attend hurricanes are very curious. One captain of a vessel states: "For nearly an hour we could not observe each other, but merely the lightning; and most astonishing, every one of our finger-nails turned quite black, and remained so nearly five weeks afterwards."

The Bore, a sudden and impetuous flow of the tide, is one of the most astonishing sights that can be witnessed on the sea-coast. At stated periods this tremendous tidal-wave comes rolling from the sea, threatening to everwhelm everything that moves on the beach. In certain parts of the Bay of Fundy the bore is more impetuous and higher—sometimes exceeding eighteen feet—than, perhaps, in any other part of the world. It comes in with such force and rapidity, with a noise resembling distant thunder, as sometimes to dash vessels on the shore. It is said also to overtake deer, swine and other beasts that feed on the beach, and swallow them up before the swiftest feet among them have time to escape. The swine, as they feed on mussels at low water, are said to sniff the bore, either by sound or smell, and generally dash up the cliffs before it rolls in. The bore is caused by the compression of the mass of advancing waters into a gradually narrowing channel. We quote from Maury's "Physical Geography of the Sea," the following incident:

"Dr. McGowan, an eye-witness, describes the bore or eagre of the Tsien-Tang River, as follows: At the upper part of the Bay, and about the mouth of the river, the eagre is scarcely observable; but, owing to the gradually descent of the shore, and the rapidity of the great flood and ebb, the tidal phenomena even here present a remarkable appearance. Vessels, which a moment before were afloat, are suddenly left high and dry on a strand nearly two miles in width, which the returning wave as quickly floods. It is not until the tide rushes beyond the mouth of the river that it becomes elevated to a lofty wave constituting the eagre, which attains its greatest magnitude opposite the City of Hang-Chow.

Generally there is nothing in its aspect, except on the third day of the second month and on the eighteenth of the eighth, or at the spring tide about the period of the vernal and autumnal equinoxes, its great intensity being at the later season. Sometimes, however, during the prevelance of easterly winds, on the third day after the sun and moon are in conjunction or in opposition, the eagre courses up the river with hardly less majesty than when paying its ordinary periodical visit. Between the river and the city walls, which are a mile distant, dense suburbs extend several miles along the bank. As the hour of flood tide approached, crowds gathered in the streets running at right angles with the Tsien-Tang, but at safe distances. My position was on a terrace in the front of the Tri-Wave Temple, which afforded a good view of the entire scene. On a sudden all traffic in the thronged mart was suspended: porters cleared the front street of every description of merchandise, boatmen ceased lading and unlading their vessels, and put out into the middle of the stream, so that a few moments sufficed to give a deserted appearance to the busiest part of one of the busiest cities of Asia. The centre of the river teemed with craft, from small boats to large barges, including the gay flower-boats. Loud shouting from the fleet annonnced the appearance of the flood, which seemed like a glittering white cable, stretched athwart the river at its mouth, as far down as the eye could reach. Its noise, compared by Chinese poets to that of thunder, speedily drowned that of the boatmen; and as it advanced with prodigious velocity—at the rate, I should judge, of twenty-five miles an hour—it assumed the appearance of an alabaster wall, or, rather, of a cataract four miles across and about thirty feet high, moving bodily onward. Soon it reached the advanced guard of the immense assemblage of vessels awaiting its approach. As the foaming wall of water dashed impetuously onward, the vast multitude were silenced, all being intensely occupied in keeping their prows toward the wave, which threatened to submerge everything afloat; but they all vaulted, as it were, to the summit with perfect safety. The spectacle was of great interest when the eagre had passed about one half way among the craft. On one side they were quietly reposing on the surface of the unruffled stream, while those on the nether portion were pitching and heaving in tumultuous confusion on the flood; others were scaling with the agility of salmon the formidable cascade. This grand and exciting scene was but of a moment's duration; it passed up the river in an instant, but from this point, with gradually diminishing force, size, and velocity, until it ceased to be perceptible, which Chinese accounts represent to be eighty miles distant from the city. From ebb to flood tide the change was almost instantaneous; a slight flood continued after the passage of the wave, but it soon began to ebb.

A very short period elapsed between the passage of the Eagre and the resumption of traffic. The vessels were soon attached to the shore again; women and children were occupied in gathering articles which the careless and unskillful had lost in the aquatic melee. The streets were drenched with spray, and a considerable volume of water splashed over the banks into the head of the grand canal, a few feet distant.

Among the most extraordinary phenomena in the universe is the volcanic action at the bottom of seas, by the upheaval of which islands are produced, the form, magnitude and character of which depend on that of the upheavėd mass. In general these islands rise suddenly, and their appearance is attended with all the phenomena that accompany eruptions: they are seen for some time, and then gradually disappear. When the dome, upheaved from the bottom of the sea, breaks at the summit so as to form a crater, a part of the circular rampart is sometimes destroyed, so that the sea enters, and an enclosed bay is formed, where innumerable tribes of coral animals build their cells.

Within the Atlantic Ocean no less than five great, and probably connected, centres of volcanic action exist. Iceland, the Azores, the Canaries, the Cape de Verd, and the West Indian Islands, besides many other points (Ascension, St. Helena, St. Paul's, etc.) at which extinct volcanic phenomena are visible. A remarkable submarine volcanic tract has been added to them, forming a belt about seventy miles from the equator on the south side. In the middle of the seventeenth century there were great and disastrous shocks in the Mediterranean basin.

A series of earthquakes and volcanic eruptions, apparently connected with each other, occurred in 1811 and 1812 in the countries surrounding the Columbian sea. The subterranean force first tried to open a vent by means of a submarine eruption in the Atlantic. Shocks of earthquake were for several days felt in the island of St. Miguel, one of the Azores, and on the 30th of January, 1811, large volumes of smoke, with which flames were observed to mingle, were seen issuing from the surface of the sea, at a distance of a few miles from the western coast of the island. They threw up mud, stones and other matter, which in a short time accumulated, so as to form a small island, which was called Sabrina; this disappeared after a few months in the sea. In 1831 an island rose out of the sea between the town of Sciacca, in Sicily, and the volcanic island of Pantellaria. Before any appearance of a change was observed in the sea, the inhabitants of Sciacca were alarmed by a number of very smart shocks of earthquake, of which two might be called severe. An Italian vessel passing near the place where, afterwards, the island rose out of the sea, observed a great disturbance of the waters at that spot. According to his statement a considerable space of the surface of the sea was seen rising to an elevation of from eighty to ninety feet above its level; the water appeared to bubble as if boiling, and the phenomenon was attended by a noise like thunder. After this agitation had lasted about ten minutes, the watery mass sank to the sea level, but after some time rose again. These risings of the water were repeated at irregular intervals of ten, fifteen and twenty minutes. A thick cloud of smoke, which enveloped the whole horizon, issued from the raised mass of water. The surface of the sea surrounding the raised mass was also considerably agitated, and a number of dead fish were floating about. For several days the atmosphere surrounding the town of Sciacca was dim and foggy, so that it was impossible to see what was going on at sea. On the 12th of July, in the morning, people were surprised at finding on the surface of the sea, in front of the town, a quantity of small porous ashes, which had been carried there by a fresh breeze from the southwest. At the same time a very unpleasant smell of sulphuric hydrogen gas incommoded the inhabitants of the town and the country near. Dead fish, recently killed, were floating in all directions.

Captain Senhouse effected a landing on the island thus formed, took possession of it, and called it Graham Island. He found the form of the crater to approach that of a perfect circle, and to be complete along its whole circumference, excepting for about two hundred and fifty yards on the south-east side, which was broken and low, apparently not above three feet high. The whole circuit of the island he conceived to be from a mile and a quarter to a mile and one-third. In the month of December following the whole island had disappeared.

The island of Santorin, one of the Greek islands called the Cyclades, is one of the most extraordinary instances of submarine volcanic action. During the last two thousand years several new islands have been formed in this locality, and singular phenomena have been exhibited even of late years.

Earthquakes are very closely related to volcanoes, although by no means confined to volcanic districts. Instances of this phenomenon at sea are frequently recorded by navigators. In most cases the motion felt on board is compared with that experienced when a ship strikes on a rock under water. During the earthquake at Lisbon, an English vessel, sailing at a distance of about fifty miles from the coast of Portugal, experienced a shock of such violence that a part of the deck was damaged. The captain, much surprised, thought that a great mistake must have crept into his reckoning, and that his vessel had got on a rock. He gave orders to put out the long boat, to save the crew, but he was soon convinced there was no danger. The source whence earthquakes originate, the power by which the ground is convulsed, is withdrawn from our investigation. The eye of the most inquisitive naturalist cannot reach it. The substances which are brought up by earthquakes bear evident signs of having endured the action of fire. It is supposed that these phenomena are produced by the efforts of accumulated elastic vapors to escape from the bowels of the earth.

Another phenomenon we may allude to is the Red-Fog or Shower-Dust, encountered by vessels at sea occasionally, and especially in the vicinity of the Cape de Verd Islands. What these showers precipitate in the Mediterranean is called "sirocco-dust," and in other parts "African dust," because the winds which accompany them are supposed to come from the Sirocco desert, or some other parched land of the continent of Africa. The dust is of a brick-red or cinnamon color, and it sometimes comes down in such quantities as to cover the sails and rigging, though the vessel may be hundreds of miles from the land. This dust, when subjected to the microscope, is found to consist for the most part of exceedingly minute animal and vegetable organisms, probably derived from some of the great river valleys of South America, being lifted up in vast clouds of impalpable sands by the fierce gales of the equinox.

CHAPTER XXV.

OCEAN STEAMSHIPS.

WE assume that there is much interest felt by all classes respecting these "ocean palaces." One cannot take up a daily paper of any pretensions without noticing the arrival or departure of many vessels of this class; and yet, though we become so familiar with their names and destinations, and so many of our friends are making them their homes for a time, we may, notwithstanding this, have but little conception of their magnitude or conveniences.

Every school-boy in the land is familiar with the name of Fulton. No school history would be complete without a mention of the *Clermont*, the first steamship, of 140 feet keel, 16½ feet beam, and rate of speed five miles an hour. A writer, in speaking of her first trip from New York to Albany in 1807, says:

"The vessel presented the most terrific appearance. The dry pine wood fuel sent up many feet above the flue a column of ignited vapor, and when the fire was stirred, tremendous showers of sparks. The wind and the tide were adverse to them, but the crowds saw with astonishment the vessel rapidly approaching them; and when it came so near that the noise of its machinery and paddles was heard, the crews of other vessels, in some instances, shrunk beneath their decks from the terrific sight; while others prostrated themselves, and besought Providence to protect them from the approach of the horrible monster which was

STEAMSHIP "CITY OF ROME."

marching on the tide, and lighting its path by the fire it vomited."

The Anchor Line of Steamships, one of the largest and best equipped that sails the ocean, comprising forty-three first-class vessels, ranging from 1,000 to 8,400 tons burthen each, has lately added to her fleet of steamships, so well and favorably known to the traveling public, a new vessel, the *City of Rome*, to ply between New York and Liverpool.

Perhaps no better evidence of our material progress during the last few decades could be suggested than that offered by a comparison of the two steamships we have mentioned. As for the *Clermont*, and other vessels of a similar character that have followed her advent by successive stages of improvement, nothing with which the reader is not already familiar could probably be said.

We submit a brief description of the *City of Rome:*

This magnificent steamer was built at Barrow, in Furness, by the Barrow Shipbuilding Company, and is the *largest passenger steamer afloat*. Her gross register tonnage is 8,415 tons; length being 560 feet, breadth of beam 52.3, and depth of holds 37 feet. And, notwithstanding her immense size, she is on all hands acknowledged to be unsurpassed for her beauty and symmetry even by the finest yachts. The vessel is rigged with four masts and has three funnels. A point calling for special notice is the large number of separate compartments formed by water-tight bulkheads, each extending to the main deck. The largest of these compartments is only about 60 feet long; and supposing that, from collision or other cause, one of these was filled with water, the trim of the vessel would not be materially affected. With a view of giving still further safety in the event of collision or stranding, the boilers are arranged in two boiler rooms, entirely separated from each other by means of a water-tight bulkhead. This reduces what in nearly all full powered steamships is a vast single compartment into two of moderate size, 60 feet in length; and in the event of either boiler room being flooded, it still leaves the vessel with half her boiler power available. There is also the usual water-tight bulkhead separating the engine room from the after boiler room.

Steam is supplied by twelve cylindrical tubular boilers. It is difficult to convey in words an adequate idea of the engine room, which is a compartment fifty feet square; but some idea as to what her engines are may be had by consideration of their dimensions: they are 47 feet, 8 inches in height from bottom of the frames to top of high-pressure cylinders, or as high as an ordinary four-story house; the engines are of 8,000 indicated horse power and capable of being worked up to 15,000 horse power.

The passenger accomodations of this floating palace are of the most complete description, affording all the comforts and luxury of a first-class New York hotel—than which there are no finer in the world. The captain and chief officers' cabins are on the promenade deck close to the ship's bridge, and immediately aft of them is the upper saloon, or drawing room, 100 feet in length, handsomely finished in white and gold, and furnished with lounges round the sides, and a grand piano. At the after end is the grand staircase to the dining saloon on main deck, a spacious apartment, 72 feet in length, 52 feet wide, 9 feet high; and in the rectangular opening to the drawing room above, the height is 17 feet, and which, being surmounted by a richly ornamented skylight, forms an effective and elegant relief to the usually heavy appearance of the ceiling. The paneling and decorations are artistic and unique, a paneled dado three feet high is carried entirely round the saloon, and the intervening space to the ceiling is filled up with rich panels of carved mahogany bordered with a margin of satinwood. The tables, of which there are three long ones in the centre, and a row of small ones along each side of the saloon, as also the revolving chairs, are of carved teakwood. At night the saloon is lighted by 32 electric lamps pendant from the ceiling, giving the whole a most brilliant appearance.

On each side of the vessel, extending aft from the saloon, are state-rooms for nearly four hundred first-class passengers, fitted in an elegant and substantial manner with everything necessary for their comfort.

The ladies' boudoir on main deck has paneled walls of figured brocaded silk, with tapestry curtains and furniture upholstered in blue velvet; there is also a magnificent ladies' cabin on the promenade deck, finished in black and gold, with furniture covered in amber-colored plush velvet and curtains of Roman cloth. There is also a comfortable reading room. Gentlemen can have their choice of two smoking-rooms; one on the upper deck is lined with cedar-wood with paneling of Japanese water-color sketches; the other, on main deck, is lined with a combination of mahogany, oak, and walnut, the floors of both being laid with parquetry.

The steamer throughout is lighted by electricity and fitted with electric call bells; then there are hot, cold, and salt water baths, lavatories, barber shops, etc., etc., conveniently situated. On the hurricane deck are twelve large life-boats, one of which is fitted as a steam launch.

CHAPTER XXVI.

THE SIGNAL SERVICE.

THE Signal Service of the Army of the United States is organized and equipped for the purpose of maintaining communication at all hours of the day and night primarily in war, but equally so in peace; such communication is effected by means of signals of any description. The modes of signalling most frequently employed by this service are by means of flags, torches, heliostats, telegraph, and telephones, by codes or ciphers.

The field-telegraph trains of the Signal Service are organized for use with armies. They are managed by soldiers who are drilled to march, manœuvre, work, and protect them. The train carries light or field-telegraph lines, which can be very quickly erected or run out at the rate of two or three miles per hour. They can be put in use for any distance, and be as rapidly taken down, repacked, and marched off with the detachment to be used elsewhere.

The Signal Service also transmits intelligence in reference to storms or approaching weather changes by the display of signals of warnings, and by reports posted in the different cities and ports of the United States. Maps showing the state of the weather over the United States are exhibited at board of trade rooms, chambers of commerce, and other places of public resort. Bulletins of data for all the stations are also prominently displayed and distributed without expense to the leading newspapers.

Signal-stations are also established in connection with the life-saving stations. These stations are connected by telegraph, and the former in addition to displaying storm-warning signals and making the usual meteorological reports, make special reports upon the temperature of the water, tempests at sea, the swells, etc. They also summon assistance to vessels in distress from the nearest life-saving stations or from the nearest port.

Stations for river-reports, to give notice of the conditions of the rivers affecting navigation and floods, are also established on the principal interior rivers and their tributaries.

The officers and men of the Signal Service are instructed for the different branches of the service at Fort Myer, Virginia, and at the Central Signal office, in Washington, D. C. They are taught signalling in all its branches, telegraphy, the use of the various meteorological instruments, the modes of observing, and the forms and duties required at stations of observation; the force is also drilled with arms, with the field-telegraph train, the construction of permanent telegraph lines, and in the usual duties of soldiers. For the duties of the observation of storms and for the display of warning signals, all stations communicate directly with the signal office in Washington over telegraphic circuits arranged with the different telegraph companies, and connecting with the office at fixed hours, each day and night. Each station is equipped with the necessary instruments.

The reading of these instruments is made five times a day at fixed hours, and the three taken at 7 A. M., 3 P. M., and 11 P. M., Washington time, are regularly telegraphed to the Central office in cipher.

The network of the Signal Service stations now extends over the continent from the Atlantic to the Pacific coasts, and the intervening territory from the Gulf (including the West Indies), to the Canadian frontier, and is in receipt, by comity of exchange, of daily telegraphic intelligence of the weather from the Canadian Dominion and its outlying posts. The office work is still hampered for want of more stations in the interior and Northwest; but it is thought provision will ere long be made for supplying them, as a new transcontinental telegraph-line is carried from Manitoba to British Columbia. These reports from one hundred and thirty-eight stations of observation are not unfrequently concentrated at the Central Office in the space of *forty-five minutes.* The stations at which cautionary signals are displayed are equipped with flags and apparatus for exhibiting the cautionary day or night signals, and also for communicating with vessels of any nationality.

The meteorological division of the United States Signal service was established in 1870, and was an additional duty imposed upon it. The progress of modern inquiry into the changes taking place in the weather, and especially into the phenomena of storms, had for many years previous, strengthened the conviction that they are not capricious, but follow certain laws. To provide, therefore, for meteorological observations, with a view to "giving notice by telegraph and signals of the approach and force of storms," was the end originally contemplated by the joint resolution of Congress, which passed February 9th of that year, authorizing the Secretary of War to carry this scheme into effect.

This was the first legislation of the United States Government inaugurating a national weather service. The great and peculiar extension of country afforded exceptional advantages for investigating and predicting storms which cross its broad area; for experience and observation had shown that they generally were more from west to east, and not frequently along the meridians. But the vast extent of the storm field, coupled with the fact that the "law of storms" was then but roughly outlined, made the execution of this task a very difficult and tedious work, calling for great caution and the most accurate observations.

At first the number of stations and the area covered by the predictions was limited. But when once the fact had been established that at any hour of the day or night, the Central Office could almost instantly call for reports from all parts of the country, and receive them from all its stations, taken at the same moment of time and revealing the actual status of the atmosphere over its whole field of inquiry, the sense of security in its scientific processes, and the confidence that the results were built upon "the solid ground of nature," gave it a powerful forward impulse.

A comparison of the simultaneous tri-daily forecasts from November 1st, 1871, to October 1st, 1872, gave an average of verification of 76.8 per cent; and during this time 354 cautionary signals were issued, with a percentage of correctness amounting to 70.

The results also offered the most complete demonstration of the laws of storms and the movements of cyclones that had ever been obtained in any country.

By act of Congress approved June 10, 1872, the Signal Service was charged with the duty of providing such stations, signals, and reports as might be found necessary for extending its research in the interests of *agriculture*. The agricultural societies over the land earnestly entered into and co-operated with the Service in this new development of its inquiries and reports. Eighty-nine such societies, thirty-eight boards of trades or chambers of commerce, numerous scientific institutions, colleges, and leading professional men put themselves in communication with the Chief Signal Officer, with a view to facilitate this branch of his work. The scientific societies at home and abroad began to take the liveliest interest in the general labors of the office, and to express the highest approval of the results attained. And, beyond the limits of the United States, numerous marine observations, which General Myer had previously desired, with the purpose of studying the atmosphere as a unit both on the ocean and the land, were forwarded regularly to his office.

The expansion of the work in 1873, under the stimulus of a world-wide favorable notice, was even more rapid than in the previous year. On March 3d, Congress authorized the establishment of Signal Service stations at the light-houses and life-saving stations on the lakes and sea-coast, and made provision for connecting the same with telegraph-lines or cables "to be constructed, maintained, and worked under the direction of the Chief Signal Officer of the Army, or the Secretary of War, and the Secretary of the Treasury." Early in this year the office also began the regular publication of a *Monthly Weather Review,* summarizing in a popular way all its data and showing the results of its investigations, as well as presenting these in graphic weather-charts adapted to the comprehension of communities it was destined to reach. The library of the Signal Office was increased to some 2,500 volumes bearing on the special scientific duties imposed upon it. The tests of meteorological instruments previously instituted enabled it to greatly improve and simplify its instrumental apparatus. The percentage of verification of its predictions for the year ending June 30th, 1873, was for each geographical division as follows:

New England	81.50
Middle States	81.17
South Atlantic States	79.92
Lower Lake region	78.90
Upper Lake region	75.25
Eastern Gulf States	77.16
Western Gulf States	74.40
Northwest	74.00

In organizing this service, the first problem that presented itself was to devise a system of observations which would, when mapped, accurately represent the aerial phenomena at the same instant of time, and in their actual relations to each other, and thus enable the investigator to discover the laws of storms and their rates of movement over the earth's surface. "The history of science," says one of its foremost representatives, "proves that unconnected, unsystematic, inaccurate observations are worth nothing." Certainly, in the domain of meteorology, no solid foundation for the science of the weather could have been laid in 1870, upon any of the then existing observational systems. The European weather-stations at that date, and long after, were engaged in making non-simultaneous reports; no two of them, unless they happened to be on the same meridian, read off their instruments at the same time.

The perfectly simple scheme of simultaneous observations aimed at the rescue of weather-research from the chaos in which for ages it had lain. Its cardinal principle of observation is to gain frequent views of the atmospheric conditions and movements over the country as they actually are, and as they would be seen, could they, so to speak, be photographed. In no other way can the bearings of the various storm-winds and their connected phenomena be detected, or the rates of their transition determined. All the predictions of the Signal Office, therefore, have from its beginning until now, been based on reports taken simultaneously.

The operations of the meteorological division of the Signal Service, popularly known as the "Weather Bureau," have been, every year since its creation, somewhat enlarged by congress, until they have become numerous and varied. The first to be specially mentioned is the daily work of weather-prediction, including storm warnings. These are issued from the Office of the Chief Signal Officer three times every day, under the title of "Indications" and "Cautionary Signals," and are based upon three series of simultaneous weather reports telegraphed to Washington from all parts of the United States and Canada; also at intermediate hours if necessary, based upon special reports. The tri-daily telegraphic observations are taken simultaneously at all stations at 7 A. M., 3 P. M., and 11 P. M., and at once put upon the wires. The number of stations from which tri-daily telegraphic reports are received at the Central Office is 138. Telegraphic reports have been also regularly received from one West-India station, and during the hurricane season from five.

Having taken their instrumental and other observations at either of the hours specified, the observers prepare their report in *ciphers*. These cipher-telegrams condensed by means of the cipher-code into five or ten words for each report, as soon as received in the Washington office, are translated from cipher, and entered on the bulletin blanks, and at the same time in their proper places on the weather-maps. This is done under the supervision of the assistant charged with the preparation of the weather predictions and the announcement of the storm warnings.

To arrive at the desired result, weather-maps are charted from the reports thus received by telegraph, this daily weather-map is one on which all the signal service stations are entered in their appropriate geographical places, and having annexed to each station the figures expressing the readings of the barometer and thermometer, the velocity of the wind, the amount of rainfall within the previous eight hours, etc.; and also symbols indicating the direction of the wind, and the form and amount of cloud, at the given time observation. The observations taken at each station are all put down on the map, and the relations between them are thus made sensible to the eye of the signal officer, by the figures and symbols, and also by lines drawn to group the geographical areas over which like conditions prevail. The weather-map is, therefore, to the meteorologist an indispensible means of obtaining a survey, and prosecuting a careful and connected study of the phenomenon he seeks to understand.

By preparing a graphic weather-map embodying the telegrahic data furnished to the Chief Signal Officer every eight hours of the day, the officer charged with formulating the storm predictions gains and retains a clear idea and mental image of the atmosphere. A great soldier has said: "There is nothing ideal in war;" and it may be said with equal force, there is no work which for its intelligent execution demands greater precision of method, more copious and circumstantial details, and closer attention to the developments of the hour, than weather-forecasting over a continent. The weather-map brings all these minutia within view, and makes the meteorologist master of the whole mass of observations, as hours consumed in the study of the numerical data could not do. Every weather-map is, therefore, a generalization in itself, as well as a record of the data. A series of weather-maps is a history of the ebb and flow, the fluctuations and tossings of the aërial ocean, and of the more subtile yet influential processes concerned in producing the weather and determining the climate of the country.

From reading in the morning newspapers the "Synopsis and Indications" for the day, no one not initiated in the method of preparing them would suspect the magnitude of the work involved in their preparation. The study pre-requisite for each of the tri-daily press-reports issued includes the draughting of seven graphic charts exhibiting the data furnished by the simultaneous reports telegraphed from all the stations.

We will not attempt to give here any formal description of these charts as such description would necessarily abound in technical terms which would not be likely to interest the general reader.

Armed with this charted material, the officer preparing the predictions proceeds first to compile the "Synopsis" and then to deduce the "Indications," and issue the necessary storm-warnings. The "Synopsis," "Indications," and cautionary signals constitute the "Press-Report" which when finished is telegraphed direct from the office of the Chief Signal Officer to all parts of the country. The average time elapsing between the simultaneous reading of the instruments at the separate stations scattered over the United States, and the issue of the "Synopsis" and "Indications" based on these readings, has been calculated at one hour and forty minutes.

An analysis of the predictions, made for the year ending June 30th, 1879, and a comparison with the weather-conditions which actually occurred within the twenty-four hours next ensuing, give the following percentages of verifications:

Percentage of verifications for the year (forecasts of barometric pressures, temperatures, wind-direction and state of weather): 86.6.

Percentage of verifications for the year (forecasts of the state of the weather only): 90.7.

These percentages of accuracy refer to predictions of barometric, thermometric, wind-direction, and general weather-changes. The average percentage of accuracy of the forecasts of the weather alone (including the state of the skies, whether clear, fair, or cloudy, and whether with or without rain) for all of the different districts is 90.7. The percentage for the Pacific Coast region is 89.3. In other words out of a hundred preannouncements of the single element of the "weather" for all parts of the country, ninety have been fulfilled by the event.

The important work of observing and reporting the fluctuations and floods of the great Western rivers was at an early period of its history undertaken by the Signal Service. The interstate commerce being necessarily much affected by the oscillations of the rivers, timely warnings of their rise and fall, and daily reports of the exact depth of water at numerous points, were eagerly asked for. These observations were found of so much importance that they have been extended over the Western, Southern, and California rivers, and deductions made from them, indicating impending changes, are daily published in the Washington weather-reports. All measurements at each river-station are made from the "benchmark," as known to the river-men of the vicinity, and the reading of the gauge is daily telegraphed to the Central Office, and all other interested stations. Knowing from such telegrams the height of the river at each station, as well as the total amount of reported rainfall higher up the river-valley, the office is thus enabled to calculate and announce the time and degree of coming changes. Thus timely premonitions of the great flood-waves that pass down the Mississippi, and also its fluctuations, are issued from this office.

The gauge used is very simple. In most cases it is a plank of pine or oak timber, two inches thick, ten inches wide, and long enough, when placed obliquely on the slope of the river-bank, to cover the extreme lowwater and high-water marks. When firmly imbedded in the earth, the "benchmark," which is generally the extreme lowwater known, is taken as the zero of the gauge, which is there carefully graduated, its subdivisions exactly corresponding to the vertical foot and subdivisions of which they are intended to be indices. A "danger-line" is marked on the gauge, showing how far the water may rise, but no farther, without danger of a flood. The reports telegraphed to the press, showing how near each stream has risen to or fallen below the "danger-line," enable the public to predetermine dangerous inundations, and furnish steamboat-men and merchants the daily information requisite for intelligently directing the movements of their craft. During the flood-months the telegraphic river-reports are especially valuable to all river-shipping, and to all interested in the traveling and transportation facilities which depend upon it, as well as giving timely warnings of ice-floods or sudden rises and falls. The levee systems of the Mississippi and other great rivers are thus guarded, and the immense agricultural interests secured, as the flood-warning comes in time to summon the State force to strengthen the imperiled works.

In connection with this service, surface and bottom water-temperatures at points upon the rivers, lakes, and sea-coasts are observed and reported for the United States Commissioner of Fish and Fisheries, with a view to ascertain the proper waters in which to plant the various food-fishes and furnish statistics desired for the development of the national system of pisciculture.

There is also to be mentioned the attention given by the office to the changes of temperature by which the canals are closed by freezing or opened by thaws for transportation. During the months when the market-rates and freight-schedules are affected by the probabilities of the canals closing, and when these water-ways are thronged with hundreds of laden barges, the daily predictions indicate thermometric conditions likely to ensue along their lines of transit. Such information protects the public from the imposition of excessive railway-rates in the shipments of grain-crops, especially in any autumn season of protracted mildness, and effects a large saving to the mercantile world.

The distribution of the tri-daily "Synopsis and Indications" over the whole Country may be understood from the following official facts. The total number of these forecasts —1,095 issued every year—are telegraphed at the moment of issue to the principal cities, and are published in some form in almost every newspaper in the country. In many public and conspicuous places, they are also bulletined for popular inspection. In order that they may reach the farming populations, an arrangement is effected with the

Post-Office Department, by which special "Farmer's Bulletins" may be distributed at an early morning hour of each day, except Sunday, along the railroads radiating from the chief cities of the Union. These bulletins contain proper selections from the matter of the "midnight" report made up in the Washington office at 1 A. M. of each day, which when it reaches the outlying stations by telegraph, is printed before daylight, and copies of it mailed to the rural postmasters for many miles around, and by them displayed in their offices. The intelligence of weather changes, with predictions and other data useful to the farmer in securing his crops or in other ways, on an average, reaches the different railway-stations, hamlets and villages throughout the country in the forenoon. As the predictions cover twenty-four hours, and often hold good for twice that period, they, therefore, reach the rural populations twelve or fourteen hours before the period to which they apply expires, and not unfrequently a day and a half or more.

The Cautionary Storm Signals, which accompany the "Synopsis and Indications" issued to the press three times each day, constitute a very important part of the Signal service work; and it was the possibility of preparing such storm-warnings for the benefit of navigation that originally gave the chief stimulous to the establishment of a Weather Bureau. The United States has a double front with over 7000 miles of sea-beaten coast, exclusive of the shore-line of its great lakes, ravaged by destructive tempests; and this vast stretch of mariginal territory needs to be environed with stations from which observations can be taken, and premonitary intelligence of cyclone and anticyclone signaled by day and by night to storm-menaced shipping. If no other duty devolved upon the Service, this alone would more than justify its whole cost, and warrant its extension. It is one of the most difficult and responsible tasks which can fall to the meteorologist, to put his science to its utmost stretch of accurate prevision (and often it must be done with a very few minutes for deliberation) to decide at what points on the coast the storm-wind will strike with dangerous effect. It is practically fatal to the value of his warnings if they are found to be superfluous, since in that case they cease to command the attention of seamen. Nor, for like reason, must they be displayed too late; nor yet too early, lest they should interfere with the movements of vessels which might run out of the dangerous vicinity before the storm can reach them. Thus the perplexing questions which spring up at every display of the signals lend to this part of the service intense interest. No such work had ever been undertaken in this country when the Signal Service was organized, and maritime storm-signaling in other countries had only been as yet rewarded by very moderate success.

On the organization of the United States Weather Service in 1870, General Myer began with great caution to prepare for this difficult and delicate part of his arduous task; and on the 24th of October, 1871, the display of signals on the sea-coasts and lakes commenced. The order regulating this display contemplated that the warning should be sent only to stations at which a wind having a velocity of twenty-five miles or more per hour would occur. As the anemometer at every station registers the wind's velocity for every hour, it is easy to ascertain whether any signal has been justified. Every such display is carefully followed up by the office, and the result—"justified" or "not justified"—is recorded, as reported by the observers hoisting the signals by telegraphic order from the Chief Signal Officer.

The cautionary signals are of two kinds: 1. Those premonishing dangerous winds to blow from any direction; 2. Those premonishing off-shore winds likely to drive vessels out to sea. Both kinds are needed by mariners as the storm-centers approach or depart from a maritime station.

A SHIP IN A STORM.

The first, distinctively termed the "*Cautionary Signal,*" consists of a *red flag* with a *black square* in the center, for warning in the day-time, and a red light by night. The second, or "Cautionary Off-shore Signal," consists of a *white flag* with a black square in the center, shown above a red flag with square black center by day, or a *white light* shown above a red light by night, indicating that, while the storm has not yet passed the station, and dangerous winds may yet be felt there, they will probably be from a northerly or westerly direction; this second signal, when displayed in the lake region in anticipation of high north to west winds, is designated the "Cautionary North-west Signal." The display of either signal, however, is always intended to be *cautionary*, and calls for great vigilance on the part of vessels within sight of it.

The total number of sea-ports and points on the lake and sea-coasts where the storm-signals are hoisted now is one hundred and eleven. The points whence storm-signals are displayed, however, are only those of the maritime margins of the field of research.

When this Weather Bureau was first proposed, the highest end thought attainable, by the most sanguine, was to give warnings of the *great* storms that traverse the lakes and sea-coast of the United States. This, however, is but a small part of the public interests it subserves. The number of persons who find that the reports and the forecasts of the Service may be utilized for every-day life is constantly increasing. Signal-observers are not unfrequently subpœnaed to bring the records of the weather into the courts, as legal evidence in cases upon which they bear. Grain and cotton merchants find the reports of value in calculations of the forthcoming crops. Emigrants consult them in the selection of favorable climatic conditions for a new abode. Physicians, sanitarians and boards of health employ the data to detect dangerous conditions of the atmosphere of the cities, and for investigating the origin and spread of diseases and epidemics, as in the case of recent yellow-fever visitations of the South. The pork-packers, fruit-importers, and fish and oyster dealers keep an eye on them to secure themselves against exposure of perishable goods to extremes of temperature or other vicissitudes of weather. They are of use to specialists in manufacturing and in hygienic interests, and are consulted by thousands planning journeys or excursions for health or pleasure. River boatmen, farmers, sugar-planters, fruit-growers, and ice-dealers find occasion to utilize them. Mechanics judge from the prognostics whether they can work outside on the morrow. The meteorological data supply engineers with information indispensable for planning economical and storm-proof architecture. Railroad officials, during snow-blockades, are kept advised by the reports, so that they are enabled to make provision for clearing the tracks; and railroad-freight officers find them useful for facilitating transportation. These are some of the daily applications made of the Signal Service work in the interior and central, not less than in the seaboard sections of the country. In every branch of agriculture and trade the deductions made from the published synopsis and indications of the weather have acknowledged value to the public when obtainable. In military operations over the vast West, the intelligence of approaching storms is highly prized, in timing movements so as to avoid heavy roads and dangerous delays. "Had we, a quarter of a century ago," says a British meteorologist, "known the rigor of the Crimean climate, who would have dared to send out an army unprepared to meet the hardships of a Black Sea winter? Ask the physician at what price he would value the power of giving timely warning of a 'cold snap' to his patients. Ask the builders of London what they have lost in the last ten years by sudden frosts or unexpected downpours of rain. Above all things, go to the farmer and ask what he would freely pay to know at seed-time what weather he might really expect in harvest. The fact is, there is not a profession, not a handicraft, not a process in animal or vegetable life, which is not influenced by meteorological changes."

We have said that this system of "simultaneous" weather reports practically began in 1870. The plan so skillfully wrought out by General Myer and his successor has become in a certain sense international. The old world has again had to take a lesson from her younger rival. Other nations have put in operation the system so successfully developed here, and not only so, but the same plan has become general all over the world, each nation as a factor contributing daily her quota of information, and the knowledge thus conveyed of the "probabilities" of so many specific localities is for the benefit of every nation. This country may justly feel proud of having been instrumental in bringing about such a desirable result.

It can be readily understood that this extension of the theatre of observation affords us much greater security by giving us a longer time for preparation, enabling us, so to speak, to erect timely barriers with which, in some sense, we may better resist the mighty forces of nature.

As perfect and desirable as this system of "international simultaneous weather reports" has become, our "wise men" are constantly improving it.

The Sea-coast Telegraph Lines are another important portion of the organization. By act of Congress, the Secretary of War was authorized to establish signal-stations at the light-houses and life-saving stations on the lakes and sea-coasts, and to connect these signal-stations with telegraph-lines, to be constructed, maintained, and worked under the direction of the Chief Signal Officer of the Army; and the use made of the life-saving stations is subject to such regulations as are fixed upon by the Chief Signal

Officer, the Secretary of War, and the Secretary of the Treasury. By this co-operative arrangement, the Signal Service has become a valuable if not an indispensible auxiliary to the sister services with which it connects, and shares very materially in the labors and responsibilities, especially of the Life-Saving Service.

The coast signal-stations aim to warn vessels within signaling distance of the approach of the storms, and to give life-saving stations quick notice of marine disasters calling for rescue, as also to furnish any intelligence to the latter, or to the light-houses, which may insure their more efficient working. Connected by wire or submarine cable, as all the signal-stations on the coast are, from Sandy Hook, N. J., to Smithville, N. C., and connected similarly with the office of the Chief Signal Officer at Washington, whence they are kept advised of any change in the meteorological status, they are thus enabled, from their full ocean view, to communicate directly any warnings from the Chief Signal Officer to passing ships, or to convey to him any facts which may be of use to the Washington office. The telegraphic wires connect each station with the Central Signal Office. The weather-reports and observations on the indications of the sea thus obtained are often of the greatest value to the Washington office in its work of preannouncing the force, direction, and velocity of the great hurricanes from the West Indies, which traverse our Atlantic seaboard.

As an illustration of this, it may suffice to note that in the summer of 1873, when the great August hurricane which so furiously assailed and wrecked seven hundred sail, was still passing over the Bermudas, its long dead swell was outrunning its centre by 600 miles, driving in the bathers at Long Branch and pouring into New York Bay. The steamer *Albemarle* encountered its fore-running wave on her voyage from Halifax to the Bermudas; and, though the morning was fair, suspecting danger, the vessel was hove to for a few hours to examine the swell. Concluding that the hurricane was advancing directly upon him, her captain changed his couɩse from southerly to westerly, and by a slight *detour* eluded the gale.

As one by one, yet all independantly, the coast-signal observers on any day telegraph to the Central Office the same significant tidings of the ocean—indications of an Atlantic gale, the intensity and direction of the swell—their concurrent observations often present unmistakable proofs of the presence, course, and progressive rate of these menacing meteors. The intelligence thus afforded is indispensable to the storm-warning and weather-prediction work of the Washington authorities. But, apart from the meteorological value of such a Coast Signal Service, its incidental contributions to the life-saving stations have already proved of the greatest assistance.

On the 20th of March, 1877, after a severe storm on the middle Atlantic coast, Sergeant William Stein of the Signal Service, in charge of the Cape Henry Station, discovered before dawn a large vessel stranded on a shoal off that station, and summoned the wreckers at Norfolk to come to the rescue. With the earliest light the sergeant displayed the "attention-flags" of the international code, with which every sea-coast signal station is supplied; and, receiving answer that she was the *Winchester* of Liverpool, with request for two steam-tugs to be sent to the vessel, he telegraphed at once to Norfolk for wrecking-steamers. Before sundown active efforts were made to save the stranded vessel. She was gotten off the shoal after some days' labor; but meantime three other vessels, in a second storm (of the 25th), were stranded within a mile of her. Sergeant Stein again telegraphed the wreckers at Norfolk for aid. He ascertained the name of the bark in greatest peril to be the *Pantzer*, a Norwegian vessel, and the crew of the Life-Saving Service a little later succeeded in firing a life-line over her deck.

The Norwegians did not comprehend its use, but after some effort, the Signal Service officer, by means of international signals, instructed her crew to "haul in on the line," and by nine o'clock all the crew of the *Pantzer* were safely landed. In the wrecks of the steamships *L'Amerique*, *Rusland*, and *Huron* (of the United States Navy), the first tidings were conveyed by the Signal Service wires, and through them succor was speedily summoned. In the case of the *Huron*, drifted ashore near Kitty Hawk, a private of the Signal Service, A. T. Sherwood, stationed at that place, received the first intelligence November 23d, and, after telegraphing to Washington, hastened to the awful scene, walking sixteen miles through the sand, and brought full reports of the situation to his station, which were instantly telegraphed to the Chief Signal Officer. The War and Navy Departments and the Life-Saving Service were thus notified, and by them steamers of the navy and wrecking companies were started to the fatal point of the shore on which the *Huron* had gone to pieces. The Kitty Hawk observer, immediately on receiving orders from the Chief Signal Officer, opened a "wreck station" abreast of the foundered vessel before daylight of the 25th, connecting it by a temporary telegraph wire with his station; and, working this improvised station on the open beach while the gale was yet raging, drew toward the spot the whole organized relief force of the government. A similar service was performed on the stormy night of January 31st, 1878, by another private soldier, of the Signal Corps, William Davis, when the steamship *Metropolis*, with 248 souls on board, became a total wreck, twenty miles from Kitty Hawk station. At 6.55 P. M. on that night, intelligence of the disaster reached Kitty Hawk, and in less than fifteen minutes, Private Davis, carrying telegraphic and signal apparatus, was riding through the night and storm to the scene. By 4 A. M. he had reached the vessel, established his telegraph-station abreast of her, opened communication, and forwarded a report to the Chief Signal Officer at Washington, and was putting in motion all the machinery of relief and succor which the country could command. The observers of the coast signal stations, whenever it is practicable, board vessels that have gone ashore, and open communication with the land. An instance of this may be cited from the action of private Harrison of the Signal Corps, at Cape Henry, when the bark *Guiseppe Masson* was wrecked near that station, February 10th, 1878. His presence prevented the crew from deserting their ship, which, by the aid of powerful wrecking steamers, was subsequently saved. Other instances of boarding vessels could be cited, as those of the Italian bark *Francesco Bellagambe* and the British steamship *Antonio*, both boarded by Signal Service men who afterwards kept up signal conversations with the shore until the ships were saved. But these cases will suffice to show the intimate alliance existing between the Coast Signal Service and the results announced by the Life-Saving Service. Without the Signal Service co-operation, the latter would often, in emergencies that arise, be powerless to command the needed help, as well as to communicate with stranded vessels. For the Signal Service, only men drilled in signaling can avail.

So arranged is the Coast Signal Service, that not only are its storm-flags and danger-warnings visible by vessels moving off the coast, but even a vessel *en voyage* (say one which is bound from the equator to New York), as she passes Cape Henlopen, may inquire by signals whether any hurricane is impending: if so, whether she has time to reach Sandy Hook before its arrival, or must take shelter behind the Delaware breakwater. Or, a vessel bound from New York or any northern port southward, on reaching the Capes of the Delaware, can make inquiry as to whether any storm is likely to strike her before she can pass Cape Hatteras, and receive full advice by telegraph from the Chief Signal

Office at Washington, in a very brief time. With adequate appropriations, this Coast Signal Service could easily be made of far greater value to all the shipping and mercantile interests.

On February 4, 1878, an order was issued by which are announced the stations of the service prepared to hold communication by the international code of flag-signals with vessels of any nation at sea coming within the proper signal distance. It is so arranged that any question as to weather changes anticipated so signaled from the vessel to the shore-station is immediately transmitted by telegraph to the central office, whence prompt reply is ordered.

This reply, on reaching the coast stations, is signaled by flags, if need be, to the inquiring vessel. It is possible thus to gain any needed information without landing a boat. It does not appear how a system of storm-warning or coast-signaling can be given a greater scope than is arrived at by this process. The instance may be imagined, for illustration, of a vessel sailing from New York for a southern port and making inquiries off the Capes of the Delaware whether it will be safe to pass Cape Hatteras, and advised from this office in reply to the inquiry transmitted to this office that a storm at the time is moving near Hatteras, and to take shelter at the Delaware Breakwater until the disturbance shall have passed northward. In the occurrences of each year there have been instances in which steamers moving along the coast have conducted their voyages from port to port upon the answers to special inquiries addressed from the ports in which they might be to the office. There have been instances in which such special inquiries have been signaled from the vessel to the station, telegraphed thence to the office, and the telegraphic answer signaled again to the vessel. With proper appliances such plans of communication might be established along the extent of the coasts of the United States.

In the annual report of the Chief Signal Officer for 1877 several disasters were described as having occurred upon the Atlantic coast, and the action of those of the Signal Corps occupying the stations upon the coast, was related at some length to illustrate by the different incidents the character of tho especial service it was expected to render. Since the date of that report several severe storms have swept the same coasts. It is since that report that the systems of display-stations established by the general office has rendered it possible to exhibit warning signals at almost every navigable inlet. There have been no disasters o. magnitude to record. The hope may be indulged that this decrease in the number of disasters has been in part due to the more extensive display of warning signals referred to, and in part to the greater attention paid by shippers and seafaring men to the warnings.

The following instances having occurred during the year, are given as illustrations of the varied uses to which the sea-coast service can be applied:

The Haytien brig *Tropic*, Captain Potter, from Marigone, Hayti, for New York, with logwood, 15 days out, came ashore near Barnegat, N. J., at 3:30 A. M., March 16, at a point 14 miles south of station, during a dense fog. Private Burrows, the repairman from the Central station, was on his way south, and, upon his arrival at the scene of wreck he immediately cut the wire and made telegraphic report to the Chief Signal Officer, via Sandy Hook. The line being down south, he then proceeded and repaired it, then returned to the wreck and reopened wreck station at 5 P. M., March 18, reporting direct to the Chief Signal Officer.

January 26, Captain Fitzgerald wrote to this office and stated that he intended taking the H. B. Plant, a small river steamer, from Wilmington, Del., to Jacksonville, Fla., by sea, and that before leaving Delaware Bay he would telegraph for the "Indications" for the next twenty-four hours; on the same date he was informed, by letter from this office, that any telegram received from him relative to the weather, would be promptly answered; at the same time he was furnished with a list of the stations of the Signal Service on the Atlantic coast from Delaware Breakwater, Del., to Jacksonville, Fla., and with the circular on the coast signal service, giving danger or distress signals. On February 9, Captain Fitzgerald telegraphed this office that steamer Plant would be at Lewis, Del., at midnight, and requested a special report as to wind and sea from Cape Henlopen to Savannah. The required information was telegraphed him at the Signal Service station at Delaware Breakwater. On the same date (February 9), Captain Fitzgerald wrote to this office from Wilmington, Del., stating that he intended to leave port at 3 P. M., and would stop at Delaware Breakwater for midnight Indications. and that in going down the coast, if he passed near enough the shore for the observers in charge of the stations of the Signal Service to see him, he would request that if bad weather be expected the "Union Jack" would be hoisted; if good weather the American flag be hoisted. Instructions were immediately telegraphed, from this office, to the observers in charge of the stations at Cape Henry, Va., Cape Hatteras, Cape Lookout, and Smithville, N. C., to look out for the steamer Plant, give all possible information, and ask for information from this office; also to signal him as desired. On the mornings of February 10 and 11, the Indications were telegraphed to Captain Fitzgerald at Delaware Breakwater. The Plant arrived off Cape Henry, Va., on the 11th. The Captain sent a boat ashore and telegraph to this office for a special report of weather. He was telegraphed in reply as follows: "Severe storm approaching from the west; it will not be safe to leave Norfolk to-night or to-morrow." Upon the receipt of this telegram the captain ran his vessel into the Hampton Roads and came to anchor. Captain Fitzgerald states in regard to this warning that it certainly saved his life and his vessel, as she could not have lived outside that night. He was telegraphed again from this office on the morning of the 12th, "Telegram of last night holds good to-day; no change in indications; if any favorable change to-day will notify you at six o'clock." And again at 6 P. M., "Remain at Norfolk until notified of safety; second storm coming from southwest." At 12:42 A. M., February 15, the observer at Cape Henry was ordered to signal Captain Fitzgerald the following message and to prevent his leaving that station: "Will not be safe for you to leave Norfolk to-night." The Plant arrived at Beaufort Harbor on the 17th and telegraphed to this office for Indications, and received the following answer: "Light southerly winds and a quiet sea expected." She then proceeded on her way and reached Jacksonville in safety.

As the Chief Signal Officer has said: "The time is not far distant when the possession of a coast not covered by sea-coast storm-signal and signal service stations, watching as sentinels each his own beat of sea and shore, and ready to summon aid by electric wires, will be held as much an evidence of semi-barbarism, as is now among civilized nations the holding of any national coast without a system of lighthouse lights." In the event of war, with a completed chain of coast signal stations, no part of our exposed sea-front could be threatened without immediate intelligence of the fact being flashed to the Washington office and all along the coast, and the defensive power of the Government concentrated at the point endangered.

In this brief description of an important subject much has necessarily been omitted that might make the chapter more readable; but the Signal Service renders such practical aid to every class and condition of man in respect to his varied employment, that a fair statement of its scope and operation will likely be deemed sufficient.

The very wise critic may tell us that this subject is hardly within the domain of the "Water World." It might seem at the first thought that this is so, but we venture to believe that a little consideration will show that there is no impropriety in treating it in this place and in this manner. Our storms come from the sea. We have some protection from phenomenal forces in nature and sudden changes of climate, if we have timely notice of the same.

Fortunately for our best interests, we generally receive such notice, and are learning day by day to pay more heed to it when received.

The "Service" is not yet perfect, but improvements are constantly being made. The Signal Stations are mainly on the coasts, and in the most likely track of the storms, and yet so situated, they are a long distance from the sources of the most destructive storms and agencies. They can tell us pretty certainly when a storm is coming, its force, and what direction it is likely to take. But this may not always be done in time, as the storm may have begun far out at sea, and notice of its exact force and direction may not be received soon enough to properly guard against its destructive tendencies.

It has been recommended that "Floating Stations" be established in mid sea and connected with the shore stations by cable. The advantages sure to result from the inauguration of such a plan are obvious, and we presume it will soon be adopted.

The "Wiggins Prophecies," concerning which there was so much ado recently, revealed two things: first, the great interest felt by all classes in this subject, and that, too, outside of and beyond any mere selfish or personal consideration; and secondly, the fact that we may safely and with entire confidence *trust* the "Weather Bureau" of the Signal Service.

CHAPTER XXVII.

THE LIFE-SAVING SERVICE.

HAS the reader spent his life within sound of the sea? Has he ever or frequently visited its coasts? There are but few more stirring and picturesque sights than some large body of water dotted here and there as far as the eye can reach, with almost unnumbered sails. Here are vessels, variously rigged and of all sizes. Whence have they come, and whither are they bound? Let us think a moment of only those the vision may take in at a single glance: what a history would a record of each of their single voyages be! The log-books of many of them could only be opened and read on the other side of the "Jasper Sea!"

What would a man give in exchange for his life? And again: "Greater love hath no man than this, that a man lay down his life for his friends." These last two sentences become appropriate texts for the theme under consideration. First, the organization of the service is an expression and evidence of what a nation will give to save its life; the second sentence finds its heroic embodiment in the labor performed by the noble-hearted officers and men of the Life-Saving Service.

We purpose in the following pages to give some portion of the history and progress of this branch of the public service, together with some of the stirring incidents illustrating its practical and far-reaching labor.

The lake and sea-coast of the United States exceeds ten

thousand miles, reaching through all varieties of climate, and containing untold features of coast danger to those sailing in its neighborhood. Probably the most dangerous of this immense stretch of coast is the long portion of barren sandy beaches lying between Cape Cod and Cape Hatteras.

To be more specific still, that portion almost within sight of the great Metropolis, the coast of New Jersey, is considered by far the worse. It was in this notoriously dangerous place, the Government, in 1848, placed a few rude huts that became the begining of what has since become the United States Life-saving Service. These little dwellings were expected to afford a shelter to the distressed sailors cast upon friendly shore, and to contain the clumsy boats and such other appliances as were then in use. These were used on occasions of shipwreck by volunteers from among the brave beachmen. Small and inadequate appropriations were occasionally made by Congress. The huts were extended to other coasts, but after all it amounted to but little, for lack of a head and coherent organization. This difficulty was finally overcome by the appointment of Mr. Sumner L. Kimball as General Superintendent in 1871, under whose able management the present excellent system was developed and perfected.

The Service now numbers about two hundred life-saving stations. Pleasant and commodious buildings have taken the place of the rude huts; here are kept ready for instant use, the surf boat and self-righting boat, the mortar-cart, loaded with wreck ordinance, lines, and appliances. Kitchen and sleeping apartments are set off. There is also a store-room and medicine chest; and here lives the keeper and his crew of six surfmen.

This band of men are chosen from among the sturdy fishermen brought up on the coast. The sea in all its moods is a familiar and not a terrible object to them. Fortunate it is, that it is into their brave and skillful hands that human life is entrusted in the many direful emergencies of those who go down into the sea in ships. From earliest youth up these surfmen become familiar with the breakers on their coasts, every eddy, bar, and current; also, though not sailors, they yet become familiar with every part and line of a ship, gained in the operation of stripping wrecks. Notwithstanding this practical apprenticeship, there must yet remain much for the surfman to learn before he can be considered as a life-saving man; his education or training is completed by officers of the Revenue Marine, who are especially qualified for this important duty.

It will be understood that these men are living apart from their families, friends, and associations, and therefore their life is sometimes monotonous, though a very busy one. Their hours of drill and practice is as regular as is that of the soldier on duty. They have their beats between stations which must be regularly and constantly patrolled. This duty is particularly severe and tedious at night-time. It is divided into three watches. Two guards set out at the commencement of each watch, one going to the right and the other to the left, ultimately meeting patrolmen from adjacent stations, with whom they exchange tokens of some character to prove to the officer in the morning that they have performed their duty faithfully. Nothing is left undone to insure the certain and perfect watching of the beach. The length of each beat is from four to five miles. It can be easily imagined that this guard or patrol duty is no light matter, especially upon a stormy night. And yet, no matter what the storm may be, it must be faced. There is no such thing as turning back. This is the very time when his duty is most imperative. Hundreds of lives may be depending upon his watchful care, and too devoted is he to duty to neglect anything of its faithful performance now.

The patrolman, discovering a vessel being driven ashore in a storm, takes the initial step in the operation of rescue.

In addition to his lantern, he carries at night a signal which when ignited gives out a brilliant red light. His senses are keenly awake, and the slightest unusual thing instantly attracts his attention. Perhaps it is a faint light in the distance, the shadowy outline of a sail or spar, or something cast at his feet that he determines must have come from some ship. Now he is all expectation, ready either for warning or rescue. But we can best convey to the reader a true picture by presenting a few incidents of actual occurrence, as follows:

The schooner *J. H. Hartzell* was wrecked about a mile south of the harbor of Frankfort, Lake Michigan, on the 16th of October, 1880. The scene on this occasion was in every respect extraordinary, and few narratives could surpass in interest the soberest recital of what took place that day abreast of, and upon the wooded steeps in the neighborhood of one of our Western towns.

The schooner was commanded by Captain William A. Jones. Her crew consisted of six men. There was also a woman on board, who acted as cook. The vessel, having made a good run, arrived off Frankfort about 3 o'clock on Saturday morning, October 16. Captain Jones concluded to wait until daylight before entering the harbor. But before daylight came a terrible storm arose. Being rather close to the shore, an attempt was made to wear ship, but in the fury of the gale the vessel would not obey her helm, and began to drift in; seeing which, her master let go both anchors and set his signal of distress. In spite of all efforts she finally went ashore on a precipitous bluff known as Little Bald Hills. The seas soon began to rush over her, and the awful staving and crashing usual in such cases began. The boats were soon carried away, the ship began to founder, and the crew had to take to the rigging. The cook was very weak from serious illness, and it took the united labor of four men to get her aloft into the cross-trees of the foremast, across which planks had been nailed. A little while after the men had got aloft, the vessel sank in sixteen feet of water, the stern resting upon the bar and the forward part in deeper water. Later the mainmast gave way and went over, remaining attached to the foundered hulk by some of the cordage, and thrashing and plunging alongside with every rush of the seas. The foremast, with the men upon it—one of them, the captain, clinging to the ratlines, about ten feet above the water; the remainder fifty feet aloft in the cross-trees, with the recumbent woman—swayed and creaked ominously, some of the wedges having become loosened, and seemed likely to go over at any moment. It was a horrible feature of this shipwreck that the vessel, now an utter ruin, had a short time before been loitering to and fro in the fresh breeze, with no anticipation of disaster, waiting only for daylight to drop into her harbor, near at hand. So suddenly and fiercely had the tempest risen that within an hour destroyed her, and placed in deadly jeopardy the lives of the wretched company that clung to her one tottering spar.

The vessel was seen from the town shortly after she struck. One of the earliest to observe her was a little boy, the son of a fisherman, who, looking through the sheeting rain and hail saw her plunging in the breakers. The lad at once told his father, who ran without delay to the village of South Frankfort with the alarm, and, accompanied by some fifteen or twenty citizens, cut across the hills and got abreast of the wreck near 8 o'clock. Other persons continued to arrive, and at length the crowd built a fire and laid pieces of driftwood along so as to form in huge rude letters, black against the white ground of the bluff, the words "LIFE-BOAT COMING." Eager signals from the sailors announced that they could read this gigantic telegram. Meanwhile, a gallant young citizen named Woodward had started on horseback for the nearest life-saving station, (No. 4, Eleventh

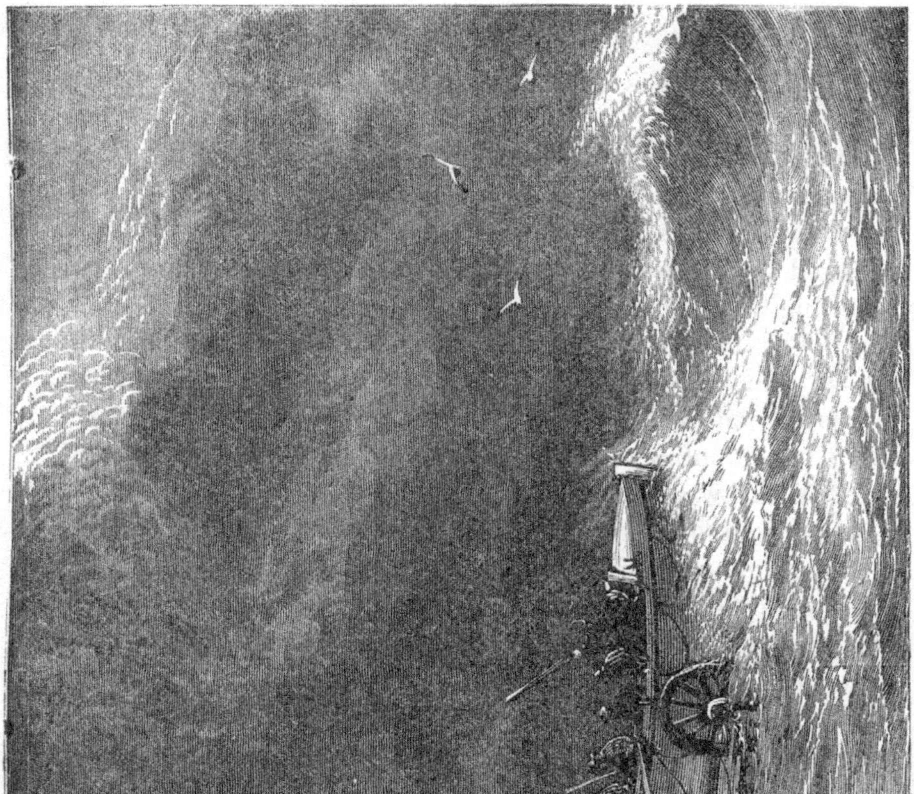

LAUNCHING THE LIFE-BOAT.

District,) ten miles distant, by a sandy and hilly road, mostly lying through woods. The young man galloped furiously through the tempest, which was constantly increasing in violence, tearing along the difficult highway to such good purpose that by half-past 8 o'clock he dashed up to the station with the news of the wreck. The keeper, Captain Thomas E. Matthews, at once ordered out the mortar-cart and beach apparatus. In a few minutes, the cart, loaded with the Lyle gun, the breeches buoy, hawser and hauling lines, and other appurtenances, left the station, dragged by the horse, which young Woodward hitched on, the hauling being also aided by himself and the station-men. One of the surfmen, was away on the south patrol when the start was made, and followed his comrades to the scene of the wreck subsequently.

The expedition had set out upon a terrible journey. The station is upon the lake shore, north of Frankfort, south of which town the wreck lay, and the intervening river and the harbor-piers making out into the lake from the town made it impossible, in any case to arrive at the wreck by following the line of the coast. The shortest route, not less than seven or eight miles long, was by a road which led off the beach to an intersecting road leading to the town, but to gain this it was necessary to travel two miles along the beach, which was now submerged by a swashing flood constantly bursting against and washing away the steep banks of the lake shore, battering the escarpment with intertangled masses of logs, stumps, and trees, and, of course, rendering the way impassable. The expedition was therefore compelled to lengthened the detour by taking an old traii or cart track. This road wandered through the woods, along winding ravines and up steep, soggy hills. Across these acclivities the way was so difficult that the men and the horse, tugging and straining at the cart together, could only make ten or fifteen yards at a pull without pausing. This violent toil was pursued amidst the roaring of the gale, which now blew almost a hurricane, until about a mile's distance from the station had been accomplished. By this time the men, despite the bitter cold, were hot and wet with their efforts, and the horse, steaming, trembled on his limbs, and could hardly draw. There were at least nine miles more of their disheartening journey before them, and the party were already sorely spent.

Fortunately some relief was at hand. The state road had been gained, and a light buggy came hurrying along with Mr. Rennie Averill, who, at the solicitation of Mr. Burmeister, the marine correspondent of the Chicago *Inter-Ocean*, had nobly undertaken the task, from which several persons had recoiled on account of the severity of the storm and the dreadful condition of the roads, of bearing the wreck alarm to the station, not knowing that this had already been done by Mr. Woodward. With the aim of getting help for the hauling, the keeper jumped into the buggy and rode on with Mr. Averill, ahead of the crew. Before long they met another brave citizen, who was also hastening to the station with a double team to give the alarm. He reported another team behind him, on the road, and, at Captain Matthews' request, which showed wise forethought, he pushed on to the station to bring up the life-car and Merriman suit, taking with him, by the keeper's order, Surfman La Cour, as he passed the cart on the way. The keeper also requested Mr. Benson to bring back Surfman La Rue, whom he judged to have reached the station by this time from his patrol. It is noticeable, and it is due to Captain Matthews, to say that his conduct of operations, from the beginning to the end of this laborious and difficult enterprise, was in the highest degree praiseworthy, no step being omitted or forgotten which could facilitate the rescue.

The life-saving crew had got on with their load some half a mile further, when they were met by another citizen with a team of stout horses, sent on by the keeper to aid the hauling. A more rapid progress was now assured. The State road upon which they were traveling was a great improvement on the trail they had left, although fearfully rough. It lay for four or five miles straight to Frankfort through dense timber, broken as it neared that place by an occasional farm clearing. The lower part of the town traversed, the road continued along the inner basin which forms the harbor, leading to a bridge spanning the river which feeds this place of anchorage. The bridge crossed, the track went on for some miles further through a dense growth of woods to the neighborhood of the wreck. Nearly the whole way was a series of steep up-grades, plentifully strown with pitch-holes. Along such a course the expedition valiantly struggled, arrived at and rushed through Frankfort, emerged again upon the rugged country road, crossed the river, plunged into the woods, and finally, about half past ten o'clock, reached the rear base of the bastion of high hills which separated them from the lake where the wreck lay. The ardor of the rushing march of this train of men and horses is shown by the fact that they conquered the rough stretch of ten miles in about two hours.

The keeper had driven on to an elevated farm, known as Greenwood's, from whence he could overlook the lake, and saw, about a quarter of a mile to the north, the wreck with the stormy water flying over her. He was returning toward the cart of apparatus, with the idea that the road to Greenwood's must be taken, when a citizen mounted on horseback, rode up to him, crying out, "Follow me and I will show you a short cut." The party followed him through a ravine about a quarter of a mile. The way then led up the overhanging hillside through the brush, and the tug with the loaded cart was terrible. So steep was the ascent that man and beast had fairly to climb, and almost to hoist the cart after them. Nothing could have been done but for the aid of a crowd of sturdy towns-folk, who had assembled there, and, anticipating the arrival of the life-saving party, had cleared away with axes and hand-spikes a great deal of the undergrowth and fallen trees. Even with these impediments removed, so precipitous was the acclivity that it took the united efforts of twenty-seven brawny men, by actual count, and a span of stout horses, to gain the summit, only about twenty feet being made at a time. By these efforts, worthy of giants, the top of the hill was reached; but the crowd were now brought up all standing by a belt of woods, as yet unpierced, which bristled along the crest of the eminence, and in which lay fallen trees half buried in the brush and dense undergrowth. The obstacle seemed to inspire all present with sudden electric energy. In an instant, all hands, citizens and crew, flung themselves upon the wood with axes, and a work began which resembled a combat. The wood seemed tumbling asunder, and its rapidly-opening depths were alive with rude figures in every variety of fiery action. In an incredibly short space of time the way through the wood was cleared, and the mortar cart, loaded with apparatus, was dragged forward to the brow of the hill. The gap cloven by this heroic onset disclosed a strange and dreadful diorama. The concourse of life-savers, fifty or sixty in number, were upon the summit of a precipitous bluff nearly three hundred feet above the sea. This bluff was composed of sand, covered near the top with a yellow sandy loam, with here and there a patch of clay upon its slanted surface. The mass not being compact, owing to the nature of its substance, yielded readily to any force brought to bear upon it, and the gale, which was now blowing with fury, beating upon the acclivity like a simoon, flung up the sand for ten or twelve feet high upon the face of the slope, so that, to the gazer looking down, the whole surface appeared in rapid and violent motion. Above the dusky layer of this sand-storm was an air thick and blurred with the snow and rain, and the crowd looking through, saw far below, looming with a sort of misty distinctness from the terrific confusion of the waters, the nearly sunken wreck, its two masts still standing, resembling grotesque dishevelled steeples made up of spar and cordage. This object had the effect of rendering all things subsidiary to itself—the immensity of livid and lowering atmosphere in which it was central—the ragged undulations of surf, bursting into foam, which flung themselves around it with furious celerity, and seemed racing toward it from the furthest sea. The hull was well smothered up in the breakers, but at intervals between seas it appeared for a moment black and streaming as the surf on the bar fell away. Standing in the spreading ladders of the lower rigging, a few feet above the water, was a diminished figure with upturned face, watching the people on the summit of the bluff. This was the captain, The monstrous waves curled and broke below his feet, and covered him with their spray. Forty feet above him could be seen, lessened by distance, a huddle of faces, peering at the crowd on shore from the swaying cross-trees. These were the faces of the crew. The foretop-mast rose above them, and the gaff-topsail, partially unfurled, bulged and flapped over them in the tempest. This frightful spectacle, seen by the crowd on the heights through the weird curtain of the tempest, amidst the uproar of the wind and sea, had something of the vivid unreality of the scenery of a vision.

What the crowd could not see, owing to the distance, was fraught with deeper elements of pity and terror. The captain of the vessel, who had but recently recovered from a fever, stood covered with frozen snow and rain in the ratlines, stiffened and discolored with exposure to the storm. High above him on their giddy and unstable perch, the six men crouched, blue in the face with cold. The fury of the wind in this tottering eyrie was such, that when one of the group had occasion to communicate with another, he could only do so by shouting through hollowed hands into his ear. Amidst all the din of the gale and sea, the unhappy men could hear the harsh creaking of the mast as the vessel swayed too and fro. They expected every moment to go over. The poor woman, carefully swathed in canvass cut from one of the sails, lay delirious and numb with the cold, and finally yielded to a state of unconsciousness in spite of the best efforts of the kind-hearted sailors.

No time was lost by the life-saving crew and citizens in commencing operations for the rescue. The prospect was discouraging in almost the last degree. Anxiously surveying the ground, Captain Matthews descried about two hundred and fifty feet beneath him, a narrow ledge or plateau, some ten or twelve feet wide, and at once determined that the cart must be lowered to this foothold as the place of operation. A portion of the whip-line was unwound from the reel and fastened to the body of the lowered vehicle as a dragrope, the other end being taken to a fallen tree as a loggerhead or snubbing-post. Surfmen Barney and Stokes and Citizen Woodward in the shafts to guide the cart, the rest of the crew and citizens seized the rope to lower away, and the perilous descent of the almost perpendicular cliff began. The descent continued steadily, without accident, to a point, when it was found that the line employed was too short to enable the cart to gain the plateau. The line was then cast off from the tree and held by the crowd, each man sitting and laying back with his feet braced in the sand, and clinging as a drag upon the burden. In this way the men slid down the bluff behind the cart. In a few minutes, panting and sweating with their effort, and looking like the dirtiest of brick-makers, they stood around the cart on the narrow ledge, the tremendous surf, thick with flood-wood, bursting in foam and spray a few feet below them.

The cart was at once unloaded, the lines made ready, and

the Lyle gun planted and fired. It was then a little before 11 o'clock. The charge was seven ounces, and at the first fire the shot, directed with great judgment on the part of the keeper, flew almost directly across the wind about two hundred yards beyond the vessel, carrying the line along her starboard broadside as she lay nearly head on to the shore, and letting it fall right upon her weather-rigging, fore and aft, where it was instantly caught by the captain. Unfortunately, the slack of the line was immediately swept by the wind and current under the head-gear of the wreck, where it fouled and could not be cleared by the people on board. The first effort to establish line communication with the wreck therefore proved a failure, and the shot-line was hauled in and faked for another trial. This time, with the view of overcoming the added weight of the line, which was wet and clogged with sand, the keeper used an eight-ounce charge. He also trained the gun a little nearer the vessel, aiming to make the line fall higher against the rigging and to prevent, if possible, its fouling with the wreckage. His calculations were superbly accurate. Before the echoes of the report of the gun had ceased along the bluff, the line, flying aloft its full length, had fallen directly across the fore-rigging, where it was caught by the men in the cross-trees.

It wanted at this time a few minutes of noon, and the shipwrecked sailors were in possession of a line from the shore. The anxious question now was, whether this line would stand the strain of hauling out the double rope, or whip, running through a tail-block, which was at once bent on to it. As allowance had to be made for the slack caused by the distance and the tremendous current, there was a vast length of this double line to be paid out between ship and shore. It was manned by at least fifty men, who strung themselves along up the face of the bluff with the aim of keeping the line as much as possible out of the sea, where it was endangered by the drift-stuff and wreckage. At times the force of the current would carry both parts of the whip far to leeward, and the sailors would fail to haul in an inch, and could only take a turn with the shot-line around the heel of the foretopmast. Then the men upon the slope of the bluff would raise and straighten out the whip as much as possible and at a signal from the keeper below, suddenly slack, giving the sailors in the cross-trees, working in concert with them, a chance to haul in a few feet at a time. These manœuvres were regulated by the keeper solely in pantomime, for such was the uproar of the gale that the voice could not have been heard beyond the distance of a few feet, even through a speaking trumpet. The strain on the slender shot-line increased as it took out more and more of the whip line, and every moment the toilers on the slant of the acclivity, timing their labors to the gestures of the grimy figure below them, felt with him the dread that the strands would part; but the tough, braided linen held, and, after more than two hours of such exertions as make the muscles tremble, they had the satisfaction of seeing the whip arrive, and the tail-block properly fastened around the lower masthead and heel of the topmast, the block hanging forward off the cross-trees.

A new obstacle, involving a terrible discouragement, had gradually been developed as the further end of the whip-line rose from the water up to the mast. The whole length of the double rope was seen to be twisted and full of turns. Every effort had been made to prevent this result; the files of men that paid out the rope had been kept widely apart, with the members of the life-saving crew judiciously stationed at certain parts among them, and two experienced surfmen had tended the reel on the cart which gave off the whip to the sea.

The ardent throng of citizens, co-workers with the life-saving crew, were reasonably enough struck with consternation at this incident. A volley of excited questions began to shower upon the keeper in regard to what he was going to do to save the men. Every other second anxious questions or expressions of dejection were shouted at him through the uproar of the storm, and for a few moments his position was exceedingly trying. The crowd, however, were good-natured and obedient in the highest degree, and presently every man rushed to his place under the keeper's orders, and all fell to work clearing the line. This was done by fastening one end to a tree on the brow of the hill and hauling it taut then untwisting or dipping the other part around it, tautening up both parts from time to time while maintaining the operation. Finally, after fully an hour's work the last of the turns were out, and the line was clear.

The breeches-buoy was at once rigged on. As the slope was constantly giving way, several small land-slides half-burying the men below, having already occurred, no sand anchor was planted, the keeper relying on the force he had under command to hold and handle the line. Surfman La Cour was stationed at the summit to tend the slack, which he did by taking a turn with the line around a fallen tree. The buoy then went out toward the wreck, urged by the eager arms of the haulers.

As the men who worked the line were compelled by the steepness of the bank to stand in constrained positions, half upright, half reclining, upon ground constantly giving way, and were also greatly hindered by the blinding sand and buffeting wind, the outward progress was slow, but at length it arrived at the mast. After some little delay, as though the people in the cross-trees hesitated, a man was seen through the dim atmosphere to get into the buoy, which was at once hauled back to the shore. The hauling was done under such difficulties that the passage of the buoy to the shore occupied seventeen minutes by the watch of one of the bystanders. As it approached, several persons rushed down the bank into the surf, and the man was pulled out and helped up to the little plateau. It was the first mate. His jaws were set, his eyes vacantly fixed, and the expression of his face dazed and frightened. A citizen gave him a draught of brandy. This seemed to revive him, and presently he said "Save the others." Two or three questions were asked him in regard to the vessel and persons on board, which he answered faintly, and he was then led away toward the town, supported on either side by two citizens.

In reply to one interrogation, he had been understood by the keeper to say that the woman in the cross-trees did not want to come ashore in the buoy, and as he left, the keeper was notified by Surfman La Cour that the tree to which the whip-line was secured was slowly giving away, and the bank coming down under the strain. This circumstance and the mate's declaration, decided the keeper to substitute the life-car for the buoy, partly because the car could be towed out like a boat until it reached the mast, thus relieving the latter of a certain amount of tension upon it, while its use also dispensed with the fallen tree and spared the pull upon the bank; partly, also, because its employment might facilitate the rescue by landing a greater number of the ship-wrecked at each trip. The car was accordingly ordered forward, and the keeper, with his own hands, attached it to the lines. Every face blazed with excitement as the hauling began. The life-car, as soon as it entered the surf, was dashed about like a cockle shell. After protracted efforts on the part of the haulers, it had at length reached the wreck, when, all at once, the jagged mainmast, which had fallen some time before, and was swinging along the side with other wreckage, rose on the summit of a huge breaker, and lunging like a battering-ram, struck the car such a blow that it tossed it a dozen feet springing into the air. Although every heart throbbed with painful anxiety, the life-savers took swift advantage of the momentary lightening of the line to haul in the slack, and raise the car up, where it hung almost straight up and down some twelve feet below the masthead. Without the least delay three of the men were successively lowered from the cross-trees by ropes around their bodies, and got in. The door of the car was closed. All hands on the hillside then fell to work and the car approached the shore. As it drew near, floundering in the surf, the keeper and several men rushed down waist-deep into the foaming flood, siezed and dragged in the car, unclasped the door and liberated the sailors. They were tended by kindly ministrations and conveyed to a place of shelter and succor which they so sorely needed.

The life-car had received some damage around the hatchway and cover from the blow of the mast and the battering wreckage. It was speedily hammered into shape and again sent out on the lines. The haulers had learned by their first experience how to handle the ropes, and the car pursued its course through the broken water without capsizing. From time to time during the strenuous hauling bursts of sand on the slope indicated the moments when the ground gave way under the feet of the files of devoted men toiling in the heart of the gale, and who could be seen on these occasions to slide and stagger as they pulled, struggling to preserve their foothold or escape engulfment. The tempest continued to scourge the escarpment with unabated violence, and the air of the waning afternoon was thicker than ever with the wind-blown rain, snow, and hail, driven in alternate gusts, and interblent with the driving substance of the hills. Amidst this continued fury the car slowly worked on toward the wreck.

It will perhaps be unnecessary for us to give further details of this thrilling example of the work of the crew of one life-saving station. All were finally brought to shore alive, except, unfortunately, the woman, and she had ceased to live long before the first car reached the vessel. It became impossible even to bring her body ashore, to the great disappointment of the anxious ones on shore.

We have detailed this incident thus minutely because of its thrilling interest, and also because it has exhibited the variety of ways in which the service employ themselves and their implements.

Imagine many similar incidents taking place annually, and it will give some idea of the immense and heroic service rendered by these crews and their keepers. Surely it is a noble and practical service to sustain liberally. Looking at it from a selfish point of view, we cannot tell but it may be our turn to need just such assistance some day.

Another fatal wreck, within life-saving limits, was that of the schooner *A. B. Goodman*, of Seaford, Delaware, bound from Baltimore, Maryland, to New Berne, North Carolina, with a cargo of guano, and having on board five men, including the captain. The wreck took place on April 4, 1881, at about half-past 6 o'clock in the evening, the vessel striking during a northwest gale, upon the outer edge of the inner shoal off Cape Hatteras, and being at once boarded by the sea, there was only time in the overwhelming rush of waters for the men to fly to the rigging; in the effort to gain which, one of them, Louis Beck, was swept overboard, and drowned.

The point at which the disaster took place was about three miles from shore, and six miles east of Life-Saving Station No. 22, Sixth District, North Carolina. This station is built upon the rise of an eminence known as Creed's Hill, and its north patrol reaches for six miles around the edge of the dreaded cape. Looking from the station, the view toward the cape presents to the eye the aspect of an immense desert of sand, strangely and fantastically sprinkled all over with gnarled and twisted trunks of black, dead trees. In winter, or during the inclement season, nothing more dismal could well be imagined than this Sahara, with its thin remnant of a former vegetation killed by the salt tides. The level is only diversified by occasional mounds of sand, and, here and there, pools of sea-water, left by some overflow in the hollows. Behind, or to the west, a forest of pines and live oaks, dense and almost impenetrable, stretches away northward to Hatteras light-house. All around the cape for two miles, in storms at flood-tides, a heavy sea swings across the low and somewhat shelving beach, in among its bordering hummocks, and back again with violence, ploughing gullies as it runs. The surf makes the sand a quag, quicksands form in the gullies, and the solitary patrolman, making his way along the top of the beach in the darkness by the dim light of his lantern, faces the chances of destruction, being liable to be swept off his feet by the rush or refluence of the surf, sucked down in the gullies by the quicksands, or struck by some fragment of wreck-stuff shot forth by the breakers. Yet his dreadful watch is made necessary by the presence off shore by a nest of shoals, range after range, which are the terror of navigators. The first, a mile wide, stretches from the point of the cape between two and three miles seaward, covered with a depth of only seven feet of water, which in storms are miles of raging foam. This formation is, in fact, a submarine prolongation of the cape. Beyond it, separated by half a mile of channel, is another formidable shoal, the Diamond, two miles long; and beyond this again, another range of shallows, the outer shoals. For six or seven miles out from shore, these terrible bottoms spread their ambush for shipping, and hence the watch in this locality for vessels in danger requires to be particularly kept around the point of the cape, no matter at what toil or hazard to the sentinel. On the evening of the disaster to the *A. B. Goodman*, the patrolman, pursuing his journey through the floods sheeting across his way, in the midst of a squall of rain and snow, saw far off, despite the distance and thick weather, the dim outlines of a vessel, and knew by this indication that there was some sort of a craft in the neighborhood of the shoals, though exactly where, or whether in danger, it was impossible to determine. The fact was reported by 10 o'clock to the keeper, B. B. Daily, who was up at dawn, and saw the schooner evidently aground, and, in fact, sunk, on the outer edge of the first range of shoals. He at once ordered out the surf-boat to the rescue.

The storm of the evening before had been brief, and the wind, blowing freshly from the north-northwest, had beaten down the surf upon the beach, the sea, therefore, was smooth for launching, but beyond, it was very heavy. Heaps of water incessantly tumbling, and thickets of bursting foam filling the offing, and the current running one way, while the wind was the other, made an ugly cross-sea. The little group of surfmen about to enter upon this stormy field had still a more serious peril before them than the chance of being overswept or capsized by the colliding waters. Their boat being light and flat-bottomed, the breeze, which was strong and off shore, might make return impossible, and force them out to sea, where they would almost certainly be lost. Nevertheless, as the stout keeper said, in his testimony, "they knew it was their duty to do what they could, so they did it." The group consisted of the keeper and six surfmen. One of the crew was left in charge at the station. It is certain none of the crew expected to return alive. The disposition of their slender effects was a part of the charge, given to the remaining surfman by his companions in case they perished. Having made this simple will as men facing the issues of life and death, they entered the boat and gave way. For some distance out the surf-boat kept the lee of the cape, where the surf was comparatively smooth. They ultimately rounded the cape, and entered the terribly rough water where the greatest circumspection was necessary in the management of the boat. Finally, at about half-past 7 o'clock, two hours after starting the life-saving crew arrived near the wrecked schooner.

She was completely sunk, her hull all under. Only her two masts stuck up from the swirling water, and perched up in the main cross-trees, wrapped in the main-gaff topsail, were huddled the four wretched survivors of her crew of five. After three or four daring and dangerous attempts to get near, baffled by the strong current and the vast commotion of the sea above the sunken hull, keeper Daily hailed the despairing group upon the mast, telling them to keep good heart and that they would be rescued as soon as possible; then dropped astern about three hundred yards and let go the anchor, having decided that it was necessary to a successful effort to wait. The efforts already made had consumed much time, and the boat anchored within an hour of noon. An hour afterward, the flood-tide somewhat smoothed the break of the sea over the sunken hull, and the life-saving crew got up their anchor, worked up to the windward of the vessel, where they again moored, and then slowly and cautiously, by slacking on the anchor line, let the boat veer down toward the mainmast of the wreck. Once within range, the keeper hove his boat-hook, by a line attached, into the rigging and held on. The fateful moment had arrived, the boat was slacked in, so that the keeper could get hold of the first man that came down from aloft, and the first mate slowly descended the rigging. As he came within reach, the keeper, standing in the stern of the boat, seized him, but the man terrified at the frightful rush and roar of waters beneath him, and doubtless unmanned by cold and hunger, and the many hours of horror he had undergone, broke from the keeper's hold and clambered up the rigging again. The boat was hauled back a little, and the keeper spoke up cheerily, encouraging the men in the cross-trees, and declaring they would all be saved. Presently, the line was again slacked, the boat veered down, and the mate once more descended. His fright again seized him, but the keeper, forwarned, got a mighty hold, and by sheer force, jerked him out of the rigging and landed him in the boat. The captain then came down, was seized by the keeper the moment he came within reach, and torn from the shrouds. The other two men, emboldened by this energetic succession of deliverance, slid down the rigging and jumped into the boat without aid. Quickly the keeper then let slack his warp, recovered his boat-hook, and gave the word to haul back to the anchor. Three of the rescued men were seated on the thwarts, the captain in the stern-sheets, the anchor was got up, and the hard work of the return was began.

By this time the wind had changed to west-southwest, blowing freshly, and so roughening the water on the south side of the shoals—which was the side on which the approach to the wreck had been made—that the keeper decided it would be safer to attempt the landing on the north side, or near Hatteras light-house. The men gave way with a will, wind and sea against them. The light keepers watching them as they toiled upon the running swells, had some time before made up their minds that they would not be able to get to land that night if they ever did. But the strenuous effort conquered, and somewhere about two o'clock the life-saving crew, dripping and exhausted, gained the beach, near the light-house tower, with the sailors they had saved.

These sailors were at once taken up to the light-house by the keepers, where a meal was set before them. No food had passed their lips since about eleven o'clock of the day previous, and they were nearly perished with cold and hunger. Their rescuers were in little better condition, having eaten nothing since four o'clock the day before. Nevertheless, without waiting to share in the repast of the sailors, they set off to their own quarters, a tramp of nearly five miles. They reached the station greatly exhausted, all of them had been out on the tempestuous patrol for some part of the night before. From this night of broken rest, they had passed abruptly to eight hours of tragic labor under the shadow of death upon the sea. Their valient rescue achieved, there still remained this long trudge, which left them finally at the station, a group of haggard, worn-out men.

Descant is unnecessary on the feat they performed in saving the four sailors. Such deeds attest themselves; and there are few scenes in human life more deeply affecting than the spectacle of this crew of poor men making their wills upon the beach, and leaving their small store of effects in charge of a comrade for the benefit of their families before entering upon a struggle of deadly peril for the lives of four unhappy creatures, who, in their dying misery, must have thought themselves abandoned forever by man. To have done this—to have quietly resigned the certainties for the chances of existence in such a case and under such circumstances—was more than noble; and there can be no hearts, however cold, that will not feel that in this action the unassuming surfmen of an obscure coast reached again, as many low-down and almost nameless men have reached, the full stature of heroism.

EDDYSTONE LIGHT-HOUSE.

CHAPTER XXVIII.

LIGHTHOUSES AND BEACONS.

NO object can be more suggestive of pleasant thoughts, of home, of peace, and security than a lighthouse. Whether placed upon a headland overlooking a wide expanse of ocean, or on a rock lashed by the foaming billows, it is a welcome sight to the traveler returning to his native country after a long absence, and it is a grateful object to those who are leaving home for distant regions, who, after leaving port, can trace for many miles the friendly light, the last visible connection that unites their thoughts to those who are left behind.

The absence of the lights that stream over the heaving waters would indeed be a calamity; indicative, probably, of war's fatal struggles, when a nation dreading hostile invasion would seek to foil their enemies by extinguishing these lights, and thus leaving them to the perils of shoals and quicksands, of breakers and the rocks. Without these glimmering lights it would be impossible to guide a ship through the perilous ocean; commerce would languish, and all the civilizing influences encouraged by trade would be lost.

The earliest allusion to lighthouses or "beacons" to guide the mariner on his sea-journey, date from a remote period. Homer, alluding to the shield of Achilles, beautifully describes the flash of a beacon-light in some solitary place, as seen by seamen leaving their friends, in verses which contain ample proof of the existence of such a provision for the mariner's safety in the poet's time.

The oldest lighthouse on record is the celebrated *Pharos*, erected on the Egyptian coast, which, being very low land, and exposed almost entirely to the west winds coming up the Mediterranean from the vast Atlantic, must of necessity have made the port of Alexandria very dangerous. This lighthouse was erected 300 years B. C. for the convenience of the Phœnician merchants who constantly traded with Egypt.

The island of Pharos, upon which the lighthouse was erected, was said, in the time of Homer, to be one day's sail from the Delta (a triangular portion of Lower Egypt comprised between the two main branches of the Nile, and so called from its resemblance to the Greek letter D, Δ); whereas, since the foundation of Alexandria, it was only a mile in distance, and was even joined to the mainland by a mole or artificial embankment having a bridge at each end.

This tower, if statements are true, was justly entitled to the honor claimed for it as one of the seven wonders of the world. It is stated to have been five hundred and fifty feet in height, and the cost of erection was equivalent, in our money, to nearly $700,000. The architect was Sostratus, the constructor of many public buildings in Alexandria.

The lighthouse consisted of several stories, one raised above the other, each decorated with columns, balustrades, and galleries made of the finest marble and of the most exquisite workmanship. It has been said that Sostratus furnished the galleries with large mirrors, by which shipping could be seen at a considerable distance. A fire was kept constantly burning on the summit, and, according to Josephus, the Jewish historian, the light was seen at the distance of three hundred stadii (about forty-two miles).

In more modern times the Turks erected two forts on the points occupied by this Pharos, one of which was the site of the far-famed lighthouse.

The celebrated colossal statue of Apollo at Rhodes is said to have supplied the purpose of a lighthouse. This figure was of bronze, and a period of twelve years is stated to have been employed in constructing it, three hundred years before the Christian era. The expenses of erection amounted to a sum estimated in our money at nearly $300,000. The gigantic size of this figure may be conceived from the account given, that few men were able to encompass one of its thumbs with their arms. The Rhodians placed this brazen sentinel at the entrance of their port. The heathen deity was represented wearing a radiant crown, holding in one hand an arrow, and in the other a brazier containing fire. This stupendous statue was overthrown by an earthquake about eighty years after its erection; and when the Saracens took Rhodes in 667, they loaded one hundred camels with the bronze that remained, and the rest was sold afterwards to a Jewish merchant for a sum equivalent to $144,000.

An edifice, called by the inhabitants of the Archipelago the "Lamp of Diogenes," stood on the shores of the Ægean sea. It was a tomb erected in memory of Diogenes, who had such a knowledge of the art of navigation that no ancient mariner would undertake a voyage without consulting him. At the angles of this building were columns, on which fires were placed for the service of vessels navigating those dangerous seas.

One of the greatest achievements of modern science in in connexion with lighthouses is that effected by the famous engineer Smeaton on the dangerous Eddystone rocks, about fourteen miles distance from Plymouth, England, which are so exposed to the heavy swells from the Bay of Biscay and the Atlantic that the waves beat upon them at times with fearful violence. In 1696, notwithstanding the many difficulties which seemed to interfere with the erection of a lighthouse on rocks so situated, Henry Winstanley succeeded in accomplishing (for a time) the desired object,

The lighthouse on the Eddystone rocks was commenced in 1696, and the first light appeared on the 14th of November, 1698. Finding that the tower was greatly exposed during the storms of winter—the waves covering the lantern at times, although at a height of sixty feet—Winstanley encompassed the building with a new work of four feet in thickness from the foundation, making all solid to the height of twenty feet; he also took down the upper part of the first building, and enlarging every part in its proportion, raised it forty feet higher than it was originally, and yet the sea in stormy weather rose one hundred feet above the vane, and at times covered half the side of the house and the lantern, as if they were under water. The building was so fantastically constructed and seemed so slight, that general opinion was against its security, but the architect himself was so firmly convinced of its stability that he frequently wished he might be there in the greatest storm that could happen. This desire was fatally gratified. In November, 1703, Winstanley went out to the rocks to superintend some repairs, and that very night a fearful tempest arose, which so increased the following day, that the lighthouse and all its inmates were swept in the foaming waters.

Nearly three years elapsed after this sad event before another attempt was made to replace the lighthouse on the Eddystone rocks. The building was commenced in 1706, and was lighted two years afterwards.

Rudyard's lighthouse enjoyed an immunity from serious accidents for a space of thirty-eight years, when, at the close of 1744, a fearful storm occurred, in which the ship *Victory* was lost close to the building, which was considerably damaged in consequence, and an opening made into the store-room. In 1755 the catastrophe occurred which destroyed the lighthouse. On the 2d December, in that year, the lighthouse keeper then on watch, about two o'clock in the morning, went into the lantern to snuff the candles. He found the whole place in a smoke, and upon opening the door of the lantern into the balcony, a flame instantly burst from the inside of the cupola. He endeavored to alarm his companions, but they being fast asleep were not able to come to his assistance so readily as the occasion required. He attempted to extinguish the flames with some water kept in a tub in the lantern, but the fire increasing, the poor fellow found himself unable to stop its progress. As he was looking upwards, a quantity of molten lead suddenly rushed like a torrent from the roof, and falling upon his head, face, and shoulders, burnt him fearfully. His companions, dismayed at the extent of the conflagration, descended to the rock, with little hope of being saved. It seems that at an early hour the flames were seen by some of the Cawsand fishermen; and on intelligence being given, a fishing-boat, manned with a strong crew, was dispatched at once. The boat arrived about eight hours after the fire had broken out, during which time the three keepers of the lighthouse had not only been driven from all the rooms and the staircase, but to avoid the fall of the timber and red-hot bolts, they were found sitting in a hole or cave, on the east side of the rock, almost in a state of stupefaction. They were conveyed safely to Plymouth.

This fire occurred in 1753, and it was then determined that the lighthouse should be constructed of stone, and that no expense should be spared to render it the most perfect of its kind. The architect chosen was the famous John Smeaton, one of the most extraordinary men of the age. The first stone of the edifice was laid the 12th June, 1757, and the whole undertaking of constructing the lighthouse was accomplished within a space of little more than three years, without the loss of life or limb to those engaged in the perilous undertaking. It was an anxious time for Smeaton and those who were engaged with him. When the weather was tempestuous the lighthouse was inaccessible. One of the keepers, Henry Edwards, after a heavy storm had prevented any boat from approaching the rocks, sent a note to the manager of the works, stating "that the sea ran over the house in such a manner that for twelve days together they could not open the door of the lantern, nor any other. The house shook as if a man had been up in a great tree. The old men were almost frightened out of their lives, wishing they had never seen the place. The fear occasioned pains in the back, but rubbing them with oil of turpentine gave them relief." The lighthouse itself bore the storm admirably, and suffered nothing from it.

The present edifice is a circular tower sweeping up with a gentle curve from the base, and gradually diminishing to the top. The upper extremity is furnished with a kind of cornice, and is surmounted by a lantern, having a gallery around it with an iron balustrade. The tower is furnished with a door, and windows, and staircase. Round the upper storeroom is the inscription: "Except the Lord build the house, they labor in vain that build it.—Psalm cxxvii." Over each side of the lantern are the words and date: "August 24th, 1759. Laus Deo."

A very great achievement in the erection of lighthouses since the days of Smeaton, is that of the "Bell-Rock" tower in Scotland, built upon a dangerous sunken reef, about eleven miles from Arbroath, on the northern side of the entrance of the great estuary or arm of the sea, called the Firth of Forth, and, as such, directly affecting the safety of all vessels entering the Firth of Tay. The "Inchcape," or Bell-Rock, had always been a perilous point to navigators, and in former times a bell was placed there by the Abbot of Aberbrothock, or Arbroath, which was put in movement by the waves. This was the only expedient that could then be devised. According to tradition, some pirates having carried off this bell, were, on a subsequent voyage, lost on the same rock. Southey's thrilling ballad of the "Inchcape Bell" is founded on this legend.

The building was commenced in 1807 (17th August), under the control of Mr. Robert Stephenson, the engineer to the Lighthouse Board, and whose plan—a tower of masonry, on the principle of the Eddystone Lighthouse—was adopted. From an account of these operations, written by that eminent engineer, we learn how severe and perilous was the undertaking, the rock being only dry for a few hours at spring-tides, and affording but little time for laying the foundations of the building with requisite security. This, however, under many difficulties, was effected, and the first stone of the lighthouse laid 10th July, 1808, at the depth of sixteen feet below high water at spring-tide. The whole of the masonry to the height of thirty feet, was completed in 1810, the light being exhibited for the first time 1st February, 1811.

Another celebrated stone lighthouse on the Scottish coast, is that placed on the "Skerryvore" rocks, which lie about twelve miles off the Isle of Tyree, in Argyllshire. These rocks were for a long period the terror of mariners, and numerous shipwrecks had occurred in their vicinity.

MONTAUK LIGHT-HOUSE.

Owing to the great difficulty of landing upon these rocks, which are worn smooth by the continual beating of the Atlantic waves, it was not until 1834 that the idea of erecting a lighthouse was seriously entertained. In such a situation as that of Skerryvore everything had to be provided before hand, and transported from a distance. The design for the building was made by Mr. Alan Stephenson, son of the celebrated architect of the Bell-Rock Lighthouse, and was an adaptation of Smeaton's Eddystone tower. Many were the vicissitudes and privations experienced in this undertaking, but this stately and noble building remains a boon to seamen, the signal of trust and confidence to the sea-bound mariner.

Other important lighthouses of stone might be mentioned, such as the "Bishop Rock," the "Needles," the "Smalls," etc. Iron lighthouses have of late been erected, and appear to be admirably suited to the purpose, and comparatively inexpensive. The "Northfleet" is built of this material, and is in open skeleton work. Iron lighthouses have been adopted in several of the British colonies.

The list so far mentioned does not include all, nor indeed any, considerable portion of the lighthouses of the world. However, these few so briefly described, with a few others to follow, that are situated on our own coasts, will probably serve to illustrate their general use and the improvements made from time to time in the different styles of architecture together with the erection of more durable structure and methods of lighting.

The earlier history of lighthouses in this country antedating the Revolution is somewhat obscure. Previous to 1789 the expense of keeping up the few then standing was borne by the commonwealths in which they were situated; after that time they were turned over by the respective States of Colonies to the General Government, and the improvement following has been commensurate with the progress of the country in other matters pertaining to its welfare.

The first light-house built by the United States was lighted in 1791. There were then but a total of nine lights on the whole North American coast. They have since multiplied in an astonishing ratio. In 1822, there were 70; in 1838, 238, and in 1875 there were 925 lighthouses and 23 lightships or beacons.

Before 1852 this matter of sea-coast illumination was under the control of a single subordinate officer of the Treasury. Since that time the Service has been reorganized and placed in the hands of the United States Lighthouse Board, consisting of eight persons, viz.: two officers of the corps of engineers, and two civilians of high scientific attainments and others. The Board is attached to the office of the Secretary of the Treasury, who is *ex officio* president of it. A chairman is chosen from among the Board who presides in the absence of the Secretary of the Treasury.

The labors of this Board have become very heavy. They now have charge of nearly 1,000 lighthouses, lightships, and beacons, and about 3,000 buoys and a great number of fog-signals and fog-bells situated all along the immense coast-line of this country. They meet in session very frequently and are continually trying difficult experiments for the betterment of the system.

Out of this great number of lighthouses on our coasts, there is, probably, none more famous than the Minot's Ledge lighthouse. The Minot's rocks are situated just east of Boston Bay, and for years they had been the terror of sailors, probably being the cause of a greater number of wrecks than any other reefs on our extensive coasts.

Captain W. H. Swift, U. S. Engineer, in 1848 began the difficult labor of putting up an iron structure on the outer or most seaward of the Minot group. Notwithstanding the many obstacles to be overcome, the work was completed in about a year. In 1851 a terrific storm arose, completely destroying this structure and all who were in it.

Every commercial and human consideration required that there should be a beacon on this ledge. Congress promptly took the matter in hand, making the necessary appropriations, and the erection of a new tower was immediately begun under the charge of the newly created Lighthouse Board. The work began in July, 1855, and was completed in September, 1860, and cost nearly $300,000. The structure is of stone and over 114 feet high. It has so far withstood all the power of the elements, and is a grand monument of modern engineering skill.

Bells, as a warning to mariners, have been and are still generally employed at light-houses as one of the best modes of signalling in times of fogs. We have already alluded to the Inchcape Bell, and to Mr. Stephenson's lighthouse on that dangerous rock. The machinery which causes the reflectors to revolve every half-minute is made to ring two large bells, each weighing about twelve hundred-weight, to warn the seaman of his danger when too near this rock.

In some lighthouses guns are fired during fogs: at those in Nova Scotia, horns are sounded during fogs, and these are heard at a distance of three miles. In a lighthouse near Holyhead, England, built on the middle of an island under a cliff, and connected with the mainland by a bridge, tamed sea-birds are employed as signals. The gulls perch on the walls of the lighthouse, and utter cries that warn the sailor. This lighthouse has a bell and a gun, but the natural signal has been judged so superior that the cannon has been removed some distance from the rock for fear lest the noise should startle the birds. In this island the young gulls run about among the white rabbits, with whom they live on the most intimate terms.

Gongs are used in some lighthouses, and are found of great use.

A steam trumpet is used, the sound of which can be heard at a considerable distance. The instrument can be tuned at will, and thus produces different notes. A small quantity of steam suffices.

In Scotland whistles are placed in the lanterns of the lighthouses, and these communicate by tubes with the rooms below, in order to summon assistance if required, and thus avoid the necessity of the keeper in charge leaving his post until relieved.

In the absence of any authority respecting the material with which the earliest lighthouses were illuminated, we may naturally conclude that wood furnished the means, with any other substance which would give intensity to the flame, or contribute to its duration.

In England, lighthouses from an early period appear to have been illuminated by coal or wood fires contained in pans. At the close of the seventeenth century an iron grate on the top of the North Foreland Lighthouse, in Kent, England, contained a coal fire open to the air; and later, in 1732, the top of the tower was covered with a sort of lantern with large sash windows, and the light of the fire was kept up by bellows, which were moved during the night by attendants. The last coal-light—that of St. Bee's in Cumberland—was only extinguished in 1822. To coal fires succeeded tallow candles, and these were fastened in wooden rods, as they are sometimes seen arranged before booths in fairs. These were in use for forty years at the Eddystone Lighthouse after it was completed by Smeaton.

The use of oil does not seem to date back beyond 1730: lamps with twisted cotton wicks were used, but these were always attended with smoke and a bad smell. The first decided improvement was made by M. Argand, a native of Geneva, who, in 1784, invented a lamp with a circular wick.

The flame thus became a hollow cylinder, with a current of air ascending through the inside, so that the burning surface was doubled. To make these lamps more effective for lighthouses, and to prevent the rays of light from escaping on all sides, a reflector was added by M. Lenoir. This threw the light forward in parallel rays toward such points of the horizon as would be useful to mariners; but M. Augustin Fresnel was the first to introduce lenticular action into lighthouse illumination by the adoption of the annular or built lens. Along with Arago he investigated the action exercised by polarized rays of light on each other, and the practical application of his theory to the improvement of the lighthouse system was of immense value, and quite abolished the old method of illuminating the "sea sentinels."

Among the later adaptations for this object are the introduction of gas, but this is attended by the uncertainty and other objections connected with its manufacture and use in remote and inaccessible places. It is, however, of great convenience and use in harbor lights.

The attention of scientific men, so far as lighthouses are concerned, is now almost confined to the discovery of the best modes of producing the light. That in ordinary use leaves little to be desired when the weather is tolerably clear; since a first-class oil light, at the height on which it is usually elevated, is visible from the masthead when the vessel comes above the horizon of the lighthouse. It is in hazy weather that a more intense light becomes desirable. There is the "Drummond" light, which consists in substituting for the Argand burner a small ball of lime ignited by the combustion of oxyogen and hydrogen; but the difficulty of obtaining a continuous combustion has hitherto prevented the adoption of this means of illumination in lighthouses. The "electric" light has undergone careful trials by the government authorities. It possesses an advantage from its not being a mere spark of very small size, as compared with an oil lamp of the same power; for this enables much smaller optical apparatus to be used, occasioning not only a saving of cost, but a saving of light—the loss of light being less when the glass is thinner. The electro-magnetic light has been seen thirty minutes after losing sight of the lower lights—that is, at the rate of a steamer's progress—a distance of about seven miles. Taking all things into consideration, it seems likely that the electric light is destined to supersede all kinds of oil for lighthouse illumination: its far-reaching property, and its power to penetrate fogs, must in the end obtain for it the victory over all other sources of light.

To these notices of the structures raised for the guidance and security of mariners, we will now add some observations on the *animated* sentinels of the seas: those men to whose care these all-important buildings are intrusted. Life in a lighthouse would seem to us a terrible hardship. Perhaps one of the worst of criminals, who might be indifferent to most kinds of penal punishment, would be terrified if he were condemned to pass the remainder of his life on a solitary rock amidst the wide ocean. It would appear, indeed, difficult to reconcile the mind to a service so lonely and peculiar, not unfrequently attended at some places with risks and privations, and requiring an amount of vigilance and responsibilty not very tempting to most men in search of employment. Yet it appears that these situations are frequently sought after, and by persons who have been engaged in far different occupations. Some of these applicants have been carpenters, blacksmiths, domestic servants, butchers, bricklayers, painters, bakers, coopers, tailors, etc. Among these individuals there are doubtless some whose taste for solitude may have induced them to forsake their occupations among the busy haunts of society, and others, perhaps, to whom the novelty of the employment might offer some attraction; but it is a very singular circumstance that almost all who have adopted a lighthouse for their home seem to have been satisfied with their choice, and some even have boasted that they would not change their solitary lot for any other.

Appointments to positions in either of these branches of the national service depends entirely upon personal and superior fitness and not upon political preferment. And we venture to say that the labor performed by the servants and the supervision of that labor by the proper Government officers is as pains-taking and conscientious as anything well can be. If the battle-scarred veteran who has fought his country's battles, be so fortunate as to return from the scenes of conflict to his home and fireside, we, from motives of gratitude and warm, patriotic sentiments tugging at our hearts, welcome him with loud plaudits and crown him with laurels. It is right and becoming that we should do this, but at the same time let us not forget these other heroes who are equally deserving, though their noble and unselfish deeds of daring are not trumpeted abroad with marshal accessories and emblazonment.

In concluding these brief notices, we trust the reader will often think of the lonely watchers to whose faithful care they are entrusted, with feelings of the deepest interest, for the lives of so many brave men, and the security of valuable vessels, depend upon their vigilance and unceasing attention.

Other books from CGR Publishing - www.CGRpublishing.com

1939 New York World's Fair: The World of Tomorrow in Photographs

San Francisco 1915 World's Fair: The Panama-Pacific International Expo.

1904 St. Louis World's Fair: The Louisiana Purchase Exposition in Photographs

Chicago 1933 World's Fair: A Century of Progress in Photographs

19th Century New York: A Dramatic Collection of Images

The American Railway: The Trains, Railroads, and People Who Ran the Rails

Magnum Skywolf #1

The World's Fair of 1893 Ultra Massive Photographic Adventure Vol. 1

The World's Fair of 1893 Ultra Massive Photographic Adventure Vol. 2

The World's Fair of 1893 Ultra Massive Photographic Adventure Vol. 3

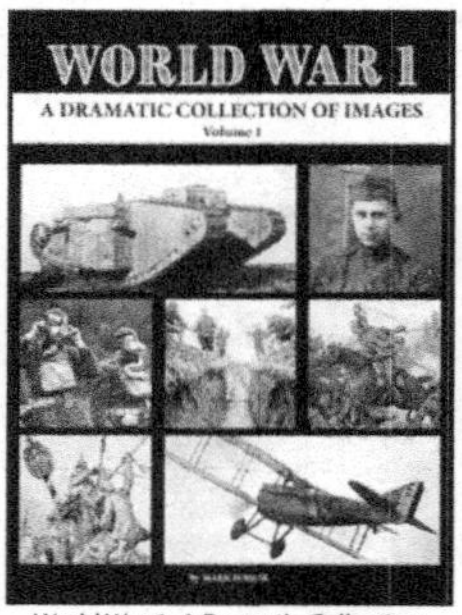

World War 1: A Dramatic Collection of Images

The White City of Color: 1893 World's Fair

Ethel the Cyborg Ninja Book 1

Ethel they Cyborg Ninja 2

How To Draw Digital by Mark Bussler

How To Draw Pandas by Mark Bussler

Other books from CGR Publishing - www.CGRpublishing.com

Ultra Massive Video Game Console Guide Volume 1

Ultra Massive Video Game Console Guide Volume 2

Ultra Massive Video Game Console Guide Volume 3

Ultra Massive Sega Genesis Guide

The Illustrated History of the 1876 Centennial Exhibition Volume 1

Chicago's White City Cookbook

Official Guide to the World's Columbian Exposition

How To Grow Mushrooms: A 19th Century Approach

The Great War Remastered WW1 Standard History Collection Vol. 1

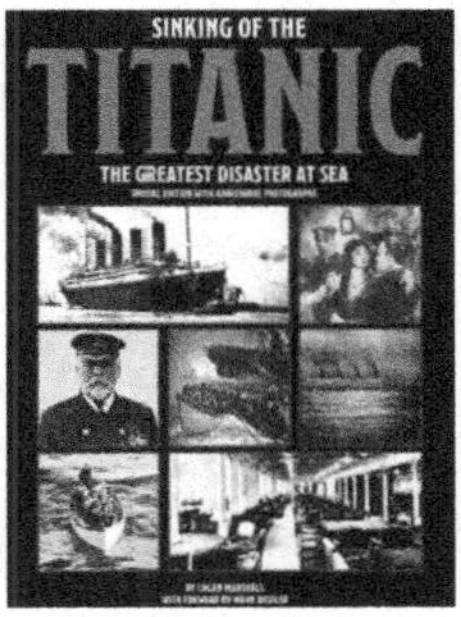

Sinking of the Titanic: The Greatest Disaster at Sea

All Hail the Vectrex Ultimate Collector's Guide

Robot Kitten Factory #1

P.T. Barnum The Greatest Showman on Earth

Electricity at the World's Fair of 1893 Columbian Exposition

Captain William Kidd and the Pirates and Buccaneers Who Ravaged the Seas

1876 Centennial Exhibition: The Illustrated Enhanced Historical Register

Oceanography and Sailing in the 19th Century:
The Water World Enlarged Illustrated Special Edition

www.CGRpublishing.com

Made in United States
North Haven, CT
10 November 2026

22106129R00285